Introduction to PDEs and Waves
for the Atmosphere and Ocean

Courant Lecture Notes in Mathematics

Executive Editor
Jalal Shatah

Managing Editor
Paul D. Monsour

Assistant Editor
Reeva Goldsmith

Copy Editor
Paul D. Monsour

Andrew Majda

Morse Professor of Arts and Sciences
Courant Institute of Mathematical Sciences and
Center for Atmosphere/Ocean Science (CAOS)

9 Introduction to PDEs and Waves for the Atmosphere and Ocean

Courant Institute of Mathematical Sciences
New York University
New York, New York

American Mathematical Society
Providence, Rhode Island

2000 *Mathematics Subject Classification.* Primary 34–XX, 35–XX, 65–XX, 76–XX, 86–XX.

Library of Congress Cataloging-in-Publication Data

Majda, Andrew, 1949–
　Introduction to PDEs and waves for the atmosphere and ocean / Andrew Majda.
　　p. cm. — (Courant lecture notes in mathematics, ISSN 1529-9031 ; 9)
　Includes bibliographical references.
　ISBN 0-8218-2954-8 (pbk. : acid-free paper)
　1. Waves.　2. Differential equations, Partial.　3. Differential equations, Nonlinear.　4. Oceanography—Mathematical models.　5. Atmosphere—Mathematical models.　6. Stratified flow. I. Title.　II. Series.

QC157.M35　2003
515′.353—dc21
　　　　　　　　　　　　　　　　　　　　　　　　　　　　　　　2002042674

Contents

Preface

These lecture notes are based on material presented by the author in graduate courses at the Courant Institute in 1995, 1997, 1999, and 2001. The lectures emphasize the serendipity between modern applied mathematics and geophysical flows in the style of modern applied mathematics where rigorous analysis, as well as asympototic, qualitative, and numerical modeling all interact. The goal was to introduce mathematicians to atmosphere/ocean science (AOS) in this fashion and conversely to develop a set of lecture notes of interest to the disciplinary community ranging from graduate students to researchers in AOS. During these courses, the beautiful applications-oriented text by Adrian Gill [11] and the well-known text by Pedlosky [29] were often used for supplementary reading material outside class. These lecture notes do not require a previous course in fluid dynamics although the texts [2, 19] are recommended for supplementary material on incompressible flow while the texts [4, 26, 33] are suggested for additional interesting topics in the mathematical physics of geophysical flows.

The author thanks Prof. Pedro Embid and his former Ph.D. student Jonathan Callet for their help with early versions of the lecture notes for Chapters 2, 4, 5, and 7. Some joint research with Professor Embid that was developed as an outgrowth of the earliest lecture courses as well as joint work with my former Courant postdocs Prof. Marcus Grote and Misha Shefter has been incorporated into the material presented here; their explicit and implicit contributions are acknowledged warmly. My current Courant postdoc, Boualem Khouider, has been a great help with the notes for Chapter 9, which were developed as part of a lecture course in spring 2001. Finally, the author acknowledges the generous support of both the National Science Foundation and the Office of Naval Research during the development of these lecture notes, including partial salary support for Professor Embid during his visit to the Courant Institute during the mid 1990s.

CHAPTER 1

Introduction

The most important features that distinguish fluid flow in the atmosphere and ocean are the effects of rotation and stratification. Here we introduce the simplest equations with both the effects of rotation and stratification and build simple exact solutions that reveal their elementary structure as well as other model equations such as the shallow water equations in other approximations.

1.1. Basic Properties of the Equations with Rotation and Stratification

In certain ranges of scales in the atmosphere and in the ocean, the fluid dynamics is controlled by the interaction of gravity and the earth's rotation with density variations about a reference state. Flow velocities are too slow to involve compressible effects. The equations at work on these scales, called the *rotating Boussinesq equations*, and the gravity waves they support, are built up in A. E. Gill's book [**11**]. We will begin with the following definitions:

Definitions.

(1.1)

$$
\begin{aligned}
\vec{x} = (x_1, x_2, x_3) &= \text{coordinates} \\
\vec{e}_3 = (0, 0, 1) &= \text{``upward'' unit vector} \\
\vec{v}(\vec{x}, t) = (v_1(\vec{x}, t), v_2(\vec{x}, t), v_3(\vec{x}, t)) &= \text{velocity} \\
\tilde{p}(\vec{x}, t) &= \text{pressure} \\
\tilde{\rho}(\vec{x}, t) &= \text{density} \\
\rho_b &= \text{reference constant density} \\
g &= \text{acceleration due to gravity} \\
&\quad \text{(points in } -\vec{e}_3 \text{ direction)} \\
\nu \geq 0 &= \text{coefficient of viscosity} \\
\kappa \geq 0 &= \text{coefficient of heat conduction} \\
f &= \text{rotation frequency} \\
\frac{D}{Dt} = \frac{\partial}{\partial t} + (\vec{v} \cdot \nabla) &= \text{advective derivative.}
\end{aligned}
$$

With this notation, we can write the *rotating Boussinesq equations*

$$
(1.2) \quad \frac{D\vec{v}}{Dt} + f\vec{e}_3 \times \vec{v} = -\nabla \tilde{p} + \nu \Delta \vec{v} - \frac{g\tilde{\rho}}{\rho_b}\vec{e}_3, \quad \text{div}\, \vec{v} = 0, \quad \frac{D\tilde{\rho}}{Dt} = \kappa \Delta \tilde{\rho}.
$$

1

These equations have an elementary exact solution with $\vec{v} \equiv 0$ with *hydrostatic balance*

$$(1.3) \qquad \frac{\partial}{\partial x_3} \tilde{p}(x_3) = -\frac{g\tilde{\rho}(x_3)}{\rho_b}.$$

Given the density as a function of x_3, we can integrate the hydrostatic balance equation to find the pressure. In fact, for many scales in the atmosphere and ocean, the x_3 pressure gradients very nearly cancel the buoyancy term in the right-hand side of (1.3). It is therefore natural to consider perturbations about a mean state in hydrostatic balance, and to let

$$(1.4) \qquad \tilde{\rho}(\vec{x}, t) = \bar{\rho}(x_3) + \rho(\vec{x}, t), \quad \tilde{p}(\vec{x}, t) = \bar{p}(x_3) + p(\vec{x}, t),$$

where $\bar{p}$ and $\bar{\rho}$ are in hydrostatic balance,

$$(1.5) \qquad \frac{\partial}{\partial x_3} \bar{p}(x_3) = -\frac{g\bar{\rho}(x_3)}{\rho_b}.$$

Then p, ρ, and $\vec{v}$ solve

$$(1.6a) \qquad \frac{D\vec{v}}{Dt} + f\vec{e}_3 \times \vec{v} = -\nabla p + \nu\Delta\vec{v} - \frac{g\rho}{\rho_b}\vec{e}_3,$$

$$(1.6b) \qquad \operatorname{div}\vec{v} = 0,$$

$$(1.6c) \qquad \frac{D\rho}{Dt} + \left(\frac{d\bar{\rho}}{dx_3}\right) v_3 = \kappa\Delta\rho + \kappa\frac{d^2\bar{\rho}}{dx_3^2}.$$

The sign of $d\bar{\rho}/dx_3$ that appears in (1.6c) is critical. For lower regions of the troposphere, sometimes lighter air is below heavier air, the sign is positive, and the situation is unstable. Think of the experiment in which a fluid in a container is heated at the bottom and cooled at the top. The situation is unstable because the hot fluid at the bottom is less dense that the fluid above it. The formation of thermal rolls in that experiment is related to cloud formation in the troposphere. Higher up in the stratosphere and in the deep ocean, the lighter fluid sits atop the heavier in a stable situation that supports waves. We will see these considerations explicitly later in the chapter.

1.2. Two-Dimensional Exact Solutions

At this point, we carry out an exercise in notation that emphasizes the difference between the horizontal and vertical components in the rotating Boussinesq equations. It will let us write a large number of two-dimensional solutions. We begin by letting $\vec{v} = (\vec{v}_H, w)$ where

$$\vec{v}_H = (v_1, v_2), \quad w = v_3, \quad (v_1, v_2)^\perp = (-v_2, v_1),$$

$$(1.7) \quad \nabla_H\psi = \left(\frac{\partial\psi}{\partial x_1}, \frac{\partial\psi}{\partial x_2}\right), \quad \Delta_H\psi = \frac{\partial^2\psi}{\partial x_1^2} + \frac{\partial^2\psi}{\partial x_2^2}, \quad \nabla_H^\perp\psi = \left(-\frac{\partial\psi}{\partial x_2}, \frac{\partial\psi}{\partial x_1}\right),$$

$$\frac{D^H}{Dt} = \frac{\partial}{\partial t} + (\vec{v}_H \cdot \nabla_H).$$

In this notation, we write the rotating Boussinesq equations in terms of the horizontal and vertical components

(1.8a)
$$\frac{D^H \vec{v}_H}{Dt} + w\frac{\partial \vec{v}_H}{\partial x_3} + f\vec{v}_H^{\perp} = -\nabla_H p + \nu\Delta\vec{v}_H \,,$$

$$\frac{D^H w}{Dt} + w\frac{\partial w}{\partial x_3} = -\frac{\partial p}{\partial x_3} - \frac{g\rho}{\rho_b} + \nu\Delta w \,,$$

(1.8b)
$$\text{div}_H\,\vec{v}_H + \frac{\partial w}{\partial x_3} = 0 \,,$$

(1.8c)
$$\frac{D^H \rho}{Dt} + w\frac{\partial \rho}{\partial x_3} + \left(\frac{d\bar{\rho}}{dx_3}\right)w = \kappa\Delta\rho = \kappa\frac{d\bar{\rho}}{dx_3^2} \,.$$

This notation obviates a form of two-dimensional solutions to the rotating Boussinesq equation. We will phrase as a proposition the observation that with $w = 0$, $\rho = 0$, and no x_3-dependence in $\vec{v}_H$, the equations of (1.8) become the *two-dimensional homogeneous fluid equation*

(1.9)
$$\frac{D^H \vec{v}_H}{Dt} + f\vec{v}_H^{\perp} = -\nabla_H p + \nu\Delta_H\vec{v}_H \,, \quad \text{div}_H\,\vec{v}_H = 0 \,.$$

PROPOSITION 1.1 *Let $\vec{v}_H(x_1, x_2)$ be any solution of the homogeneous two-dimensional Navier-Stokes equation (1.9). Then $\rho = 0$ and $\vec{v} = (\vec{v}_H, 0)$ generate special solutions to the rotating Boussinesq equations; for $\kappa = 0$, $\bar{\rho}(x_3)$ is arbitrary, while for $\kappa > 0$, it must have the form $\bar{\rho}(x_3) = \rho_b + cx_3$ for some constant c.*

An enormous number of exact solutions are known for the much studied two-dimensional homogeneous fluid equations. The solutions of (1.9) do not depend on depth. They give the bulk averaged properties of fluid motion at large scales and are called the *equations for barotropic flow*. These equations are important in their own right and provide an important example of dispersive waves where the differential effects of rotation are important.

Barotropic Equations. Let's take a more detailed look at these barotropic equations (neglecting viscosity),

(1.10)
$$\frac{D\vec{v}}{Dt}H + f(y)\vec{v}_H^{\perp} = -\nabla p \,, \quad \text{div}\,\vec{v}_H = 0 \,.$$

Since div $\vec{v}_H = 0$, there is a stream function ψ so that

(1.11)
$$\vec{v}_H = \begin{pmatrix} -\frac{\partial\psi}{\partial y} \\ \frac{\partial\psi}{\partial x} \end{pmatrix} \,.$$

Introducing the two-dimensional curl of a vector field $\vec{v} = (u, v)$ by

(1.12)
$$\text{curl}\,\vec{v} = -\frac{\partial u}{\partial y} + \frac{\partial v}{\partial x} \,,$$

the curl of the fluid velocity is called the *vorticity*. Thus, curl $\vec{v}_H = \omega$, which from (1.11) means that

$$\text{(1.13)} \qquad \Delta_H \psi = \omega = \text{curl } \vec{v}_H \quad \text{where} \quad \Delta_H = \frac{\partial^2}{\partial x^2} + \frac{\partial^2}{\partial y^2}.$$

Using a stream function automatically satisfies the divergence condition, while the pressure can be eliminated in (1.10) by taking the curl of (1.10) and deriving an equation for the vorticity, ω, alone. Elementary calculations show that

$$\text{curl}\left(\frac{D\vec{v}_H}{Dt}\right) = \frac{D\omega}{Dt} = \frac{\partial \omega}{\partial t} + \vec{v}_H \cdot \nabla\omega$$

while

$$\text{(1.14)} \qquad \text{curl}\left(f(y)\begin{pmatrix} -v \\ u \end{pmatrix}\right) = \text{curl}\left(f(y)\begin{pmatrix} -\frac{\partial \psi}{\partial x} \\ -\frac{\partial \psi}{\partial y} \end{pmatrix}\right) = \frac{\partial f}{\partial y}\frac{\partial \psi}{\partial x}.$$

Furthermore, for any function g, from (1.11)

$$\text{(1.15)} \qquad \vec{v}_H \cdot \nabla g = -\frac{\partial \psi}{\partial y}\frac{\partial g}{\partial x} + \frac{\partial \psi}{\partial x}\frac{\partial g}{\partial y} = \det\begin{pmatrix} \nabla\psi \\ \nabla g \end{pmatrix} = J(\psi, g)$$

where $J(\psi, g)$ denotes the Jacobian determinant of the transformation $\begin{pmatrix} x \\ y \end{pmatrix} \longmapsto \begin{pmatrix} \psi \\ g \end{pmatrix}$. Combining (1.14) and (1.15), we see that the barotropic equations in (1.10) are equivalent to the barotropic equations in vorticity stream form.

Barotropic Equations in Vorticity Stream Form.

$$\text{(1.16)} \qquad \frac{\partial \omega}{\partial t} + J(\psi, \omega) + \frac{\partial f}{\partial y}\psi_x = 0, \quad \omega = \Delta\psi.$$

Note that if there are no differential effects of rotation, i.e., $\frac{\partial f}{\partial y} \equiv 0$, the equations in (1.16) become the vorticity stream form of two-dimensional fluid flow (see Majda and Bertozzi [**19**]). In geophysical flows, the β-plane tangent approximation uses the value

$$\text{(1.17)} \qquad f = f_0 + \beta y, \quad \beta \neq 0,$$

and (1.16) becomes

$$\text{(1.18)} \qquad \frac{\partial \Delta\psi}{\partial t} + J(\psi, \Delta\psi) + \beta\psi_x = 0.$$

For $\beta \neq 0$, this is the simplest dispersive equation with wave solutions called *Rossby waves*, which are important for planetary wave propagation. We will see this later on in these lecture notes. In general, in (1.16) we can introduce the quantity Q, called the *potential vorticity*, given by

$$\text{(1.19)} \qquad Q = \omega + f(y)$$

so that (1.16) can be rewritten elegantly as

$$\text{(1.20)} \qquad \frac{\partial Q}{\partial t} + J(\psi, Q) = 0.$$

1.3. Buoyancy and Stratification

Next, we build elementary exact solutions of the Boussinesq equations in (1.6) with $\partial\bar\rho/\partial x_3$ constant that illustrate the important effect of gravity. These special solutions involve only vertical motion and density changes that both vary only in the horizontal; i.e., we seek exact solutions of (1.8) with the form

$$(1.21) \qquad \vec v_H = 0\,, \quad p = 0\,, \quad w = w(\vec x_H, t)\,, \quad \rho = \rho(\vec x_H, t)\,.$$

Inserting this ansatz into (1.8) shows that (w, ρ) satisfy the simple equations

$$(1.22) \qquad \frac{\partial w}{\partial t} = -\frac{g}{\rho_b}\rho\,, \quad \frac{\partial \rho}{\partial t} = -\left(\frac{d\bar\rho}{dx_3}\right)w\,.$$

These equations are independent of position and are simple constant-coefficient ODEs for each $\vec x_H$. In fact, both w and ρ satisfy the second-order equations

$$(1.23) \qquad w_{tt} = -\omega_b^2 w\,, \quad \rho_{tt} = -\omega_b^2 \rho\,,$$

with

$$(1.24) \qquad \omega_b^2 = \left(-\frac{g}{\rho_b}\frac{d\bar\rho}{dx_3}\right).$$

When do these solutions exhibit pure stable oscillation? Clearly, when ω_b^2 is a positive quantity, i.e.,

$$(1.25) \qquad \omega_b = \left(-\frac{g}{\rho_b}\frac{d\bar\rho}{dx_3}\right)^{1/2} = N > 0$$

is a real frequency, called the *buoyancy* or *Brunt-Väisälä frequency*, N then w and ρ have solutions of the form

$$(1.26) \qquad w = A\cos\omega_b t + B\sin\omega_b t\,, \quad \rho = \tilde A\cos\omega_b t + \tilde B\sin\omega_b t\,,$$

and vertical motions simply are restored to their original locations on average and oscillate about them. From (1.25) we see that ω_b from (1.25) is a real frequency precisely when we have

$$(1.27) \qquad \frac{d\bar\rho}{dx_3} < 0 \qquad \begin{array}{l}\text{the density decreases with height and}\\ \text{heavier fluid is below lighter fluid.}\end{array}$$

This is the situation satisfied typically in both the atmosphere and the ocean once one moves a sufficient distance from the atmosphere/ocean/land boundary layer. For obvious reasons, it is called *stable stratification*.

On the other hand,

$$(1.28) \qquad \text{for } \frac{d\bar\rho}{dx_3} > 0 \qquad \begin{array}{l}\text{the density increases with height and}\\ \text{heavier fluid is above lighter fluid.}\end{array}$$

This is the situation of *unstable* stratification, and the exact solutions in (1.23) have the form

$$(1.29) \qquad w = Ae^{|\omega_c|t} + Be^{-|\omega_c|t}\,, \quad \rho = \tilde A e^{|\omega_c|t} + \tilde B e^{-|\omega_c|t}\,.$$

The exponential growth reflected in these exact solutions is a very simple manifestation of unstable stratification.

1.4. Jet Flows with Rotation and Stratification

A simple fluid velocity

$$(1.30) \qquad \vec{v} = \begin{pmatrix} u(y,z) \\ 0 \\ 0 \end{pmatrix}$$

looks like a simple jet flow. Such flows are among the simplest basic model solutions for the atmosphere and ocean.

First, let's only consider the effect of stratification, i.e., set $f \equiv 0$ in (1.8) so that rotation is ignored. Then the jet flows in (1.30) are automatically exact solutions of (1.8) provided that

$$(1.31) \qquad \text{the pressure } p \text{ vanishes identically, i.e., } p \equiv 0.$$

Thus, these solutions have hydrostatic balance in the vertical.

What happens to these solutions when we include the effects of rotation? Below, we assume $f(y) \neq 0$ in the region of interest so, in particular, a discussion of flows near the equator is eliminated. The effects of rotation are dramatic here and the pressure is necessarily a nontrivial function of (y, z). In fact, it is best to let the pressure, $P(y, z)$, define the velocity in (1.30) for these exact solutions. Looking back at (1.8) we have an exact solution of the form in (1.30) only if

$$(1.32) \qquad u(y,z) = \frac{-\frac{\partial P}{\partial y}(y,z)}{f(y)}, \qquad \rho(y,z) = -\frac{\rho_b}{g}\frac{\partial P}{\partial z}(y,z).$$

The first condition in (1.32) is called *geostrophic balance* where rotation induces a balance between horizontal pressure gradients and the fluid velocity. Such effects are very important at large scales in rotating geophysical flows.

1.5. From Vertical Stratification to Shallow Water

We will be considering the shallow water equations, both at midlatitudes and near the equator in subsequent lectures. We should say how these are related to the real governing equations that contain vertical stratification. We begin with the rotating Boussinesq equations. In fact, the density properties in the atmosphere change so dramatically with height that the Boussinesq approximation is not used, but this distinction is not crucial for our purposes since the more general equations have similar behavior.

Rotating Boussinesq Equations.

$$(1.33) \qquad \begin{aligned} \frac{D}{Dt}\vec{v}_H + f(y)\vec{v}_H^{\perp} &= -\nabla_H p, \\ \frac{D}{Dt}w &= -\frac{\partial p}{\partial z} - g\rho, \\ \frac{D}{Dt}\rho + \frac{\partial \bar{\rho}}{\partial z}w &= 0, \\ \text{div}_H\, \vec{v}_H + \frac{\partial w}{\partial z} &= 0. \end{aligned}$$

The mean density $\bar{\rho}$ is a known function of z. Typically, the assumption is immediately made that because the atmosphere is very thin, the vertical waves equilibrate quickly and the term $\frac{Dw}{Dt}$ can be dropped. (It would be interesting to understand the accuracy of this assumption in a PDE context.) In the so-called *primitive equations*, the second equation above is replaced by the hydrostatic balance condition

$$(1.34) \qquad -\frac{\partial p}{\partial z} = g\rho.$$

Mathematically, the primitive equations are rather different from the stratified equations. They essentially evolve the pressure in time with the vertical velocity determined by the incompressibility condition. It is not clear that the primitive equations are always well posed. It might be that viscosity effects in numerical weather simulations mask ill-posedness. Nevertheless, these are the standard equations utilized.

We introduce $\theta = -g\rho$ and the buoyancy frequency $N(z) = (-g\bar{\rho}_z)^{1/2}$ and linearize.

Linearized Primitive Equations.

$$(1.35a) \qquad \frac{\partial \vec{v}_H}{\partial t} + f(y)\vec{v}_H^{\perp} = -\nabla_H p,$$

$$(1.35b) \qquad \frac{\partial p}{\partial z} = \theta,$$

$$(1.35c) \qquad \frac{\partial \theta}{\partial t} = -N^2 w,$$

$$(1.35d) \qquad \mathrm{div}_H \vec{v} + \frac{\partial w}{\partial z} = 0.$$

We claim that solutions can be written as infinite combinations of eigenmodes of the shallow water equations. We assume the rigid lid boundary conditions,

$$w|_{z=0,H} = 0.$$

To see the connection with the shallow water equations, we separate variables. Let

$$\begin{pmatrix} u \\ v \\ p \end{pmatrix} = \begin{pmatrix} U(x,y,t) \\ V(x,y,t) \\ P(x,y,t) \end{pmatrix} G(z).$$

Immediately we use (1.35b) and (1.35c) to find

$$\theta = P(x,y,t)G'(z), \qquad w = -N^{-2}\frac{\partial P}{\partial t}G'.$$

We continue by inserting our ansatz into (1.35d)

$$(1.36) \qquad \frac{\partial U}{\partial x} + \frac{\partial V}{\partial y} = \frac{\partial P}{\partial t}N^{-2}G^{-1}G''.$$

The separation condition is found by noting that the left-hand side is a function of (x,y,t) only, and so the right-hand side must also be independent of z; that is, for m constant,

$$(1.37) \qquad G'' = -m^2 N^2 G,$$

while the rigid lid boundary conditions yield $G'|_{z=0,H} = 0$. We rewrite (1.36) along with (1.35a),

(1.38)
$$\frac{\partial U}{\partial t} - f(y)V = -\frac{\partial P}{\partial x}, \quad \frac{\partial V}{\partial t} + f(y)U = -\frac{\partial P}{\partial y},$$
$$\frac{\partial P}{\partial t} + \frac{1}{m^2}\left(\frac{\partial U}{\partial x} + \frac{\partial V}{\partial y}\right) = 0.$$

For $m \neq 0$, the equations in (1.38) are the linearized rotating shallow water equations. Thus, we have derived the rotating shallow water equations where the equivalent height is given by $H_m = 1/m^2$. For instance, with a constant buoyancy frequency $N = 1$ and boundary conditions applied at $z = 0$ and $z = H$, we find

$$G(z) = \cos\left(\frac{\pi k z}{H}\right) \quad \text{for } k = 1, 2, 3, \ldots$$

and the equivalent depth is

$$H_k = \frac{H^2}{\pi^2 k^2}$$

so that the effective gravity wave speed $c = \sqrt{gH_k}$ is proportional to inverse wavenumber $|k|^{-1}$. Higher wavenumbers are slower.

What about the special mode for (1.37) and (1.38) with $G \equiv 1$? Clearly the solutions of (1.38) do not involve the depth at all, and in fact, the equations in (1.38) are a linearized version of the barotropic equations discussed earlier in (1.10)–(1.20).

CHAPTER 2

Some Remarkable Features of Stratified Flow

In this chapter, we will see some remarkable properties of stably and unstably stratified flows including energy and vorticity propagation principles. We will examine the local structure of a divergence free velocity field, and build a large class of solutions to the Boussinesq equations that reflect this local analysis. After that we consider specific examples of these solutions and nonlinear plane waves to

(1) illustrate some key properties of the local structure,
(2) provide models for observed atmospheric and oceanic features, and
(3) provide an arena for studying stability properties of two-dimensional solutions to three-dimensional perturbations and of a stably stratified fluid to the introduction of vorticity.

We begin with a discussion of the energetics of stratified fluid.

2.1. Energy Principle

Recall the Boussinesq equations for a rotating stratified fluid:

$$(2.1) \quad \frac{D\vec{v}}{Dt} + f(y)\vec{v}_H^{\perp} = -\nabla p - \frac{g\rho}{\rho_b}\vec{e}_3, \quad \operatorname{div}\vec{v} = 0, \quad \frac{D\rho}{Dt} + \left(\frac{d\bar{\rho}}{dx_3}\right)v_3 = 0.$$

The full density $\tilde{\rho}$ consists of perturbations ρ about a density $\bar{\rho}$ in hydrostatic balance, which itself creates only small deviations from the baseline constant ρ_b,

$$\tilde{\rho}(\vec{x}, t) = \rho_b + \bar{\rho}(x_3) + \rho(\vec{x}, t).$$

For the following, we make the usual assumption valid for local considerations that $d\bar{\rho}/dx_3$ is constant. We look for local kinetic energy of the fluid of the standard form,

$$(2.2) \quad \text{K.E.} = \frac{1}{2}\rho_b\vec{v}\cdot\vec{v}$$

and compute its advective derivative

$$(2.3) \quad \frac{D}{Dt}\left(\frac{1}{2}\rho_b\vec{v}\cdot\vec{v}\right) = \rho_b\vec{v}\cdot\frac{D\vec{v}}{Dt} = -\rho_b\vec{v}\cdot\nabla p - g\rho v_3 + f(y)\vec{v}_H^{\perp}\cdot\vec{v}$$

$$= \rho_b \operatorname{div}(-\vec{v}p) - g\rho v_3.$$

The incompressibility of $\vec{v}$ was used in the last step. To obtain an energy principle we wish to cancel the part of (2.3) that is not a perfect divergence, and noting the

9

ρv_3 that appears, we try the computation

$$\frac{D}{Dt}\left(\frac{A}{2}\rho^2\right) = A\rho\frac{D\rho}{Dt} = -A\rho\left(\frac{d\bar{\rho}}{dx_3}\right)v_3\,.$$

The choice of constant A is made to cancel the last term of (2.3):

$$(2.4) \qquad\qquad A = \left(-\frac{d\bar{\rho}}{dx_3}\right)^{-1}g\,.$$

Then we have the following result for the infinitesimal propagation of energy:

$$(2.5) \qquad\qquad \frac{D}{Dt}\left(\frac{1}{2}\rho_b\vec{v}\cdot\vec{v} + \frac{1}{2}A\rho^2\right) = -\operatorname{div}(\rho_b\vec{v}p)\,.$$

It is an easy exercise to check that the effect of rotation does not change the energy considerations derived here, since the Coriolis force acts perpendicular to the velocity. We'll see by the following general considerations that our local result, combined with certain boundary conditions, yields global energy conservation.

Energy Conservation. Let the quantity E by advected by an incompressible velocity $\vec{v}$ according to

$$\frac{DE}{Dt} = -\operatorname{div}(\vec{v}F)\,.$$

Since $\operatorname{div}\vec{v} = 0$, we can write the advective derivative

$$\frac{DE}{Dt} = \frac{\partial E}{\partial t} + \vec{v}\cdot\nabla E = \frac{\partial E}{\partial t} + \operatorname{div}(\vec{v}E)\,;$$

therefore,

$$\frac{\partial E}{\partial t} = -\operatorname{div}(\vec{v}(E+F))\,.$$

We are now in a position to convert this to a global result by integrating in both space and time

$$\int_\Omega E(t_1,\vec{x})d\vec{x} - \int_\Omega E(t_2,\vec{x})d\vec{x} = \int_{t_1}^{t_2}\int_\Omega \operatorname{div}\left(\vec{v}(\vec{x})(E(\tau,\vec{x})+F(\tau,\vec{x}))\right)d\vec{x}\,d\tau\,.$$

According to Gauss' law, we can convert the last space integral into an surface integral at the boundary

$$\int_\Omega E(t_1,\vec{x})d\vec{x} - \int_\Omega E(t_2,\vec{x})d\vec{x} = \int_{t_1}^{t_2}\int_{\partial\Omega} \vec{v}(\vec{x})\cdot\hat{n}(\vec{x})(E(\tau,\vec{x})+F(\tau,\vec{x}))d\vec{x}\,d\tau$$

where $\hat{n}$ is the unit normal to the boundary surface $\partial\Omega$. In words, the right-hand side is the time integral of the flux of the quantity $E + F$ into or out of our region. We might be interested in the limit that the boundary is located far away, where the fluid velocity is negligible, or we might concern ourselves with periodic boundary conditions where the boundary terms on opposite ends cancel out. These are two cases where we have no fluxes bringing energy into our system. We can conclude

$$(2.6) \qquad\qquad \frac{d}{dt}\int_\Omega E(t,\vec{x})d\vec{x} = 0\,.$$

With one of these boundary conditions, and in the absence of forcing and dissipation, we find a conservation law for a global quantity, the integral of the local energy.

Returning to stratified Boussinesq flows, the local energy is given by

$$(2.7) \qquad E = \frac{1}{2}\rho_b\vec{v}\cdot\vec{v} + \frac{1}{2}A\rho^2.$$

As mentioned before, we associate the first term with the kinetic energy. The second term is potential energy. Kinetic energy can be converted into potential energy, and vice versa, so it is only their sum that is conserved. To interpret potential energy physically, begin with a stratified fluid $\tilde{\rho}^0 = \bar{\rho}(x_3)$. To introduce a small perturbation ρ at height z is equivalent to transporting fluid along the density gradient (from either above or below depending on the sign of ρ) from height h where $\tilde{\rho}^0(z) + \rho(z) = \tilde{\rho}^0(z + h)$. For small h we Taylor-expand the right-hand side of this equation and find the correspondence

$$\rho(z) \approx h\frac{d\bar{\rho}}{dx_3}.$$

The energy change due to gravity to work against the gradient is

$$\int_0^h g\tilde{\rho}^0(x_3)dx_3 = g\frac{d\bar{\rho}}{dx_3}\frac{h^2}{2}.$$

Then, making the association for h, we get the form of local potential energy $A\rho^2/2$ with A given by (2.4).

The tilt of the density slope, the sign of $d\bar{\rho}/dx_3$, is critical in our interpretation of the conservation of energy. A negative value (light fluid on top of heavy) gives an energy that is positive definite. The kinetic energy part is therefore bounded for all time by the total energy. Recall the gravity wave solutions from (1.3) of the previous chapter; with the energy principle, we could have predicted that these solutions could not have blown up in this stably stratified case. In the opposite case, heavy fluid above light, the possibility exists that both the kinetic and potential energy terms could grow while the total energy stays constant, and this is just what we saw in the unstably stratified example from Section 1.3 where the wave solutions grew exponentially. For gravity waves, we can relate the constant A with the Brunt-Väisälä frequency N by

$$A = \frac{g^2}{\rho_b N^2}.$$

In many applications there is strong stratification so that $N \gg 1$, and these fast oscillations contain little density variation.

2.2. Vorticity in Stratified Fluids and Exact Solutions Motivated by Local Analysis

What do fluid flows look like locally? To answer this question, we Taylor-expand a general incompressible flow about some point $\vec{x}_0$,

$$(2.8) \qquad \vec{v}(\vec{x}, t) = \vec{v}(\vec{x}_0, t) + (\nabla\vec{v})\big|_{(\vec{x}_0, t)}(\vec{x} - \vec{x}_0) + O(|\vec{x} - \vec{x}_0|^2)$$

where the tensor $\nabla \vec{v}$ is the 3×3 matrix with (i, j) component given by $\partial v_i / \partial x_j$. The local structure for the scalar density is also linear:

$$(2.9) \qquad \tilde{\rho}(\vec{x}, t) = \rho_b + (\nabla \tilde{\rho})\big|_{(\vec{x}_0, t)} (\vec{x} - \vec{x}_0) + O(|\vec{x} - \vec{x}_0|^2).$$

To understand $\nabla \vec{v}$, we decompose it into symmetric and antisymmetric parts. Any antisymmetric matrix multiplying a vector acts as a cross product. Here,

$$\nabla \vec{v}\big|_{(\vec{x}_0, t)} = \left(\frac{^{\mathsf{T}}\nabla \vec{v} + \nabla \vec{v}}{2}\right) \vec{h} + \left(\frac{^{\mathsf{T}}\nabla \vec{v} - \nabla \vec{v}}{2}\right) \vec{h}$$

$$(2.10)$$

$$= \mathcal{D}(\vec{x}_0, t)\vec{h} + \frac{1}{2}\vec{\omega}(\vec{x}_0, t) \times \vec{h}.$$

The symmetric part of $\nabla \vec{v}$ is called the *deformation matrix* and labeled $\mathcal{D}$. It has the property that its trace equals the divergence of $\vec{v}$,

$$\operatorname{tr} \mathcal{D}(\vec{x}_0, t) = 0,$$

and $\vec{\omega}$, called the *vorticity*, is the curl of the vector field $\vec{v}$. The physical interpretation is that locally, every incompressible velocity field looks like a combination of translation, stretching, and rotation. In fact, the translation part is trivial and can be removed by a Galilean transformation, so we will set $\vec{v}(\vec{x}_0, t) = 0$. For the moment we ignore the effects of rotation.

We now seek to use this local form to write exact solutions to the Boussinesq equations. We begin by writing down the Boussinesq equations for the gradient of velocity. I will write this component-wise first and then move into tensor notation

$$\frac{\partial}{\partial t}\frac{\partial v_i}{\partial x_j} + \sum_k v_k \frac{\partial}{\partial x_k}\frac{\partial v_i}{\partial v_j} + \sum_k \frac{\partial v_i}{\partial x_k}\frac{\partial v_k}{\partial x_j} = -\frac{\partial^2 p}{\partial x_i \partial x_j} - \frac{g}{\rho_b}\frac{\partial \rho}{\partial x_j}\delta_{i3}.$$

In terms of the matrix $V = \nabla \vec{v}$, this is just

$$(2.11) \qquad \frac{DV}{Dt} + V^2 = -\widehat{P} - \frac{g}{\rho_b}(\vec{e}_3 \otimes \nabla \rho).$$

The new notation in (2.11) is defined in component form above; $\widehat{P}$ is the 3×3 Hessian matrix of pressure, i.e., $\widehat{P} = (\frac{\partial^2 p}{\partial x_i \partial x_j})$. The equation for the gradient of density is

$$(2.12) \qquad \frac{D}{Dt}\nabla \rho + {}^{\mathsf{T}}V \nabla \rho + \left(\frac{d\bar{\rho}}{dx_3}\right)\nabla v_3 = 0.$$

We now examine the symmetric and antisymmetric parts of V

$$(2.13) \qquad \mathcal{D} = \frac{^{\mathsf{T}}\nabla \vec{v} + \nabla \vec{v}}{2}, \quad \Omega = \frac{\nabla \vec{v} - {}^{\mathsf{T}}\nabla \vec{v}}{2}.$$

Using that $V^2 = \mathcal{D}^2 + \Omega^2 + \mathcal{D}\Omega + \Omega\mathcal{D}$ where the first two terms are symmetric while the latter two form the antisymmetric part, we can break up the equation for the velocity gradient (2.11). The symmetric part is given by

$$(2.14) \qquad \frac{D}{Dt}\mathcal{D} + \mathcal{D}^2 + \Omega^2 = -\widehat{P} - \frac{g}{\rho_b}\left(\frac{\vec{e}_3 \otimes \nabla \rho + \nabla \rho \otimes \vec{e}_3}{2}\right).$$

The more complicated antisymmetric part will be stated as a proposition that uses a lemma.

PROPOSITION 2.1 *The antisymmetric part of (2.11) can be written via an equation for $\vec{\omega} = \text{curl } \vec{v}$ that acts for the antisymmetric part of the velocity gradient through the equation $\Omega \vec{h} = \frac{1}{2}\vec{\omega} \times \vec{h}$. The result is*

$$(2.15) \qquad \frac{D\vec{\omega}}{Dt} = \vec{\omega} \cdot \nabla \vec{v} + \frac{g}{\rho_b}\begin{pmatrix} \nabla_H^\perp \rho \\ 0 \end{pmatrix}$$

where $\nabla_H^\perp \rho = (-\partial\rho/\partial x_2, \partial\rho/\partial x_1)$.

The first term in (2.15) is the *vortex stretching term*, while the second one is called the *baroclinic production term*.

LEMMA 2.2 *For Ω the antisymmetric part of $\nabla \vec{v}$ and $\Omega \vec{h} = \frac{1}{2}\vec{\omega} \times \vec{h}$, then $\vec{\omega} \cdot \nabla \vec{v} = \vec{\omega} \cdot {}^T\nabla \vec{v}$.*

PROOF OF LEMMA 2.2: For any $\vec{h}$,

$$0 = \frac{1}{2}\vec{\omega} \cdot (\vec{\omega} \times \vec{h}) = \vec{\omega} \cdot \Omega \vec{h} = \frac{1}{2}\vec{\omega} \cdot (\nabla \vec{v} - {}^T\nabla \vec{v})\vec{h} \,.$$

Since $\vec{h}$ was arbitrary, this gives us the lemma. $\qquad\square$

PROOF OF PROPOSITION 2.1: The antisymmetric part of (2.11) is

$$(2.16) \qquad \frac{D}{Dt}\Omega + \mathcal{D}\Omega + \Omega\mathcal{D} = -\frac{g}{2\rho_b}\begin{pmatrix} 0 & 0 & -\frac{\partial\rho}{\partial x_1} \\ 0 & 0 & -\frac{\partial\rho}{\partial x_2} \\ \frac{\partial\rho}{\partial x_1} & \frac{\partial\rho}{\partial x_2} & 0 \end{pmatrix}.$$

We multiply this equation by an arbitrary vector $\vec{h}$

$$\frac{1}{2}\frac{D}{Dt}\vec{\omega} \times \vec{h} + (\mathcal{D}\Omega + \Omega\mathcal{D})\vec{h} = -\frac{g}{2\rho_b}\begin{pmatrix} -\frac{\partial\rho}{\partial x_1}h_3 \\ -\frac{\partial\rho}{\partial x_2}h_3 \\ \frac{\partial\rho}{\partial x_1}h_1 + \frac{\partial\rho}{\partial x_2}h_2 \end{pmatrix} = \frac{g}{2\rho_b}\begin{pmatrix} -\frac{\partial\rho}{\partial x_2} \\ \frac{\partial\rho}{\partial x_1} \\ 0 \end{pmatrix} \times \vec{h} \,.$$

Then for

$$\Omega = \frac{1}{2}\begin{pmatrix} 0 & -\omega_3 & \omega_2 \\ \omega_3 & 0 & -\omega_1 \\ -\omega_2 & \omega_1 & 0 \end{pmatrix} \quad \text{and for} \quad \mathcal{D} = \begin{pmatrix} d_{11} & d_{12} & d_{13} \\ d_{12} & d_{22} & d_{23} \\ d_{13} & d_{23} & d_{33} \end{pmatrix}$$

with $d_{11} + d_{22} + d_{33} = 0$, it is left to the reader to show that

$$\mathcal{D}\Omega + \Omega\mathcal{D} = \frac{1}{2}\begin{pmatrix} 0 & -C_{12} & C_{13} \\ C_{12} & 0 & -C_{23} \\ -C_{13} & C_{23} & 0 \end{pmatrix}$$

or

$$(\mathcal{D}\Omega + \Omega\mathcal{D})\vec{h} = \frac{1}{2}\vec{C} \times \vec{h}$$

where

$$\vec{C} = \begin{pmatrix} C_{23} \\ C_{13} \\ C_{12} \end{pmatrix} = \begin{pmatrix} -\omega_1 d_{11} - \omega_2 d_{12} - \omega_3 d_{13} \\ -\omega_1 d_{12} - \omega_2 d_{22} - \omega_3 d_{23} \\ -\omega_1 d_{13} - \omega_2 d_{23} - \omega_3 d_{33} \end{pmatrix} = -\vec{\omega} \cdot \mathcal{D} \,.$$

Summarizing, we have

$$\frac{1}{2} \frac{D}{Dt} \vec{\omega} \times \vec{h} - \frac{1}{2} \vec{\omega} \cdot \mathcal{D} \times \vec{h} = \frac{g}{2\rho_b} \begin{pmatrix} \nabla^\perp \rho \\ 0 \end{pmatrix} \times \vec{h} \,.$$

Finally, it remains to note from Lemma 2.2 that $\vec{\omega} \cdot \Omega = 0$ implies that $\vec{\omega} \cdot \mathcal{D} = \vec{\omega} \cdot \nabla \vec{v}$. Since $\vec{h}$ was arbitrary, we can conclude (2.15). $\qquad\square$

Equation (2.15) tells us what happens to infinitesimal vorticity elements. They are advected with the fluid, amplified by interaction with velocity gradients (the so-called *tornado mechanism*), and amplified by particular gradients of density. We are now in position to prove the remarkable identity that the component of vorticity along the density gradient cannot be generated or destroyed, but is merely advected with the fluid.

THEOREM 2.3 (Ertel's Theorem) *In the absence of forcing and dissipation,*

(2.17)
$$\frac{D}{Dt}(\vec{\omega} \cdot \nabla \tilde{\rho}) = 0 \,.$$

PROOF OF ERTEL'S THEOREM: Since $\tilde{\rho}$ satisfies $\frac{D}{Dt}\tilde{\rho} = 0$, as in (2.12) we have the identity

(2.18)
$$\frac{D\nabla}{Dt}\tilde{\rho} = (^\mathsf{T}\nabla\vec{v})\nabla\tilde{\rho}$$

together with (2.15),

$$\frac{D\vec{\omega}}{Dt} = (\nabla\vec{v})\vec{\omega} + \frac{g}{\rho_b} \begin{pmatrix} \nabla_H^\perp \tilde{\rho} \\ 0 \end{pmatrix} \,.$$

Note that for $\tilde{\rho} = \rho_b + \bar{\rho}(x_3) + \rho$, $\nabla_H^\perp \tilde{\rho} = \nabla_H^\mathsf{T}\rho$. Thus,

$$\frac{D}{Dt}(\vec{\omega} \cdot \nabla \tilde{\rho}) = \frac{D\vec{\omega}}{Dt} \cdot \nabla\tilde{\rho} + \vec{\omega} \cdot \frac{D\nabla}{Dt}\tilde{\rho}$$

$$= \left\{ ((\nabla\vec{v})\vec{\omega}) \cdot \nabla\tilde{\rho} - \vec{\omega} \cdot ((^\mathsf{T}\nabla\vec{v})\nabla\tilde{\rho}) \right\} + \left\{ \frac{g}{\rho_b} \nabla\tilde{\rho} \cdot \begin{pmatrix} \nabla_H^\perp \tilde{\rho} \\ 0 \end{pmatrix} \right\} \,.$$

Both of the terms in braces are separately zero, and Ertel's theorem is proved. $\quad\square$

We claimed at the beginning of this chapter that we would find exact solutions to the Boussinesq equations, and the following theorem provides just that. Looking ahead to the form of these solutions, (2.19), we see that they are linear functions of $\vec{x}$, with coefficients that are functions of time only. By this clever choice, the PDEs of the Boussinesq equations reduce to ODEs. Physically, the behavior far from the origin where the velocities and fluid densities are proportional to $|\vec{x}|$ as $|\vec{x}| \to \infty$ is not meaningful. But locally these solutions are very interesting because they allow us to take the local features shown by Taylor's theorem and see how they evolve and interact in time. They provide an arena for us to build intuition about this structure in a rigorous setting.

Note further that the form of the density is perturbations about a constant state. The built-in linear term is removed: $\bar{\rho} = 0$. We will analyze stable versus unstable stratification directly in the perturbation term $\rho = \vec{b} \cdot \vec{x}$.

We have the following:

THEOREM 2.4 *There are special solutions to the Boussinesq equations* (2.1) *of the form*

$$(2.19) \quad \vec{v}(\vec{x}, t) = \mathcal{D}(t)\vec{x} + \frac{1}{2}\vec{\omega}(t) \times \vec{x}, \quad \tilde{\rho} = \rho_b + \vec{b}(t) \cdot \vec{x}, \quad p = \frac{1}{2}\widehat{P}(t)\vec{x} \cdot \vec{x},$$

where $\mathcal{D}$ is an arbitrary 3×3, symmetric matrix with zero trace, when $\vec{\omega}(t) = $ curl $\vec{v}$ and $\vec{b}$ satisfy the ODEs

$$(2.20) \qquad \frac{d\vec{\omega}}{dt} = \mathcal{D}(t)\vec{\omega}(t) + \frac{g}{\rho_b} \begin{pmatrix} -b_2(t) \\ b_1(t) \\ 0 \end{pmatrix},$$

$$(2.21) \qquad \frac{d\vec{b}}{dt} = -\mathcal{D}(t)\vec{b}(t) + \frac{1}{2}\vec{\omega}(t) \times \vec{b}(t),$$

and the matrix $\widehat{P}(t)$ is given by

$$(2.22) \qquad -\widehat{P} = \frac{d\mathcal{D}}{dt} + \mathcal{D}^2 + \Omega^2 + \frac{g}{2\rho_b}(\vec{e}_3 \otimes \vec{b} + \vec{b} \otimes \vec{e}_3).$$

Finally, Ω is defined by

$$(2.23) \qquad \Omega\vec{h} = \frac{1}{2}\vec{\omega} \times \vec{h}.$$

Equations (2.20) *and* (2.21) *are simply equations* (2.15) *and* (2.18) *for the special solutions with the form in* (2.19).

The interested reader can compare these exact solutions with stratification discussed here and below to those in the book by Majda and Bertozzi [19, chap. 1].

PROOF: We need to show that $\vec{v}$, p, and ρ satisfy the Boussinesq equations (2.1). The condition div $\vec{v} = 0$ is automatically satisfied by having a traceless $\mathcal{D}$. For the momentum equation, note that we have defined $\vec{v} = V\vec{x}$ where $V = \mathcal{D} + \Omega$ is only a function of time, not space. The reader may confirm that for this form, the advection term is given by

$$(\vec{v} \cdot \nabla)\vec{v} = V^2\vec{x}$$

so that

$$\frac{D}{Dt}\vec{v} = \frac{d}{dt}\mathcal{D}\vec{x} + \frac{d}{dt}\Omega\vec{x} + V^2\vec{x}.$$

From the proof of Proposition 2.1, which allowed an arbitrary traceless $\mathcal{D}$, we know that with (2.23), the equation for $\vec{\omega}$ (2.20) is equivalent to the equation for Ω

(2.16), with the advective derivative returning to the normal time derivative:

$$\frac{d}{dt}\Omega + \mathcal{D}\Omega + \Omega\mathcal{D} = -\frac{g}{2\rho_b}\begin{pmatrix} 0 & 0 & -\frac{\partial\rho}{\partial x_1} \\ 0 & 0 & -\frac{\partial\rho}{\partial x_2} \\ \frac{\partial\rho}{\partial x_1} & \frac{\partial\rho}{\partial x_2} & 0 \end{pmatrix}.$$

Inserting this and substituting for $d\mathcal{D}/dt$ by using (2.22), we find

$$\frac{D}{Dt}\vec{v} = -\widehat{P}\vec{x} - \mathcal{D}^2\vec{x} - \Omega^2\vec{x} - \frac{g}{\rho_b}\begin{pmatrix} 0 & 0 & b_1/2 \\ 0 & 0 & b_2/2 \\ b_1/2 & b_2/2 & b_3 \end{pmatrix}\vec{x}$$

$$- (\mathcal{D}\Omega + \Omega\mathcal{D})\vec{x} + \frac{g}{2\rho_b}\begin{pmatrix} 0 & 0 & b_1/2 \\ 0 & 0 & b_2/2 \\ -b_1/2 & -b_2/2 & 0 \end{pmatrix}\vec{x} + V^2\vec{x}.$$

The $(V^2 - \mathcal{D}^2 - \Omega^2 - \mathcal{D}\Omega - \Omega\mathcal{D})\vec{x}$ goes away, and we are left with

$$\frac{D}{Dt}\vec{v} = -\widehat{P}\vec{x} - \frac{g}{\rho_b}\begin{pmatrix} 0 \\ 0 \\ \vec{b}\cdot\vec{x} \end{pmatrix}.$$

Using the definition of pressure in (2.19), where $\nabla\rho = \widehat{P}\vec{x}$, we arrive at

$$\frac{D}{Dt}\vec{v} = -\nabla p - \frac{g\rho}{\rho_b}\vec{e}_3.$$

It remains to verify the Boussinesq equation for density

$$\frac{D}{Dt}\rho = \frac{D}{Dt}(\vec{b}(t)\cdot\vec{x})$$

$$= \frac{db}{dt}\cdot\vec{x} + \vec{v}\cdot\vec{b}$$

$$= -(\mathcal{D}\vec{b})\cdot\vec{x} + \frac{1}{2}(\omega\times\vec{b})\cdot\vec{x} + (\mathcal{D}\vec{x})\cdot\vec{b} + \frac{1}{2}(\omega\times\vec{x})\cdot\vec{b} = 0.$$

Recall that $\bar{\rho} = 0$ for these solutions. $\square$

2.3. Use of Theorem 2.4: Exact Two-Dimensional Solutions

EXAMPLE 2.1. This solution is of the type mentioned in the last chapter, confined to the horizontal plane. We introduce it to see the stretching and rotation structure of a velocity field. It also serves as a simple test problem to study the stability of two-dimensional flows to three-dimensional perturbations; however, that will not be part of this chapter.

As mentioned, the flow is confined to the horizontal, so the vorticity $\vec{\omega} = (0, 0, \overline{\omega})$. The vector $\vec{b}$ is also in the x_3-direction, $\vec{b} = (0, 0, b_3)$, so the density is hydrostatically balanced. The deformation matrix for this example will be in the horizontal:

$$\mathcal{D} = \begin{pmatrix} \lambda & 0 & 0 \\ 0 & -\lambda & 0 \\ 0 & 0 & 0 \end{pmatrix}.$$

The reader may verify that this time-independent solution obeys the conditions of the theorem. The velocity is then given by

$$(2.24) \qquad \vec{v} = \begin{pmatrix} \tilde{v}_H \\ 0 \end{pmatrix}, \quad \vec{v}_H = \begin{pmatrix} \lambda & \frac{1}{2}\overline{\omega} \\ -\frac{1}{2}\overline{\omega} & -\lambda \end{pmatrix} \begin{pmatrix} x_1 \\ x_2 \end{pmatrix}.$$

The velocity field is a combination of a pure rotation, like a turntable, a stretching in the x_1-direction, and a compression in the x_2-direction (Majda and Bertozzi [19]).

EXAMPLE 2.2. This solution has flow pointing in one direction but depending in magnitude on another direction and is called a *shear flow*. It provides a local model for the atmospheric jet stream. One might also consider the nonlinear stability of this solution,

$$\vec{b} = \begin{pmatrix} 0 \\ 0 \\ b_3 \end{pmatrix}, \quad \vec{\omega} = \begin{pmatrix} -a_2 \\ a_1 \\ 0 \end{pmatrix}, \quad \mathcal{D} = \frac{1}{2} \begin{pmatrix} 0 & 0 & a_1 \\ 0 & 0 & a_2 \\ a_1 & a_2 & 0 \end{pmatrix}.$$

The velocity is then given by

$$(2.25) \qquad \vec{v} = \begin{pmatrix} a_1 x_3 \\ a_2 x_3 \\ 0 \end{pmatrix}.$$

These two solutions were cooked up to be time independent, but more general solutions involve the interaction of stretching and vorticity, as well as dissipation, in striking ways. The interested reader is directed to Andrew Majda's 1986 paper [17], which treats the subject in the context of the Navier-Stokes equations, or the new book by Majda and Bertozzi [19].

EXAMPLE 2.3. Can a vortex make a stably stratified fluid locally unstable? We are motivated to ask this question by the observation that vorticity with a projection in the (x_1, x_2)-plane in a stably stratified fluid can advect heavier fluid from below and lighter fluid from above to create the unstable situation of a positive x_3 density gradient.

We seek to understand this issue with a simple model that has no stretching ($\mathcal{D} = 0$). Vorticity is in the x_2-direction, $\vec{\omega} = (0, \overline{\omega}(t), 0)$, and density slope initially in the x_3-direction. We see that the density slope and velocity will stay in the (x_1, x_3)-plane

$$\vec{b} = \begin{pmatrix} b_1(t) \\ 0 \\ b_3(t) \end{pmatrix}, \quad \vec{b}(t=0) = \begin{pmatrix} 0 \\ 0 \\ -B_0 \end{pmatrix}, \quad B > 0,$$

$$\vec{v} = \begin{pmatrix} 0 & 0 & \frac{1}{2}\overline{\omega}(t) \\ 0 & 0 & 0 \\ -\frac{1}{2}\overline{\omega}(t) & 0 & 0 \end{pmatrix} \begin{pmatrix} x_1 \\ x_2 \\ x_3 \end{pmatrix}.$$

The ODEs solved by the vorticity and pressure are given by Theorem 2.4,

$$(2.26a) \qquad \frac{d\overline{\omega}}{dt} = \frac{g}{\rho_b} b_1(t),$$

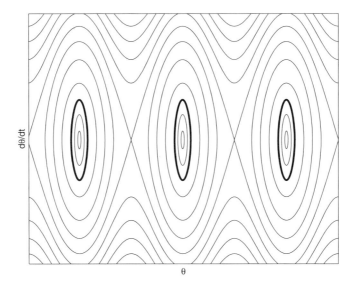

FIGURE 2.1. The phase portrait for the pendulum equation. The critical orbits with $\omega_0^2 = 4N^2$ is denoted with a dark line. For motions with $\omega_0^2 < 4N^2$ inside this orbit, there are no overturnings. From *Journal of Fluid Mechanics*, vol. 376 (1998), p. 326, fig. 1. Reprinted with the permission of Cambridge University Press.

$$(2.26\text{b}) \qquad \frac{d}{dt} \begin{pmatrix} b_1 \\ 0 \\ b_3 \end{pmatrix} = \begin{pmatrix} 0 & 0 & \frac{1}{2}\overline{\omega}(t) \\ 0 & 0 & 0 \\ -\frac{1}{2}\overline{\omega}(t) & 0 & 0 \end{pmatrix} \begin{pmatrix} b_1 \\ 0 \\ b_3 \end{pmatrix},$$

$$b_1(t=0) = 0, \quad b_3(t=0) = -B_0, \quad \overline{\omega}(t=0) = \overline{\omega}_0.$$

To solve these ODEs, we write

$$\vec{b} = B(t) \begin{pmatrix} \sin\theta(t) \\ 0 \\ -\cos\theta(t) \end{pmatrix}.$$

Equation (2.26b) tells us that $\frac{dB}{dt} = 0$ and $\frac{d\theta}{dt} = -\overline{\omega}/2$ with the initial conditions $\theta(t=0) = 0$, $B = B_0$. Taking one more derivative, we get

$$(2.27) \qquad \frac{d^2\theta}{dt^2} = -\frac{g B_0}{2\rho_b} \sin\theta(t).$$

This is the pendulum equation, and is of Hamiltonian form. The conserved energy for the system is

$$H = \left(\frac{d\theta}{dt}\right)^2 - \frac{g B_0}{\rho_b} \cos\theta(t).$$

We wish to inquire whether the fluid can go unstable by overturning the density gradient. The condition for this is that the vertical density slope (b_3) be positive, or $\cos\theta < 0$. To answer the inquiry, we plot the lines of constant H in phase space in Figure 2.1. The initial point will be at $\theta = 0$ and will have $\frac{d\theta}{dt}$ determined by the initial vorticity according to

$$\left.\frac{d\theta}{dt}\right|_{t=0} = -\frac{1}{2}\overline{\omega}_0 \,. \tag{2.28}$$

Once we find the initial point in phase space, the solution will move forward on the level sets. The question is, does this level set enter the region where the fluid is unstably stratified, $\cos\theta < 0$ ($\frac{\pi}{2} < \theta < \frac{3\pi}{2}$). From Figure 2.1, we see that this will only happen if the magnitude of the initial vorticity is large enough. Small $\overline{\omega}_0$ will create only small perturbations that stay within the stable region, such as the first loop around $(0,0)$ in Figure 2.1. Much higher up the $\theta = 0$ axis, the level sets clearly take you into the unstable region, for example, on or above the separatrix, the line connecting the saddle points. But even the first loop inside the separatrix in Figure 2.1 takes you to $\theta > \frac{\pi}{2}$, which is unstable. We then ask, what is the critical value for the initial vorticity that leads to instability? To answer this, we look for the level that just reaches $\theta = \frac{\pi}{2}$ when $\frac{d\theta}{dt} = 0$. This point will have $H = 0$ because $\cos(\frac{\pi}{2}) = 0$ and $\frac{d\theta}{dt} = 0$. To find the strength of the initial vortex, we then look along the $\theta = 0$ axis for the initial value of $\frac{d\theta}{dt}$ that will have the same H:

$$0 = \left(\frac{d\theta}{dt}\right)^2 = \frac{g B_0}{\rho_b}\cos(0) \,,$$

or since $\left.\frac{d\theta}{dt}\right|_{t=0} = -\frac{\omega_0}{2}$,

$$\overline{\omega}_0^c = \pm 2\left(\frac{g B_0}{\rho_b}\right)^{1/2} = \pm 2N \,. \tag{2.29}$$

Thus, the initial strength for the vorticity to cause overturning, $\overline{\omega}_0^c$, is exactly twice the buoyancy or Brunt-Väisälä frequency. For initial vortices of weaker strength, the fluid does not overturn with an unstable density gradient; i.e., for $\omega_0^2 < 4N^2$ there is never overturning of the fluid and always overturning otherwise.

Physically, we recall the sketch at the beginning of this example and realize that the destabilizing effect of vorticity must act faster than the natural restoring frequency of the stably stratified fluid in order to drive the heavy fluid from below on top of the lighter fluid. The oscillating vortical flows for $\omega_0^2 < 4N^2$ admit the following physical interpretation: The initial vortex deflects the initially horizontal isopycnal surfaces (*isopycnal* is the oceanographer's name for the level set of constant density) to a maximum angle and converts all of its kinetic energy into potential energy in the process; baroclinic vorticity production then converts this potential energy back into kinetic energy and restarts the next phase of the oscillation cycle described by the pendulum solutions. In the next chapter, we discuss the linear and nonlinear stability of these exact solutions.

EXAMPLE 2.4. The reader is asked to consider the more general case, to observe that a strong jet can inhibit overturning

$$\vec{v} = \begin{pmatrix} d_{11} & \frac{1}{2}\overline{\omega}(t) + d_{12} \\ -\frac{1}{2}\overline{\omega}(t) + d_{21} & d_{22} \end{pmatrix}\begin{pmatrix} x_1 \\ x_3 \end{pmatrix} \,. \tag{2.30}$$

Will the vorticity again produce instability?

We can further our intuition of stably stratified flows by solving numerically more complicated examples using Theorem 2.4. This is an exercise for the interested reader.

2.4. Nonlinear Plane Waves in Stratified Flow: Internal Gravity Waves

Here we build special exact solutions of the stratified Boussinesq equation,

$$(2.31) \qquad \frac{D\vec{v}}{Dt} = -\nabla p - \frac{g\rho}{\rho_b}\vec{e}_3, \quad \text{div } \vec{v} = 0, \quad \frac{D\rho}{Dt} + \left(\frac{d\bar{\rho}}{dx_3}\right)v_3 = 0.$$

We assume that $d\bar{\rho}/dx_3$ is constant in the discussion here. Here we show that the effects of stratification induce nonlinear wave motion.

PROPOSITION 2.5 *The Boussinesq equations in* (2.31) *have exact solutions of the form*

$$\vec{v} = \vec{A}(t)F(\vec{\alpha}(t)\cdot\vec{x}), \quad \rho = B(t)F(\vec{\alpha}(t)\cdot\vec{x}), \quad p = P(t)G(\vec{\alpha}(t)\cdot\vec{x}),$$

with $G'(s) = F(s)$ *provided that* $\vec{\alpha}(t)$, $\vec{A}(t)$, $B(t)$, *and* $P(t)$ *satisfy the ODEs* (2.32a), (2.32d), *and* (2.32e) *and the constant equations in* (2.32b) *and* (2.32c),

$$(2.32a) \qquad \frac{d\vec{\alpha}}{dt} = 0,$$

$$(2.32b) \qquad \vec{A}(t)\cdot\vec{\alpha} = 0,$$

$$(2.32c) \qquad P(t) = -\frac{g}{\rho_b}B(t)\frac{\alpha_3}{|\vec{\alpha}|^2},$$

$$(2.32d) \qquad \frac{d\vec{A}}{dt} = \frac{g}{\rho_b}\left(\frac{\alpha_3}{|\vec{\alpha}|^2}\vec{\alpha} - \vec{e}_3\right)B(t),$$

$$(2.32e) \qquad \frac{dB}{dt} + \frac{d\bar{\rho}}{dx_3}A_3 = 0.$$

The following lemma will show that the nonlinear equations simplify greatly for our special solutions:

LEMMA 2.6 *For* $\vec{v}$ *of the form* $\vec{v} = \vec{A}(t)F(\vec{\alpha}(t)\cdot\vec{x})$, *div* $\vec{v} = 0$ *implies*

    (i) $\vec{A}(t)\cdot\vec{\alpha}(t) = 0$ *and*
    (ii) $\vec{v}\cdot\nabla W(\vec{\alpha}(t)\cdot\vec{x}) = 0$

for arbitrary W.

PROOF OF LEMMA 2.6(i):

$$\text{div }\vec{v} = \sum_i A_i(t)\frac{\partial}{\partial x_i}F(\vec{\alpha}(t)\cdot\vec{x}) = \sum_i A_i(t)\alpha_i(t)F'(\vec{\alpha}(t)\cdot\vec{x})$$

$$= F'\vec{A}(t)\cdot\vec{\alpha}(t) = 0.$$

$\square$

PROOF OF LEMMA 2.6(ii):

$$\vec{v} \cdot \nabla W(\vec{\alpha}(t) \cdot \vec{x}) = F(\vec{\alpha}(t) \cdot \vec{x}) \vec{A}(t) \cdot \nabla W(\vec{\alpha}(t) \cdot \vec{x})$$

$$= F \sum_i A_i(t) \frac{\partial}{\partial x_i} W(\vec{\alpha}(t) \cdot \vec{x})$$

$$= F \sum_i A_i \alpha_i W'(\vec{\alpha}(t) \cdot \vec{x})$$

$$= F W' \vec{A} \cdot \vec{\alpha} = 0$$

by part (i). $\qquad\square$

PROOF OF PROPOSITION 2.5: Lemma 2.6 applies and part (ii) tells us that when we substitute our plane wave forms into the Boussinesq equation, the non-linear part of the advective derivative will drop out, and we are left with

(2.33a) $$\frac{\partial \vec{v}}{\partial t} = -\nabla p - \frac{g\rho}{\partial_b} \vec{e}_3 ,$$

(2.33b) $$\mathrm{div}\, \vec{v} = 0 ,$$

(2.33c) $$\frac{\partial \rho}{\partial t} + \left(\frac{d\bar{\rho}}{dx_3}\right) v_3 = 0 .$$

Inserting our plane wave forms into (2.33a) and using $G' = F$, we find

$$\frac{d\vec{A}}{dt} F + \vec{A}(t) \left(\frac{d\vec{\alpha}}{dt} \cdot \vec{x}\right) F' = -\vec{\alpha}(t) P(t) F - \frac{g}{\rho_b} B(t) \vec{e}_3 F .$$

Since F and F' are arbitrary functions, they must be treated as independent terms, which forces condition (2.32a) of Proposition 2.5, $\frac{d\vec{\alpha}}{dt} = 0$. Condition (2.32b) $\vec{A} \cdot \vec{\alpha} = 0$ comes from Lemma 2.6(i). Taking the time derivative of this result yields

$$\frac{d\vec{A}}{dt} \cdot \vec{\alpha} = 0 .$$

We can now take the dot product of the equation above for $d\vec{A}/dt$ with $\vec{\alpha}$ to yield condition (2.32c),

$$-|\vec{\alpha}|^2 P(t) - \frac{g\alpha_3}{\rho_b} B(t) = 0 .$$

This result can be plugged back into the $\vec{A}$ equation to remove $P(t)$, yielding equation (2.32d) of Proposition 2.5. Finally, we plug the plane waves into (2.33c) to find the equation for $B(t)$, condition (2.32e). $\qquad\square$

Let us rewrite the equations for our wave amplitudes:

(2.34)
$$\frac{dA_1}{dt} = \frac{g}{\rho_b} \frac{\alpha_3 \alpha_1}{|\vec{\alpha}|^2} B(t) , \qquad \frac{dA_2}{dt} = \frac{g}{\rho_b} \frac{\alpha_3 \alpha_2}{|\vec{\alpha}|^2} B(t) ,$$

$$\frac{dA_3}{dt} = \frac{g}{\rho_b} \left(\frac{\alpha_3^2}{|\vec{\alpha}|^2} - 1\right) B(t) , \qquad \frac{dB}{dt} = -\frac{d\bar{\rho}}{dx_3} A_3(t) .$$

We see that B and A_3 are coupled independently of the others, and we can solve by taking another derivative

$$(2.35) \qquad \frac{d^2 B}{dt^2} = \frac{g}{\rho_b} \frac{d\bar{\rho}}{dx_3} \left(1 - \frac{\alpha_3^2}{|\vec{\alpha}|^2} \right) B = -\omega^2 B .$$

The behavior here depends on the sign of ω^2. Since the angular term in parentheses is always positive, the overall sign depends on the sign of the density gradient.

Case 1: $d\bar{\rho}/dx_3 > 0$ (heavier fluid on top). This case will have exponentially growing solutions of the form $e^{|\omega|t}$. We conclude that the steady state is unstable.

Case 2: $d\bar{\rho}/dx_3 < 0$ (heavier fluid on bottom). This case supports oscillating solutions, and thus we refer to it as stable stratification. The fundamental frequency of oscillation that falls out of this analysis is the same frequency that we discussed in Chapter 1, the buoyancy frequency or Brunt-Väisälä frequency:

$$(2.36) \qquad N = \left(-\frac{g}{\rho_b} \frac{d\bar{\rho}}{dx_3} \right)^{1/2} .$$

For special motions with $\alpha_3 = 0$, we recover plane wave versions of the special exact solutions involving only vertical motion and density perturbations with stable stratification that were discussed in Chapter 1. The nonlinear plane waves with $\alpha_3 \neq 0$ supported by stable stratification are called *internal gravity waves*. We direct your attention to sections 6.4 and 8.4 of A. E. Gill's textbook [**11**] for a reference. Changing the notation for the general parameter $\vec{\alpha}$ to the wavenumber $\vec{k}$ to correspond to more standard notation, the frequency defined in (2.35) is given by

$$(2.37) \qquad \omega(\vec{k}) = N \frac{|\vec{k}_H|}{|\vec{k}|}$$

for the wave vector $\vec{k} = (\vec{k}_H, k_3)$. This is the dispersion relation for internal gravity waves. It remains to write the complete solutions. The general solution to (2.35) is

$$B(t) = c_1 \sin \omega(\vec{k})t + c_2 \cos \omega(\vec{k})t .$$

$P(t)$ is immediately obtained via equation (2.32c),

$$P(t) = -\frac{g}{\rho_b} \frac{k_3}{|\vec{k}|^2} \left(c_1 \sin \omega(\vec{k})t + c_2 \cos \omega(\vec{k})t \right) .$$

A_3 is obtained using the last equation of (2.34) and the definition of the Brunt-Väisälä frequency (2.36) to get

$$A_3(t) = \frac{g}{\rho_b N} \frac{|\vec{k}_H|}{|\vec{k}|} \left(c_1 \cos \omega(\vec{k})t - c_2 \sin \omega(\vec{k})t \right) .$$

A_1 and A_2 are obtained by integrating the first two parts of (2.34),

$$A_1(t) = -\frac{g}{\rho_b N}\frac{k_3 k_1}{|\vec{k}|\,|\vec{k}_H|}\left(c_1 \cos\omega(\vec{k})t - c_2 \sin\omega(\vec{k})t\right),$$

$$A_2(t) = -\frac{g}{\rho_b N}\frac{k_3 k_2}{|\vec{k}|\,|\vec{k}_H|}\left(c_1 \cos\omega(\vec{k})t - c_2 \sin\omega(\vec{k})t\right).$$

In order to write the physical variables, we must merely remember their definitions in Proposition 2.5, recalling that $G'(s) = F(s)$,

$$\rho = \left(c_1 \sin\omega(\vec{k})t + c_2 \cos\omega(\vec{k})t\right)F(\vec{k}\cdot\vec{x}),$$

$$p = -\frac{g}{\rho_b}\frac{k_3}{|\vec{k}|^2}\left(c_1 \sin\omega(\vec{k})t + c_2 \cos\omega(\vec{k})t\right)G(\vec{k}\cdot\vec{x}),$$

(2.38)
$$\vec{v} = -\frac{g}{\rho_b N}\frac{1}{|\vec{k}|\,|\vec{k}_H|}\left(c_1 \cos\omega(\vec{k})t - c_2 \sin\omega(\vec{k})t\right)$$

$$F(\vec{k}\cdot\vec{x})(k_3 k_1, k_3 k_2, -|\vec{k}_H|^2),$$

$$= -\frac{g}{\rho_b N}\left(c_1 \cos\omega(\vec{k})t - c_2 \sin\omega(\vec{k})t\right)F(\vec{k}\cdot\vec{x})\frac{(\vec{k}_H^\perp, 0)\times\vec{k}}{|\vec{k}|\,|\vec{k}_H|}.$$

The last result shows us that the fluid velocities $\vec{v}$ are perpendicular to the direction of wave propagation $\vec{k}$.

Sinusoidal Waveforms.

(2.39)
$$F(\vec{k}\cdot\vec{x}) = \sin(\vec{k}\cdot\vec{x}).$$

Writing out the density in this case gives

(2.40)
$$\rho = \frac{c_1}{2}\left(\cos(\omega(\vec{k})t - \vec{k}\cdot\vec{x}) - \cos(\omega(\vec{k})t + \vec{k}\cdot\vec{x})\right)$$
$$+ \frac{c_2}{2}\left(\sin(\omega(\vec{k})t + \vec{k}\cdot\vec{x}) - \sin(\omega(\vec{k})t - \vec{k}\cdot\vec{x})\right).$$

This manipulation illustrates that there are waves moving in different directions corresponding to the two branches of the dispersion relation. Let's simplify to the case $c_2 = 0$ and write the solutions

(2.41)
$$\rho = \frac{c_1}{2}\left(\cos(\omega(\vec{k})t - \vec{k}\cdot\vec{x}) - \cos(\omega(\vec{k})t + \vec{k}\cdot\vec{x})\right),$$

$$p = \frac{c_1}{2}\frac{g}{\rho_b}\frac{k_3}{|\vec{k}|^2}\left(\sin(\omega(\vec{k})t - \vec{k}\cdot\vec{x}) + \sin(\omega(\vec{k})t + \vec{k}\cdot\vec{x})\right),$$

$$\vec{v} = \frac{c_1}{2}\frac{g}{\rho_b N}\frac{(\vec{k}_H^\perp, 0\times\vec{k})}{|\vec{k}|\,|\vec{k}_H|}\left(\sin(\omega(\vec{k})t - \vec{k}\cdot\vec{x}) - \sin(\omega(\vec{k})t + \vec{k}\cdot\vec{x})\right).$$

See Gill's book [**11**, sect. 6.5, fig. 6.6] for a picture of these waves. Fluid velocities lie along wave crests, and are in phase with the pressure perturbations but out of phase with the density variations. The buoyancy, that final term in the momentum equation (2.33a), takes its sign from the density perturbation, and has components both along and perpendicular to the direction of propagation. The buoyancy's projection along $\vec{k}$ cancels the pressure gradient, while the perpendicular component provides the restoring force for the fluid velocities. The last result means that internal gravity waves cannot propagate purely in the x_3-direction, as the buoyancy has no component perpendicular to $\vec{e}_3$. The dispersion relation confirms that $\omega \to 0$ in this case. The restoring force for the density variations is the vertical advection of the background, hydrostatic density. Thus, we observe that as the pattern advances, the downward pointing fluid velocity will bring the less dense fluid from above into the currently "heavy" region, and the "light" region will be made denser with fluid from below. Meanwhile, the heavy region works with gravity to generate a downward velocity, while the light region generates oppositely directed buoyancy.

Result (2.41) shows that for phase propagation in the opposite direction, the diagram would need to keep, for instance, the velocities and pressures fixed, while interchanging the heavy and light density markings. The reader can verify the restoring dynamics in this case.

This dispersion relation for internal gravity waves (2.37) is rather peculiar and creates some remarkable effects for internal gravity waves. The frequency is independent of the wave vector amplitude and depends instead on the angle between the wave vector and the horizontal plane. The waves with greater tilt from horizontal propagate more slowly. The group velocity, defined by

$$\vec{c}^g = \left(\frac{\partial \omega}{\partial k_1}, \frac{\partial \omega}{\partial k_2}, \frac{\partial \omega}{\partial k_3} \right)$$

for internal gravity waves, is

$$\vec{c}^g = \frac{Nk_3}{|\vec{k}_H| |\vec{k}|^3} \left(-k_3 k_1, -k_3 k_2, k_1^2 + k_2^2 \right) = -\frac{Nk_3}{|\vec{k}_H| |\vec{k}|^3} \frac{(\vec{k}_H^\perp, 0) \times \vec{k}}{|\vec{k}| |\vec{k}_H|} \cdot$$

We observe that group velocities lie on the same axis as fluid particle velocities, perpendicular to the velocity of individual wave crests. Energy is propagated in the direction of the group velocity for wave packets as will be discussed in Chapter 5 below; thus, internal gravity waves propagate energy in a direction orthogonal to the phase variations! Gill's discussion of gravity wave dispersion can be found in section 6.6 of his textbook.

2.5. Exact Solutions with Large-Scale Motion and Nonlinear Plane Waves

Here we build exact solutions that combine the large-scale structure from Theorem 2.4 with the superimposed nonlinear plane waves as illustrated by the internal gravity waves discussed in Proposition 2.5. We have the following:

THEOREM 2.7 *There are exact solutions to the Boussinesq equations of the form*

$$\vec{v}(\vec{x}, t) = \mathcal{D}(t)\vec{x} + \frac{1}{2}\vec{\omega}(t) \times \vec{x} + \vec{A}(t) F(\vec{\alpha}(t) \cdot \vec{x}),$$

(2.42)
$$\tilde{\rho} = \rho_b + \vec{b}(t) \cdot \vec{x} = B(t) F(\vec{\alpha}(t) \cdot \vec{x}),$$

$$p = \frac{1}{2}\widehat{P}(t)\vec{x} \cdot \vec{x} + P(t) G(\vec{\alpha}(t) \cdot \vec{x}),$$

where $G'(s) = F(s)$ and $\mathcal{D}(t)$ is an arbitrary 3×3, traceless, symmetric matrix, provided $\vec{\omega}(t) = \text{curl } \vec{v}$ and $\vec{b}$ satisfy the ODEs

$$(2.43) \qquad \frac{d\vec{\omega}}{dt} = \mathcal{D}(t)\vec{\omega}(t) + \frac{g}{\rho_b} \begin{pmatrix} -b_2(t) \\ b_1(t) \\ 0 \end{pmatrix},$$

$$(2.44) \qquad \frac{d\vec{b}}{dt} = -\mathcal{D}(t)\vec{b}(t) + \frac{1}{2}\vec{\omega}(t) \times \vec{b}(t),$$

and provided the wave phase and amplitudes satisfy the ODEs

$$(2.45a) \qquad \frac{d\vec{\alpha}}{dt} = -^{\mathsf{T}}V(t)\vec{\alpha}(t),$$

$$(2.45b) \qquad \frac{d\vec{A}}{dt} = -V(t)\vec{A}(t) + \vec{\alpha}\frac{2(^{\mathsf{T}}V(t)\vec{\alpha}(t) \cdot \vec{A}(t))}{|\vec{\alpha}|^2} + \frac{g}{\rho_b}\left[\frac{\alpha_3}{|\vec{\alpha}|^2}\vec{\alpha} - \vec{e}_3\right]B(t),$$

$$(2.45c) \qquad \frac{dB}{dt} = -\vec{A}(t) \cdot \vec{b}(t),$$

where $V(t)$ acts on vectors by $V\vec{h} = (\mathcal{D} + \Omega)\vec{h} = \mathcal{D}\vec{h} + \frac{1}{2}\vec{\omega} \times \vec{h}$. The initial conditions are arbitrary, except that we require

$$(2.46) \qquad \vec{\alpha} \cdot \vec{A}\big|_{t=0} = 0.$$

The pressure is then determined via the matrix

$$(2.47) \qquad -\widehat{P}(t) = \frac{d\mathcal{D}}{dt} + \mathcal{D}^2 + \Omega^2 + \frac{g}{2\rho_b}(\vec{e}_3 \otimes \vec{b} + \vec{b} \otimes e_3)$$

and the scalar

$$(2.48) \qquad -P(t) = \frac{2(^{\mathsf{T}}V(t)\vec{\alpha}(t) \cdot \vec{A}(t))}{|\vec{\alpha}|^2} + \frac{g}{\rho_b}\frac{B\alpha_3}{|\vec{\alpha}|^3}.$$

SKETCH OF THE PROOF OF THEOREM 2.7: We seek to verify the Boussinesq equations. We compute

$$\text{div } \vec{v} = \text{tr } \mathcal{D}(t) + \vec{A}(t) \cdot \vec{\alpha}(t)F'(\vec{\alpha}(t) \cdot \vec{x}).$$

$\vec{A} \cdot \vec{\alpha}$ satisfies

$$\frac{d}{dt}(\vec{A} \cdot \vec{\alpha}) = \frac{d\vec{A}}{dt} \cdot \vec{\alpha} + \vec{A} \cdot \frac{d\vec{\alpha}}{dt} = 0$$

by using ODEs (2.45a) and (2.45b). Together with the initial condition (2.46), we get

$$(2.49) \qquad \vec{A}(t) \cdot \vec{\alpha}(t) = 0$$

at all times. With $\mathcal{D}$ having trace 0, the defined velocity is divergence free. We now compute the advective derivative

$$\frac{\partial}{\partial t}\vec{v} = \frac{dV}{dt}\vec{x} + \frac{d\vec{A}}{dt}F + \vec{A}\frac{d\vec{\alpha}}{dt}\cdot \vec{x}F',$$

$$\vec{v}\cdot\nabla\vec{v} = \vec{v}\cdot V + \vec{A}\cdot(\vec{v}\cdot\vec{\alpha})F'$$

$$= V^2\vec{x} + \vec{A}\cdot VF + \vec{A}(V\vec{x}\cdot\vec{\alpha})F' + \vec{A}(\vec{A}\cdot\vec{\alpha})FF'.$$

The last term will go away with (2.49). Using $G'(s) = F(s)$,

$$\nabla p = \widehat{P}\vec{x} + P\vec{\alpha}F.$$

The next result, which was also true in Theorem 2.4, uses the definitions for V and $\widehat{P}$ and the equation for $d\vec{\omega}/dt$,

$$\frac{dV}{dt}\vec{x} + V^2\vec{x} = -\widehat{P}\vec{x} - \frac{g}{\rho_b}(\vec{b}\otimes\vec{e}_3)\vec{x},$$

so that

$$\frac{D}{Dt}\vec{v} + \nabla p + \frac{g}{\rho_b}\rho\vec{e}_3 = \left(\frac{d\vec{A}}{dt} + \vec{A}\cdot V + P\vec{\alpha} + \frac{gB}{\rho_b}\vec{e}_3\right)F$$

$$+ \vec{A}\left(\frac{d\vec{\alpha}}{dt}\cdot\vec{x} + {}^\mathsf{T}V\vec{\alpha}\cdot\vec{x}\right)F'.$$

The equations for $d\vec{A}/dt$, P, and $d\vec{\alpha}/dt$ will make the right-hand side zero. The particular form of the pressure is chosen to guarantee that $d(\vec{A}\cdot\vec{\alpha})/dt = 0$. Finally, we need to consider density

$$\frac{D\tilde{\rho}}{Dt} = \frac{d\vec{b}}{dt}\cdot\vec{x} + \frac{dB}{dt}F + B\frac{d\vec{\alpha}}{dt}\cdot\vec{x}F' + \vec{v}\cdot\vec{b} + B\vec{v}\cdot\vec{\alpha}F'$$

$$= \left(\frac{d\vec{b}}{dt} + {}^\mathsf{T}V\vec{b}\right)\cdot\vec{x} + \left(\frac{dB}{dt} + \vec{A}\cdot\vec{b}\right)F + \left(\frac{d\vec{\alpha}}{dt} + {}^\mathsf{T}V\vec{\alpha}\right)\cdot\vec{x}F'$$

$$= 0.$$

More details of the derivation can be found in the next section. $\square$

2.6. More Details for Theorem 2.7 on Special Exact Solutions for the Boussinesq Equations Including Plane Waves

We are seeking exact solutions to the Boussinesq equations

$$(2.50)\qquad \frac{D}{Dt}\vec{v} = -\nabla p - \frac{g}{\rho_b}\rho\vec{e}_3,\quad \mathrm{div}\,\vec{v} = 0,\quad \frac{D}{Dt}\tilde{\rho} = 0$$

(where $\tilde{\rho} = \rho_b + \rho$), of the form

$$\vec{v}(\vec{x}, t) = \mathcal{D}(t)\vec{x} + \frac{1}{2}\vec{\omega}(t) \times \vec{x} + \vec{A}(t)F(\vec{\alpha}(t) \cdot \vec{x}),$$

(2.51)
$$\tilde{\rho} = \rho_b + \vec{b}(t) \cdot \vec{x} + B(t)\widetilde{F}(\vec{\alpha}(t) \cdot \vec{x}),$$

$$p = \frac{1}{2}\widehat{P}(t)\vec{x} \cdot \vec{x} + P(t)G(\vec{\alpha}(t) \cdot \vec{x}),$$

where the relationships among the waveform functions $F(s)$, $\widetilde{F}(s)$, and $G(s)$ are to be determined, and where $\mathcal{D}(t)$ is an arbitrary 3×3 symmetric matrix, so that $\vec{\omega}(t) = \operatorname{curl} \vec{v}$. Let the antisymmetric matrix Ω act on vectors by $\Omega \vec{h} = \frac{1}{2}\vec{\omega} \times \vec{h}$, and let the matrix $V = \mathcal{D} + \Omega$ so that

$$\vec{v}(\vec{x}, t) = V(t)\vec{x} + \vec{A}(t)F(\vec{\alpha}(t) \cdot \vec{x}),$$

which is in fact the more general starting point. We begin with the requirements for an incompressible fluid:

$$\operatorname{div} \vec{v} = \operatorname{tr} \mathcal{D} + \vec{A} \cdot \vec{\alpha}F'.$$

Since F is arbitrary, we see that we must require a traceless strain matrix

(C1)
$$\operatorname{tr} \mathcal{D}(t) = 0,$$

and also that we must guarantee

(2.52)
$$\vec{A}(t) \cdot \vec{\alpha}(t) = 0$$

at all times. This will not be an imposed condition, but rather enforced from separate conditions. We use it now and will return to show that it is true. We turn to the density equation

$$\frac{D\tilde{\rho}}{Dt} = \frac{d\vec{b}}{dt} \cdot \vec{x} + \frac{dB}{dt}\widetilde{F} + B\frac{d\vec{\alpha}}{dt} \cdot \vec{x}\widetilde{F}' + \vec{v} \cdot \vec{b} + B\vec{v} \cdot \vec{\alpha}\widetilde{F}'$$

$$= \left(\frac{d\vec{b}}{dt} + {}^{\mathsf{T}}V\vec{b}\right) \cdot \vec{x} + \left(\frac{dB}{dt} + \vec{A} \cdot \vec{b}\right)\widetilde{F}$$

$$+ B\left(\frac{d\vec{\alpha}}{dt} + {}^{\mathsf{T}}V\vec{\alpha}\right) \cdot \vec{x}\widetilde{F}' + B(\vec{A} \cdot \vec{\alpha})\widetilde{F}\widetilde{F}'.$$

The last term is zero once we have (2.52). Since each of the other terms must be zeroed independently, we must enforce the following ODEs for the density slope

(C2)
$$\frac{d\vec{b}}{dt} = -\mathcal{D}(t)\vec{b}(t) + \frac{1}{2}\vec{\omega}(t) \times \vec{b}(t),$$

the density wave part

(C3)
$$\frac{dB}{dt} = -\vec{A}(t) \cdot \vec{b}(t),$$

and the wave vector

(C4)
$$\frac{d\vec{\alpha}}{dt} + {}^{\mathsf{T}}V(t)\vec{\alpha}(t) = 0.$$

We now compute the terms in the momentum equation of (2.50)

$$\frac{\partial}{\partial t}\vec{v} = \frac{dV}{dt}\vec{x} + \frac{d\vec{A}}{dt}F + \vec{A}\frac{d\vec{\alpha}}{dt}\cdot\vec{x}F',$$

$$\vec{v}\cdot\nabla\vec{v} = \vec{v}\cdot V + \vec{A}\cdot(\vec{v}\cdot\vec{\alpha})F'$$

$$= V^2\vec{x} + \vec{A}\cdot VF + \vec{A}(V\vec{x}\cdot\vec{\alpha})F' + \vec{A}(\vec{A}\cdot\vec{\alpha})FF'.$$

Again, the last term is zero given (2.52),

$$\nabla p = \widehat{P}\vec{x} + P\vec{\alpha}G', \quad \text{and} \quad \frac{g\rho}{\rho_b}\vec{e}_3 = \frac{g}{\rho_b}(\vec{b}\otimes\vec{e}_3)\vec{x} + \frac{g}{\rho_b}B\vec{e}_3\widetilde{F}.$$

We combine these terms to see that we must satisfy the following:

(2.53)
$$\left(\frac{dV}{dt} + V^2 + \widehat{P} + \vec{b}\otimes\vec{e}_3\right)\vec{x} + \left(\frac{d\vec{A}}{dt} + \vec{A}\cdot V\right)F$$

$$+ P\vec{\alpha}G' + \frac{gB}{\rho_b}\vec{e}_3\widetilde{F} + \vec{A}\left(\frac{d\vec{\alpha}}{dt} + {}^\mathsf{T}V\vec{\alpha}\right)\cdot\vec{x}F' = 0.$$

The term multiplying F' is zero by condition (C3). Equation (2.53) must hold for $G = F = F' = 0$, where we need

(2.54)
$$\frac{dV}{dt} + V^2 + \widehat{P} + \vec{b}\otimes\vec{e}_3 = 0.$$

For this case, $\widehat{P}$ is just the Hessian matrix for pressure,

$$\widehat{P} = \left(\frac{\partial^2 p}{\partial x_i\partial x_j}\right),$$

which we want to be symmetric. It then does not enter into the antisymmetric part of (2.54), which is

$$\frac{d\Omega}{dt} + \mathcal{D}\Omega + \Omega\mathcal{D} + \frac{g}{2\rho_b}\left((\vec{b}\otimes\vec{e}_3) - (\vec{e}_3\otimes\vec{b})\right) = 0.$$

In terms of $\vec{\omega}$, this condition is

(C5)
$$\frac{d\vec{\omega}}{dt} = \mathcal{D}(t)\vec{\omega}(t) + \frac{g}{\rho_b}\begin{pmatrix}-b_2(t)\\b_1(t)\\0\end{pmatrix}.$$

This prescribes the ODE satisfied by $\vec{\omega}$ or, equivalently, Ω. In taking the symmetric part of (2.54), we see that the time dependence of $\mathcal{D}$ is arbitrary and is absorbed into the pressure according to

(C6)
$$-\widehat{P}(t) = \frac{d\mathcal{D}}{dt} + \mathcal{D}(t)^2 + \Omega(t)^2 + \frac{g}{2\rho_b}\left(\vec{e}_3\otimes\vec{b}(t) + \vec{b}(t)\otimes\vec{e}_3\right).$$

Returning to (2.53), we are left with

$$\left(\frac{d\vec{A}}{dt} + \vec{A}\cdot V\right)F + P\vec{\alpha}G' + \frac{gB}{\rho_b}\vec{e}_3\widetilde{F} = 0.$$

We see here that the necessary relationship among the waveforms is

(C7) $$F(s) = \widetilde{F}(s) = G'(s).$$

Then

(2.55) $$\frac{d\vec{A}}{dt} + \vec{A} \cdot V + P\vec{\alpha} + \frac{gB}{\rho_b}\vec{e}_3 = 0.$$

We are now in a position to enforce condition (2.52) for a divergence-free field. We do so by requiring the initial condition

(C8) $$\vec{\alpha} \cdot \vec{A}\Big|_{t=0} = 0$$

together with

$$\frac{d}{dt}(\vec{A} \cdot \vec{\alpha}) = 0.$$

For this, we use (2.55) and (C4),

$$-\left(\frac{d\vec{A}}{dt} \cdot \vec{\alpha} + \vec{A} \cdot \frac{d\vec{\alpha}}{dt}\right) = (V\vec{A}) \cdot \vec{\alpha} + P|\vec{\alpha}|^2 + \frac{gB}{\rho_b}\alpha_3 + \vec{A} \cdot ({}^{T}V\vec{\alpha}).$$

We can make this zero by prescribing the pressure wave amplitude

(C9) $$-P(t) = \frac{2({}^{T}V(t)\vec{\alpha}(t) \cdot \vec{A}(t))}{|\vec{\alpha}(t)|^2} + \frac{g}{\rho_b}\frac{B(t)\alpha_3(t)}{|\vec{\alpha}(t)|^2}.$$

Lastly, this is inserted into (2.55) to give the final form for the velocity wave amplitude equation

(C10) $$\frac{d\vec{A}}{dt} = -V(t)\vec{A}(t) + \vec{\alpha}\frac{2({}^{T}V(t)\vec{\alpha}(t) \cdot \vec{A}(t))}{|\vec{\alpha}(t)|^2} + \frac{g}{\rho_b}\left[\frac{\alpha_3(t)}{|\vec{\alpha}(t)|^2}\vec{\alpha}(t) - \vec{e}_3\right]B(t).$$

Conditions (C1) through (C10) form the conditions of Theorem 2.7.

EXERCISE 2.1 (Effect of Vertical Jets on Stratification). The simplest elementary three-dimensional flow with vertical motion but no vorticity is the two-dimensional strain flow

(2.56) $$\vec{V} = \begin{pmatrix} 0 \\ -\gamma(t)y \\ \gamma(t)z \end{pmatrix}$$

where $\gamma(t)$ is a prescribed time-dependent strain rate.

(1) Graph the flow at an instant of time for $\gamma(t) > 0$ or $\gamma(t) < 0$ to convince yourself this looks like a jet flow.
(2) How does such a jet flow alter the stratification?

Consider an initial stratification

$$\tilde{\rho} = b_0 z$$

and find the exact solution of Theorem 2.4 with the velocity in (2.56).

Write down the solution carefully for a general $\gamma(t)$ and discuss the effect of the strain rate, $\gamma(t)$, on stratification. Consider a simple example like $\gamma(t) = \gamma_0 \cos \omega t$ to illustrate your points.

(3) While the flow in (1) and (2) involves the y- and z-variables, it is possible to build nonlinear plane waves that are three-dimensional about these two-dimensional flows. Use Theorem 2.7 and write the *explicit equation* for these three-dimensional nonlinear plane waves for a general initial direction $\vec{\alpha}_0 = (\alpha_1, \alpha_2, \alpha_3)$ for the plane wave for a general $\gamma(t)$. Specialize these equations to the special strain flows, $\gamma(t) = \gamma_0 \cos \omega t$. Do these linear ODEs have time-periodic coefficients in this case? Read your favorite book on ODEs, regarding Floquet analysis for linear equations with periodic coefficients.

Show a lot of detail.

Such solutions from this exercise and their Floquet analysis could actually give some insight into current research in stratified flows motivated by small-scale mixing in the ocean. This will be discussed at the end of the next chapter.

CHAPTER 3

Linear and Nonlinear Instability of Stratified Flows with Strong Stratification

In the last chapter, we saw that the effects of stable stratification are remarkable in the sense that stable stratification can produce waves called *internal gravity waves*. The two-dimensional pendulum flows in Section 2.3 are especially interesting elementary flows since they illustrate how the effects of density gradients from gravity can balance and dynamically interact with a vortex attempting to overturn the fluid and nevertheless produce nonlinear oscillations in the fluid where the density gradient is never overturned by the vortex and remains stably stratified throughout the evolution. Recall from Chapter 2 that these exact solutions of the Boussinesq equations have the forms for the density and velocity

$$\tilde{\rho} = \rho_b + \sin\theta(t)x - \cos\theta(t)z = \rho_b + \bar{\rho},$$

(3.1)
$$\vec{V} = \left(\frac{\omega(t)}{2}z, -\frac{\omega(t)}{2}x \right), \quad \omega(t) = -2\frac{d\theta}{dt},$$

where $\omega(t)$ is the time-dependent vorticity. In (3.1) we have picked the unit of time as the reciprocal of the buoyancy frequency or Brunt-Väisälä frequency; thus, $N \equiv 1$ for the initial stable stratification. As established in Section 2.3, in these units (3.1) is an exact solution of the two-dimensional Boussinesq equations provided $\theta(t)$ satisfies the pendulum equation

(3.2)
$$2\frac{d^2\theta}{dt^2} = -\sin\theta, \quad \omega(t) = -2\frac{d\theta}{dt},$$

with the initial data

(3.3)
$$\theta(t)\big|_{t=0} = 0, \quad \omega(t)\big|_{t=0} = \text{Fr}.$$

Here Fr is a nondimensional number called the *Froude number*, which we will encounter throughout the remaining chapters. Here is how the Froude number is defined:

(3.4)
$$\text{Fr} = \frac{\text{buoyancy time}}{\text{eddy turnover time}}.$$

The eddy turnover time measures the strength of the velocity field while the buoyancy time measures the strength of the stratification. Clearly, the buoyancy time is measured by the reciprocal of the buoyancy frequency, N^{-1}, with $N^2 = -g\frac{d\rho}{dz}/\rho_b$. The eddy turnover time is measured by the initial vorticity that has units $(\text{time})^{-1}$.

Thus, the initial condition in (3.3) emerges. Furthermore,

(3.5)   for Fr ≪ 1 there is strong stratification relative to fluid motion.

These conditions are often met in the atmosphere and ocean away from the boundary layers created by the air/sea/land interface. As established in the last chapter, there is no overturning of the density, and stable stratification persists for the pendulum flows for

$$\text{(3.6)} \qquad\qquad\qquad\qquad \text{Fr} < 2 \,.$$

The problem that we investigate here in this chapter is the fundamental issue of whether the basic pendulum flows are unstable to perturbations and whether these perturbations can create local unstable overturnings in the fluid flow, even with very strong initial stratification so that Fr ≪ 1. We will find out that, somewhat surprisingly, the answer is yes. Our approach will be an example of modern applied mathematics where mathematical theory, scientific computing, and physical reasoning all interact simultaneously. To understand these issues, we will rewrite the Boussinesq equations first in vorticity stream form and then use a novel mean Lagrangian coordinate to study perturbations. In this mean Lagrangian coordinate system both linear stability analysis and numerical computations become very elegant and clear. In fact, the linear stability analysis reduces to studying the instability of an infinite number of decoupled Hill's equations through Floquet theory.

To put the work in this chapter into a historical context, we mention important classical results on linearized stability of the simple steady shear flows, $\vec{V} = (v(z), 0, 0)$, mentioned earlier in Chapter 1. Richardson introduced a classical nondimensional number, the Richardson number Ri, defined by

$$\text{(3.7)} \qquad\qquad\qquad\qquad \text{Ri} = \frac{N^2}{\left(\frac{\partial v}{\partial z}\right)^2}$$

with N the buoyancy frequency. Note that

(3.8)   for Ri ≫ 1 there is strong stratification relative to the shear motion.

Will steady shear flows under strong enough stratification become linearly stable? There is a celebrated theorem of Miles and Howard that states that with suitable boundary conditions,

(3.9)   the steady shear flows are stable for Richardson numbers Ri $> \frac{1}{4}$.

The proof can be found in the book by Cushman-Roisin [4]. This criterion for stability is often interpreted and applied literally for time-dependent flow fields in both theoretical and numerical modeling of the atmosphere and ocean. For example, a popular turbulent eddy diffusivity for numerical models in the atmosphere/ocean community is the Lilly-Smagorinsky eddy diffusivity where the turbulent diffusivity is completely switched off and set to zero for Ri $\geq \overline{\text{Ri}} \geq \frac{1}{4}$ with $\overline{\text{Ri}}$ of order unity. In this chapter, we use the elementary time-dependent pendulum flows in (3.1)–(3.3) to demonstrate that such reasoning can be violated in dramatic fashion for time-dependent strongly stratified flows. For the pendulum flows, we identify the Richardson number in (3.7) with the Froude number in (3.3) and (3.4) through the obvious formula

$$(3.10) \qquad\qquad \text{Fr} = \text{Ri}^{-1/2}$$

and use either Fr or sometimes Ri below for historical reasons in the discussion below. Thus, there is no overturning for the vortical flows for $\text{Ri} \geq \frac{1}{4}$.

The material presented in this chapter is based on recent work of Majda and Shefter in [**23, 24, 25**]. Many more references to work on the important topic of overturning and mixing in stratified fluids can be found in the bibliographies of these papers.

3.1. Boussinesq Equations and Vorticity Stream Formulation

In the units for this chapter, the nondimensional Boussinesq equations are

$$(3.11) \qquad \frac{D\vec{v}}{Dt} = -\nabla p - \tilde{\rho}\vec{e}_3 \,, \quad \text{div}\,\vec{v} = 0\,, \quad \frac{D\tilde{\rho}}{Dt} = 0\,.$$

Here, we restrict our consideration to purely two-dimensional flows where $\vec{v}(x, z, t) = (u(x, z, t), w(x, z, t))^{\mathsf{T}}$ are two-dimensional velocity fields, with no dependence on the horizontal direction y. In (3.11), $\tilde{\rho}(x, z, t)$ is the total nondimensional flow density and p is the nondimensional hydrodynamic pressure. For such two-dimensional flows, the Boussinesq equations in (3.11) allow a simple vorticity-stream formulation similar to the one we used in Chapter 1 in discussing barotropic flow. With the vorticity

$$\Omega(x, z, t) = \frac{\partial u}{\partial z} - \frac{\partial w}{\partial x}$$

the stream function $\psi(x, z, t)$ is introduced via

$$(3.12) \qquad \vec{v} = \nabla^{\perp}\psi = \begin{pmatrix} \partial_z \psi \\ -\partial_x \psi \end{pmatrix}.$$

The incompressibility constraint $u_x + w_z = 0$ is satisfied trivially, provided that (3.12) holds. Taking the curl of the velocity equations in (3.11), a simple calculation shows the equations in (3.11) take the vorticity-stream form

$$(3.13) \qquad \begin{aligned} \Omega_t + \nabla^{\perp}\psi \cdot \nabla\Omega &= \tilde{\rho}_x \,, \\ \tilde{\rho}_t + \nabla^{\perp}\psi \cdot \nabla\tilde{\rho} &= 0 \,, \\ \Delta_2 \psi &= \Omega \,. \end{aligned}$$

With the above nondimensionalization, the initial conditions have the form

$$\Omega(0) = \text{Fr} + \omega_0'(x, z)\,, \quad \tilde{\rho}(0) = -z + \rho_0'(x, z)\,,$$

where ω_0' and ρ_0' are perturbations of the initial data for the pendulum flows. Now, we consider nonlinear perturbations about the basic pendulum solutions from (3.1) in the form

$$\Omega(x, z, t) = \omega(t) + \omega'(x, z, t),$$
(3.14)
$$\vec{v}(x, z, t) = \vec{V}(x, z, t) + \vec{v}(x, z, t),$$
$$\tilde{\rho}(x, z, t) = \bar{\rho}(x, z, t) + \rho'(x, z, t).$$

The equations of motion in (3.13) become

$$\omega'_t + \vec{V} \cdot \nabla \omega' + (\nabla^\perp \psi' \cdot \nabla \omega') = \rho'_x,$$
(3.15)
$$\rho'_t + \vec{V} \cdot \nabla \rho' + (\nabla^\perp \psi' \cdot \nabla \rho') + (\nabla^\perp \psi' \cdot \nabla \bar{\rho}) = 0,$$
$$\Delta \psi' = \omega'.$$

Equations in Mean Lagrangian Coordinates. The equations in (3.15) for perturbations look very complicated to analyze. Next we show how to simplify them by looking at these equations in a new reference frame. In order to remove the large-scale piece from the equation of motion (3.15), we introduce Lagrangian coordinates $\vec{\xi} = (\xi_1, \xi_3)$ moving along with the large-scale flow by the following equations:

$$(3.16) \qquad \vec{x} = \begin{pmatrix} X(\vec{\xi}, t) \\ Z(\vec{\xi}, t) \end{pmatrix}, \quad \frac{d\vec{x}}{dt} = \begin{pmatrix} \frac{\omega(t)}{2} Z(\vec{\xi}, t) \\ \frac{-\omega(t)}{2} X(\vec{\xi}, t) \end{pmatrix}.$$

The incompressibility constraint in Lagrangian coordinates takes the following form:

$$(3.17) \qquad \det\left(\frac{d\vec{x}}{d\vec{\xi}}\right) = 1.$$

The condition above can be easily obtained in an explicit form by showing that $\frac{d}{dt}(\det \frac{d\vec{x}}{d\vec{\xi}}) = 0$ and noting that $\frac{d\vec{x}}{d\vec{\xi}}|_{t=0} = I$. For a change of variables from $\vec{x}$ to $\vec{x}$, $\begin{pmatrix} \partial_{\xi_1} \\ \partial_{\xi_2} \end{pmatrix} = (\frac{d\vec{x}}{d\vec{\xi}})^\top \begin{pmatrix} \partial_x \\ \partial_z \end{pmatrix}$. Thus, for any two functions ψ and f under a Lagrangian coordinate transformation the following property is satisfied:

$$(3.18) \qquad \det(\nabla_\xi \psi, \nabla_\xi f) = \det\left(\frac{d\vec{x}}{d\vec{\xi}}\right)^\top \det(\nabla_x \psi, \nabla_x f).$$

Due to the incompressibility constraint in (3.17),

$$(3.19) \quad \nabla_x^\perp \psi \cdot \nabla_x f = -\det(\nabla_x \psi, \nabla_x f) = -\det(\nabla_\xi \psi, \nabla_\xi f) = \nabla_\xi^\perp \psi \cdot \nabla_\xi f,$$

so that the nonlinear advective term in Lagrangian coordinates preserves its Eulerian form. See the book by Majda and Bertozzi [19, chap. 1] for more extensive discussion of such Lagrangian particle trajectory equations.

For the pendulum flows, the Lagrangian transformation assumes an explicit form in terms of the pendulum equation solution,

$$(3.20) \quad \vec{x}(\vec{\xi}, t) = |\vec{\xi}| \begin{pmatrix} \sin(\theta(t) + \theta_0) \\ -\cos(\theta(t) + \theta_0) \end{pmatrix}, \quad \vec{\xi} = |\vec{\xi}| \begin{pmatrix} \sin \theta_0 \\ -\cos \theta_0 \end{pmatrix}, \quad \theta(0) = 0,$$

and therefore,

$$(3.21) \qquad \vec{x} = U(t)\vec{\xi} = \begin{pmatrix} \cos\theta(t) & -\sin\theta(t) \\ \sin\theta(t) & \cos\theta(t) \end{pmatrix} \vec{\xi} \,.$$

Since the Lagrangian transformation in this case is linear and the matrix $d\vec{x}/d\vec{\xi} = U(t)$ is unitary, that is, $U^{\mathsf{T}}(t) = U^{-1}(t)$, the following relation is true:

$$(3.22) \qquad \begin{pmatrix} \partial_{\xi_1} \\ \partial_{\xi_3} \end{pmatrix} = \left(\frac{d\vec{x}}{d\vec{\xi}}\right)^{\mathsf{T}} \begin{pmatrix} \partial_x \\ \partial_z \end{pmatrix}, \quad \begin{pmatrix} \partial_x \\ \partial_z \end{pmatrix} = \left(\frac{d\vec{x}}{d\vec{\xi}}\right) \begin{pmatrix} \partial_{\xi_1} \\ \partial_{\xi_3} \end{pmatrix}.$$

To finish the transformation to Lagrangian coordinates, we compute the two remaining operators, namely Δ_x and ∂_x, in the $\vec{\xi}$-coordinates,

$$(3.23) \qquad \partial_x = \cos\theta(t)\partial_{\xi_1} - \sin\theta(t)\partial_{\xi_3}, \quad \Delta_x = \Delta_\xi\,.$$

Using (3.17), (3.19), (3.23), and the expression for a substantial derivative in Lagrangian coordinates moving with the large-scale flow velocity $\vec{V}$, we rewrite the equations in (3.15) in the form that follows (we also drop the primes for simplicity).

Perturbation Equations in Mean Lagrangian Coordinates.

$$(3.24) \qquad \begin{aligned} \omega_t + (\nabla_\xi^\perp \psi \cdot \nabla_\xi \omega) &= (\cos\theta(t)\partial_{\xi_1} - \sin\theta(t)\partial_{\xi_3})\rho\,, \\ \rho_t + (\nabla_\xi^\perp \psi \cdot \nabla_\xi \rho) + \frac{\partial\psi}{\partial\xi_1} &= 0\,, \quad \Delta_\xi \psi = \omega\,. \end{aligned}$$

The $\partial\psi/\partial\xi_1$ term in the second equation of the system above originates from the cross product $\nabla^\perp \psi' \cdot \nabla\bar{\rho}$ if we observe that in Lagrangian coordinates $\bar{\rho} = -\xi_3$ and use (3.19). Recall that $\partial\psi/\partial\xi_x = -w$, the vertical velocity. The appropriate initial conditions are

$$(3.25) \qquad \omega(\vec{\xi}, t) = \mathrm{Fr}\omega_0(\vec{\xi})\,, \quad \rho(\vec{\xi}, t) = \rho_0(\vec{\xi})\,.$$

The equations for perturbations of pendulum solutions in mean Lagrangian coordinates look like the standard Boussinesq equations with the direction of gravity oscillating like a pendulum!

Linear Stability Analysis in Lagrangian Coordinates and Hill's Equation.
Consider plane wave solutions of the sinusoidal shape in Lagrangian coordinates,

$$\psi(\vec{\xi}, t) = \widehat{\psi}_k(t)\sin(\vec{k} \cdot \vec{\xi}) = -\frac{\widehat{\omega}_k(t)}{|\vec{k}|^2}\sin(\vec{k} \cdot \vec{\xi})\,,$$

$$\omega(\vec{\xi}, t) = \widehat{\omega}_k(t)\sin(\vec{k} \cdot \vec{\xi})\,,$$

$$(3.26) \qquad \vec{v}(\vec{\xi}, t) = \begin{pmatrix} k_3\widehat{\psi}_k(t) \\ -k_1\widehat{\psi}_k(t) \end{pmatrix}\cos(\vec{k} \cdot \vec{\xi}) = \begin{pmatrix} -\frac{k_3}{|\vec{k}|^2}\widehat{\omega}_k(t) \\ \frac{k_1}{|\vec{k}|^2}\widehat{\omega}_k(t) \end{pmatrix}\cos(\vec{k} \cdot \vec{\xi})\,,$$

$$\rho(\vec{\xi}, t) = \hat{\rho}_k(t)\cos(\vec{k} \cdot \vec{\xi})\,,$$

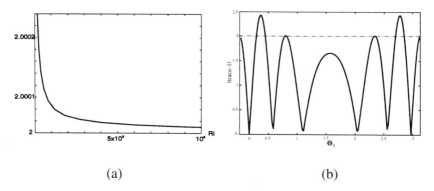

(a)   (b)

FIGURE 3.1. Maximum trace of Floquet matrix at large values of
Richardson number (a) and the trace of Floquet matrix for different inclination angles for fixed Richardson number Ri $= 5$ (b).

where $\vec{k}$ is an arbitrary fixed wave vector with integer coordinates in Lagrangian
space. Since all the nonlinear terms in (3.24) are of the form

$$(3.27) \qquad \nabla^{\perp}\psi(\vec{k}\cdot\vec{\xi},t)\cdot\nabla f(\vec{k}\cdot\vec{\xi},t)\,,$$

they vanish and the system in (3.24) assumes the following linear structure:

$$(3.28) \qquad \omega_t = (\cos\theta(t)\partial_{\xi_1} - \sin\theta(t)\partial_{\xi_3})\rho\,, \quad \rho_t = -\frac{\partial\psi}{\partial\xi_1}\,, \quad \Delta_{\xi}\psi = \omega\,.$$

Substituting the plane wave solutions in (3.26) into the system above reduces it to
a system of two linear ODEs with the coefficients periodic in time,

$$(3.29) \qquad \frac{d\widehat{\omega}_k}{dt} = -(\cos\theta(t)k_1 - \sin\theta(t)k_3)\hat{\rho}_k(t)\,, \quad \frac{d\hat{\rho}_k}{dt} = \frac{k_1}{|\vec{k}|^2}\widehat{\omega}_k(t)\,.$$

These equations are equivalent to the classic Hill's equation [12]

$$(3.30) \qquad \frac{d^2}{dt^2}\hat{\rho}_k = -P(t)\hat{\rho}_k(t)\,, \quad P(t) = \frac{k_1}{|\mathbf{k}|^2}[k_1\cos\theta(t) - k_3\sin\theta(t)]\,.$$

To investigate whether perturbations grow in time and have instability, we use
elementary Floquet theory for the time-periodic equations in (3.30). A good reference on Floquet theory is Hochstadt's ODE book [12]. The basic facts are listed
at the end of this chapter. This theory yields the criterion that for instability, we
need to check only that the trace of the Floquet matrix exceeds 2 in magnitude.
Clearly, the periodic function $P(t)$ in (3.30) depends on the direction of $\mathbf{k}$ but not
on the magnitude of $\mathbf{k}$. Thus, we set $\mathbf{k} = (k_1, k_3) = (\sin\theta_0, -\cos\theta_0)|\mathbf{k}|$, where the
angle θ_0 parameterizes this variation in (3.30). In Figure 3.1, we give an explicit
demonstration of instability at arbitrarily large Richardson numbers.

In Figure 3.1(a), through numerical evaluation, we graph the value of the Floquet trace, maximized over all angles θ_0 for Richardson numbers $1 \leq$ Ri $\leq 10^6$.

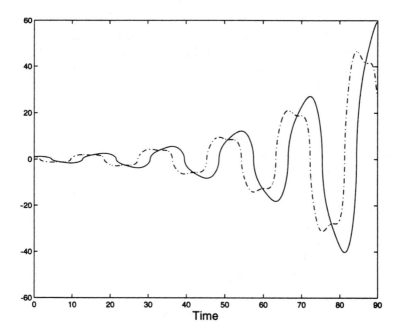

FIGURE 3.2. Amplitude of linearized perturbation vorticity (solid line) and perturbation density (broken line) undergoing exponential growth at twice the period of the underlying pendulum flow. From *Chaos*, vol. 10 (2000), no. 1, p. 7, fig. 1. Copyright © American Institute of Physics.

Even though the growth rates become extremely small at these large Richardson numbers, nevertheless, there is unambiguous evidence for instability. In Figure 3.1(b), we plot the Floquet trace as a function of the angle θ_0 for the representative moderately large Richardson number Ri $= 5$. The angle θ_0^* with the two peaks in the growth rate correspond to a parametric instability where the instability grows with twice the period of the underlying fluid flow. See Figure 3.2. The qualitative features displayed here for Ri $= 5$ in Figure 3.1(b) in fact persist for all Richardson numbers investigated for Ri $> \frac{1}{4}$. Thus, linear instability theory predicts the spontaneous formation of density stratified shears aligned in preferred directions of variation near the direction θ_0^*. How do these instabilities alter the nonlinear dynamics in the stably stratified flow? We address this issue next.

3.2. Nonlinear Instability of Stratified Flows

Numerical Method and Diagnostics. Natural quantities that measure the intensity of evolving perturbations are the kinetic and potential energy defined as

$$E(t) = \text{K.E.}(t) + \text{P.E.}(t),$$

(3.31)
$$\text{K.E.}(t) = \frac{1}{2} \iint |\vec{v}(\vec{\xi}, t)|^2 \, d\xi_1 \, d\xi_3, \quad \text{P.E.}(t) = \frac{1}{2} \iint \rho^2(\vec{\xi}, t) d\xi_1 \, d\xi_3.$$

Here, $\vec{v}$ and ρ are the perturbation Eulerian velocity and density fields. We point out that none of the energies above is conserved, as they can exchange energy with each other and also with the mean flow. Large portions of energy extracted from the mean flow may signify the onset of instability. In this paper we will consider perturbations with periodic structure in space, so that the integrations in the formulas in (3.31) above should be understood as integrals over the doubly periodic domain in Lagrangian coordinates.

The nonlinear stability of elementary flows in two space dimensions is studied here. With the nonlinear equations for perturbations in mean Lagrangian coordinates, we utilize a standard filtered pseudospectral method with fourth-order Runge-Kutta time differencing. Both for the physical interpretation of results as well as a numerical check on the accuracy of the time-stepping procedure, we monitor the kinetic and potential energy of perturbations from (3.31). All of the simulations reported below utilize $(128)^2$ Fourier modes, and the initial data for perturbations is concentrated uniformly on the low wavenumber band $|k_1| + |k_3| \leq 10$ with random phases and total energy fluctuations representing 10% of the mean flow value.

Both to monitor the accuracy of the simulations with the given spatial resolution as well as to give physical insight into the nonlinear transfer of energy between scales, we monitored the energy in the six bands,

$$|k_1| + |k_3| \leq 10,$$
$$10 < |k_1| + |k_3| \leq 20,$$
$$20 < |k_1| + |k_3| \leq 30,$$
$$30 < |k_1| + |k_3| \leq 40,$$
$$40 < |k_1| + |k_3| \leq 50,$$
$$50 < |k_1| + |k_3| \leq 64.$$

In all the simulations reported below, through this diagnostics, we concluded that the energy of perturbations remained confined essentially to wavenumbers $0 < |k_1| + |k_3| \leq 30$, so that the spatial resolution of $(128)^2$ Fourier modes is completely justified.

Pendulum Flows and Overturning. To investigate whether perturbations to vortical flows, given by (3.15), grow in time in the fully nonlinear regime, we integrate the equations in (3.21) with the standard pseudospectral method mentioned above on a rectangular grid with a resolution of $(128)^2$ Fourier modes. We chose random large-scale initial data by assigning random amplitudes and random phases to Fourier modes with wavenumbers restricted by $|k_1| + |k_3| \leq 10$. The values of the random amplitudes were selected in such a way that the initial energy fluctuations are 10% of the mean flow energy. The main results will be summarized in this section for Ri $= 3$.

We illustrate the development of perturbations with plots of perturbation energy and some snapshots of representative density contours, velocity field, spectral energy distribution, and the regions of overturning. Local overturning takes place when the total density gradient satisfies $d\tilde{\rho}/dz > 0$. It is convenient to take the snapshots at the whole periods of the mean vortical flow in (3.1), when the Eulerian and Lagrangian coordinate systems coincide exactly. As an illustrative example we selected a vortical mean flow with Ri $= 3$, with the period roughly equal to 9 buoyancy times. We note, as a side remark, that value 3 of the Richardson number corresponds to Ri $= 12$ when the conventional definition of Richardson number is used and thus is 48 times greater than the critical value given by the Miles-Howard theorem.

Figures 3.3 through 3.5 show flow parameters after four, six, and eight periods of the mean flow. The slight perturbations at four periods shown in Figure 3.3 present only very limited overturning, depicted by shaded regions in Figure 3.3(b), and are strongly aligned along the direction with the inclination angle $\alpha_0 \approx 2.75$. This direction is easily identifiable with the direction of fastest growth predicted by the linear theory and shown in Figure 3.1. Note that at this stage most of the perturbation energy is concentrated in large scales, as shown in Figure 3.3(c). At the later times, presented in Figures 3.4 and 3.5, the perturbations cause violent overturning throughout the domain of integration. The velocity field in Figure 3.4(d) and overturning regions in Figure 3.4(b) are still largely organized along the preferred direction predicted by the linear theory. The regions of strong shear are visible in Figure 3.4(d). According to conventional expectations, they become subject to shear instability, whose typical signature is obvious in Figure 3.5. The overturning regions shown in Figure 3.5(b) overtake the whole integration domain, with well-formed intensive structures resembling Kelvin-Helmholtz billows. Coherent same-sign vortices corresponding to the overturning regions appear in Figure 3.5(d). We point out that the spectrum of perturbation energy undergoes considerable spreading, with substantial amounts transferring to the smaller scales, as seen in Figures 3.3(c), 3.4(c), and 3.5(c).

Plots of the kinetic and potential energy budgets of the perturbation, shown in Figure 3.6, give a very revealing picture of the instability development. At early stages, when perturbation amplitudes are relatively small, parametric instability of the linear theory, shown in Figure 3.2, is clearly the main driving mechanism of amplification. Near $t = 80$ nonlinear interactions start dominating the flow and change the picture completely. The time of transition to the nonlinear regime of instability development correlates very well with the emergence of the direct cascade to small scales in the spectrum and formation of the Kelvin-Helmholtz-like billows shown in Figures 3.3(c), 3.4(c), and 3.5(c). Although the energy spectrum spreads over three bands with $|k_1| + |k_3| \leq 30$, we note that our numerical procedure produces reliable answers, since the bands of the spectrum with $64 \geq |k_1| + |k_3| > 30$ remain very weakly excited. Qualitatively similar results were observed in numerical experiments with vortical flows at Ri $= 1, 5, 10$, which are not presented here. As one may naturally expect, instabilities take a longer time to develop as the Richardson numbers grow, but strong overturning with nonlinear saturation and structures resembling Kelvin-Helmholtz instability were observed at all Richardson numbers we studied [23, 24, 25].

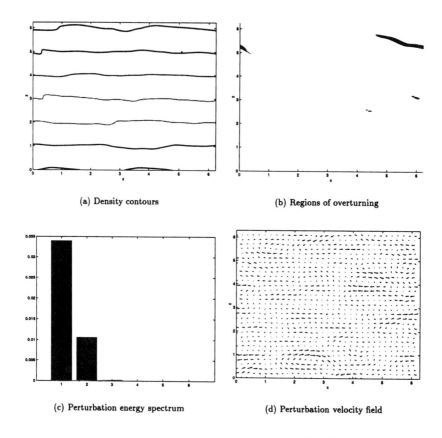

(a) Density contours   (b) Regions of overturning

(c) Perturbation energy spectrum   (d) Perturbation velocity field

FIGURE 3.3. Density contours, regions of local overturning, spectral distribution of energy and velocity field of perturbation to the vortical flow with Ri $= 3$, at $t = 36$. Heights of the six bars in the energy distribution plot indicate the amount of energy allocated in Fourier modes with wavenumbers $|k_1| + |k_3| \leq 10$, $10 < |k_1| + |k_3| \leq 20$, $20 < |k_1| + |k_3| \leq 30$, $30 < |k_1| + |k_3| \leq 40$, $40 < |k_1| + |k_3| \leq 50$, and $50 < |k_1| + |k_3| \leq 64$. From *Chaos*, vol. 10 (2000), no. 1, p. 14, fig. 9. Copyright © American Institute of Physics.

3.3. Shear Flows

Here we also investigate the instability properties of elementary linear shear flows, given by (2.25) with $a_2 = 0$, utilizing the pseudospectral code in mean Lagrangian coordinates. We are testing for a numerical confirmation of the Miles-Howard theorem in (3.9) in the nonlinear regime. The preliminary runs with random large-scale initial conditions, such as described in Section 3.2, revealed no significant amplification in the perturbation amplitudes for all flows tested, with Ri $= 1, 3, 5$. The results we report here were produced by integrating similar Lagrangian equations for perturbations of shear flows as in (3.15) (see [**23, 24, 25**])

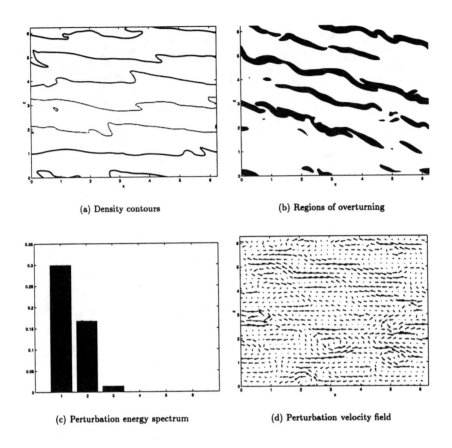

(a) Density contours

(b) Regions of overturning

(c) Perturbation energy spectrum

(d) Perturbation velocity field

FIGURE 3.4. As for Figure 3.3, at $t = 54$. From *Chaos*, vol. 10 (2000), no. 1, p. 15, fig. 10. Copyright © American Institute of Physics.

with the initial conditions of the following form. We select the plane wave whose direction provides the fastest transient growth within the linear theory as described in Section 3.2 for pendulum flows, and added a random large-scale perturbation to this wave in order to include the nonlinear interactions. The amplitudes of initial conditions were selected in such a way that the initial energy of fluctuations totaled 10%, with 90% of that energy assigned to the preferred wave, and the remaining 10% distributed uniformly with random phase among the other large-scale modes. Here we deliberately chose a flow with the smallest value of the Richardson number, $Ri = 1$, to illuminate the robustness of stability demonstrated by this study.

The evolution of perturbations is illustrated by the characteristic density contours, velocity field, and the spectral distribution of perturbation energy plotted at the initial moment and at $t = 4.5$, $t = 9$, and $t = 18$ buoyancy times, shown in Figures 3.7 through 3.10. The density contours shown in Figures 3.7(a), 3.8(a), 3.9(a), and 3.10(a) are plotted in the Eulerian frame to emphasize that no overturning takes place for this flow. The Eulerian components of the velocity field, shown

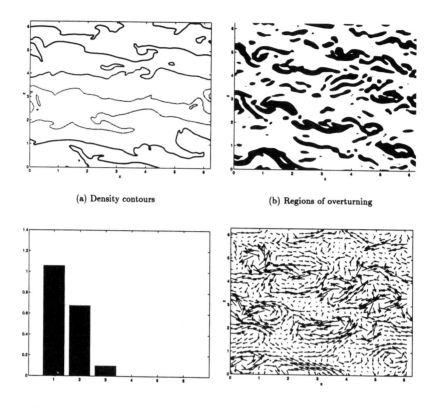

(a) Density contours   (b) Regions of overturning

FIGURE 3.5. As for Figure 3.3, at $t = 72$. From *Chaos*, vol. 10 (2000), no. 1, p. 16, fig. 11. Copyright © American Institute of Physics.

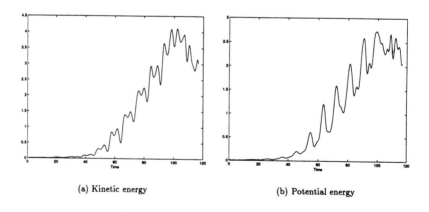

(a) Kinetic energy   (b) Potential energy

FIGURE 3.6. Kinetic and potential energy of perturbation for vortical flow with Ri $= 3$. From *Chaos*, vol. 10 (2000), no. 1, p. 17, fig. 12. Copyright © American Institute of Physics.

in Figures 3.7(c), 3.8(c), 3.9(c), and 3.10(c), are stripped from the mean shear piece and are mapped back into the Lagrangian frame to show the detailed structure of the perturbation field.

A general picture of perturbation evolution is given by the plots of potential and kinetic energy in Figure 3.11. At an early time $t = 4.5$, presented in Figure 3.8, the perturbation velocity field is growing and is visibly organized along the preferred direction. At this time, the density contours are only slightly deflected from their unperturbed horizontal locations. At a later time, $t = 9$, the density contours are visibly deflected (see Figure 3.9(a)) due to the strong shearing effect of the mean flow; however, no overturning is produced. The velocity field shown in Figure 3.9(c) at the same moment is strongly amplified and has the structure of a quasi-traveling gravity wave going through the domain rather than a shear; the arrows reflecting streamlines are directed towards the wave fronts rather than along them. In Figure 3.10 we observe the complete decay of the perturbations occurring at $t = 18$. We remark on the spectral distribution of perturbation energy. In contrast to an intensive direct energy cascade observed for purely vortical flows and discussed in Section 3.2, there is no apparent energy propagation to the smaller scales in Figures 3.8(b), 3.9(b), and 3.10(b). Perturbation motion remains largely confined to the large scales, which strongly correlates with the absence of small-scale secondary vortices, which are a signature of Kelvin-Helmholtz instability; in fact, as time progresses, there is an inverse cascade nearly back to the mean flow.

Finally, we present the plots of kinetic and potential energy of perturbations in Figure 3.11. Note that intermittent behavior of perturbations for shears differs greatly from parametric instability observed for the purely vortical pendulum flows. Instead, we see the perturbation amplitudes grow and peak very rapidly at $t \approx 10$ at values comparable with instability levels for vortical flows, followed by a rapid decay to the level of numerical noise at $t \approx 20$. Despite relatively large transient perturbation energy, no overturning is generated. We believe that this property is reflected in the fact that only kinetic energy amplifies strongly, while potential energy, linked directly by its definition in (3.31) to density perturbations, reaches only moderate values of roughly 0.5.

As a final comment, we mention the results of numerical simulations for shear flows at even smaller Richardson numbers. In this range, nonlinear stability becomes susceptible to the magnitude of initial perturbations. We discovered that the shear flow with Ri $= 0.5$ was stable to perturbations with the initial energy equal to 5% of the mean flow energy. At the same time the shear flow with Ri $= 0.2$ generated overturning when we perturbed it with fluctuations of the same strength. A more precise estimate of the stability threshold can be obtained by using weaker initial perturbations.

These results show that linear shear flows are stable to perturbations at large Richardson numbers as expected from the Miles-Howard theorem. It would be very interesting to have rigorous nonlinear stability theorems supporting this observed behavior. However, through a combination of mathematical theory and numerical experiments, we have shown that the pendulum flows are unstable even with very

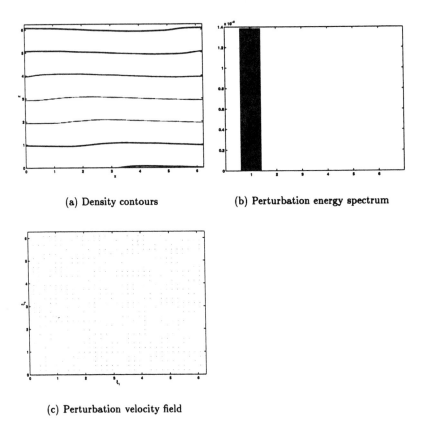

(a) Density contours   (b) Perturbation energy spectrum

(c) Perturbation velocity field

FIGURE 3.7. Density contours, spectral distribution of energy, and velocity field of perturbation to the shear flow with Ri = 1, at initial time $t = 0$. Bars in the plot of spectral distribution of energy are as in Figure 3.3. From *Chaos*, vol. 10 (2000), no. 1, p. 17, fig. 13. Copyright © American Institute of Physics.

strong stratification. Furthermore, this instability leads to local overturning of the density contours.

3.4. Some Background Facts on ODEs

Fundamental Matrix. For any nonautonomous linear system of ordinary differential equations $\frac{d\mathbf{x}}{dt} = \mathcal{A}(t)\mathbf{x}$, one can define a fundamental matrix $\mathcal{R}(t)$ that solves

$$(3.32) \qquad\qquad \mathcal{R}' = \mathcal{A}(t)\mathcal{R}, \quad \mathcal{R}(0) = \mathbf{I}.$$

Then any solution of a general initial value problem

$$(3.33) \qquad\qquad \mathbf{x}' = \mathcal{A}(t)\mathbf{x}, \quad \mathbf{x}(0) = \mathbf{x}_0$$

can be written as $\mathbf{x}(t) = \mathcal{R}(t)\mathbf{x}_0$.

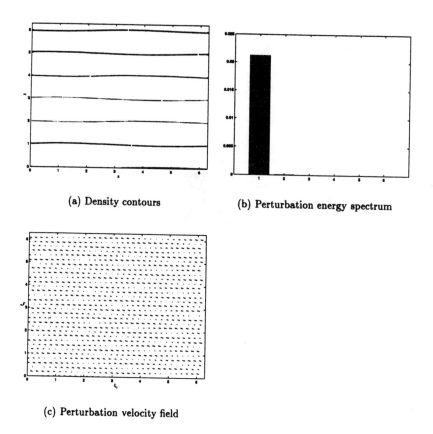

(a) Density contours   (b) Perturbation energy spectrum

(c) Perturbation velocity field

FIGURE 3.8. As in Figure 3.7, at $t = 4.5$. From *Chaos*, vol. 10
(2000), no. 1, p. 18, fig. 14. Copyright © American Institute of
Physics.

Liouville's Theorem. Liouville's theorem claims that for any flow described
by a linear system with time-dependent coefficients, the determinant of the fundamental matrix can be expressed in terms of the trace of the system matrix,

$$(3.34) \qquad \det \mathcal{R}(t) = \exp \int_0^t \operatorname{tr} \mathcal{A}(s)ds .$$

Periodicity. Floquet theory says that for any T-periodic matrix $\mathcal{A}(t)$ there
exists a matrix $\mathcal{P}(t)$ and constant matrix $\mathcal{F}$ such that

$$(3.35) \qquad \mathcal{R}(t) = \mathcal{P}(t)e^{\mathcal{F}t}, \quad \mathcal{P}(t+T) = \mathcal{P}(t) .$$

Since $\mathcal{P}(t+T) = \mathcal{P}(t)$ and $\mathcal{R}(T) = \mathcal{P}(T)e^{\mathcal{F}T} = e^{\mathcal{F}T}$, it suffices to show that all
eigenvalues of $e^{\mathcal{F}T}$ are not greater than 1 to prove stability and to find an eigenvalue
larger than 1 to have instability.

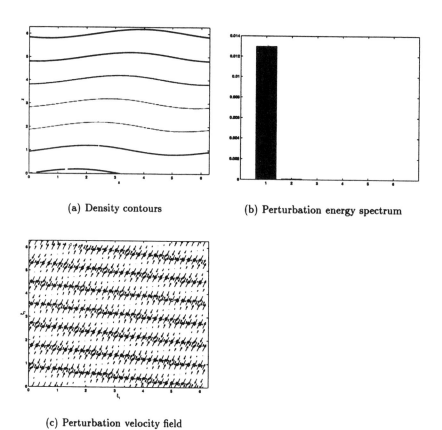

(a) Density contours   (b) Perturbation energy spectrum

(c) Perturbation velocity field

FIGURE 3.9. As in Figure 3.7, at $t = 9$. From *Chaos*, vol. 10 (2000), no. 1, p. 19, fig. 15. Copyright © American Institute of Physics.

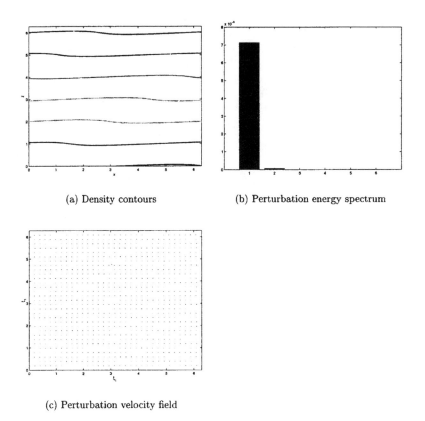

(a) Density contours (b) Perturbation energy spectrum

(c) Perturbation velocity field

FIGURE 3.10. As in Figure 3.7, at $t = 18$. From *Chaos*, vol. 10 (2000), no. 1, p. 20, fig. 16. Copyright © American Institute of Physics.

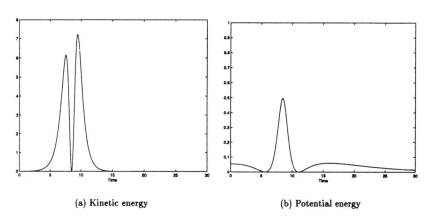

(a) Kinetic energy (b) Potential energy

FIGURE 3.11. Kinetic and potential energy of perturbation for shear flow with Ri = 1. From *Chaos*, vol. 10 (2000), no. 1, p. 20, fig. 17. Copyright © American Institute of Physics.

CHAPTER 4

Rotating Shallow Water Theory

Rotating shallow water theory is an appropriate approximation for atmospheric and oceanic motions in the midlatitudes with relatively large length and time scales. For example, the study of large-scale motions of atmospheric weather patterns involves scales of horizontal motion of the order of 1000 km, whereas the scale of vertical motion is only of the order of 10 km. Clearly, the shallow water approximation is then appropriate. On the other hand, the effects of the earth's rotation are important in situations where the fluid motions evolve on a time scale t_U that is comparable or longer than the time scale of rotation t_R (this corresponds to small Rossby number Ro $= t_R/t_U$). For example, in the midlatitudes the (Coriolis) rotation frequency is of the order of 10^{-4} s^{-1}. With scales of motion of the order of 1000 km and wind velocities of the order of 10 ms^{-1}, the time scale of the fluid motion dynamic is of the order of days ($\approx 10^5$ s). In this case the Rossby number is Ro ≈ 0.1 and the effects of the earth's rotation are relevant.

The rotating shallow water equations are capable of describing important aspects of atmospheric and oceanic motions, but as we will see later, these equations do not include the effects of density stratification discussed in the models of the previous chapters. Therefore, rotating shallow water theory applies to phenomena that do not depend crucially on temporal changes of the density stratification. Recall from Chapter 1 that we derived the linearized shallow water equations from the primitive equations by expanding the vertical structure of stratification in time-independent eigenmodes. Here we begin by summarizing the more traditional route for the derivation.

4.1. Rotating Shallow Water Equations

The conventional formulation of the rotating shallow water equations is based on the following assumptions:

Assumption 1: The scale L of horizontal motion is much larger than the scale D of vertical motion, so that $D/L \ll 1$.

Assumption 2: The three-dimensional fluid is incompressible and homogeneous, with the constant density normalized to 1 for convenience.

Assumption 3: The external force is due to gravity, and we assume the hydrostatic approximation, where $\partial p/\partial x_3 = -g$.

Assumption 4: The axis of rotation of the fluid coincides with the vertical x_3-axis, and the (Coriolis) frequency of rotation is f.

Assumption 5: The bottom surface topography is given by $x_3 = h_B(x_1, x_2)$. In this situation, $x_3 = h(x_1, x_2, t)$ defines the top fluid surface and $H = h(x_1, x_2, t) - h_B(x_1, x_2)$ is the total depth of the fluid.

With these assumptions we have the following:

Rotating Shallow Water Equations. The rotating shallow water equations are

$$(4.1) \qquad \frac{D\vec{v}}{Dt} + f\vec{v}^{\perp} = -g\nabla h,$$

$$(4.2) \qquad \frac{DH}{Dt} + H \operatorname{div} \vec{v} = 0,$$

where $\vec{v} = (v_1(x_1, x_2, t), v_2(x_1, x_2, t))$ is the horizontal component of the fluid velocity, $\vec{v}^{\perp} = (-v_2, v_1)$ is the "orthogonal velocity," $H = h - h_B$ is the total depth, f is the rotation frequency, g is the acceleration of gravity, and $\frac{D}{Dt} = \frac{\partial}{\partial t} + \vec{v} \cdot \nabla$ is the material derivative.

For the complete derivation of the rotating shallow water equations starting with the three-dimensional rotating incompressible Euler equations, the reader is referred to Pedlosky [**29**, pp. 59–63].

Here we remark that the rotating shallow water equations are a two-dimensional system, with the variables $\vec{v}$ and H depending only on (x_1, x_2, t). The momentum equation in (4.1) does not include vertical motions, and the forcing effects are due to rotation and the hydrostatic approximation. Equation (4.2) arises from integrating the three-dimensional incompressibility condition

$$\frac{\partial v_1}{\partial x_1} + \frac{\partial v_2}{\partial x_2} + \frac{\partial v_3}{\partial x_3} = 0$$

in the vertical variable x_3 and imposing the top surface kinematic boundary condition and the bottom surface topography. As a result, the incompressibility condition becomes an evolution equation balancing changes in the total depth H and the horizontal divergence $\operatorname{div} \vec{v}$. Equation (4.2) clearly shows that, for the incompressible fluid, a positive horizontal divergence produces a decrease in H, and conversely, a negative horizontal divergence will make H grow.

Compressible Isentropic Fluid Flow. The rotating shallow water equations are similar in form to the equations of compressible fluid flow. More precisely, the equations for compressible isentropic gas dynamics are given by

$$(4.3) \qquad \frac{D\vec{v}}{Dt} + \frac{1}{\rho}\nabla p = 0,$$

$$(4.4) \qquad \frac{D\rho}{Dt} + \rho \operatorname{div} \vec{v} = 0,$$

where the pressure p and the density ρ are related by the isentropic ideal gas law equation

$$(4.5) \qquad p = A\rho^{\gamma},$$

with A and γ constants, and $\gamma > 1$. Then we observe that in the absence of rotation effects ($f = 0$) and with flat bottom topography ($h_B = 0$), the rotating shallow water equations reduce to

$$(4.6) \qquad \frac{D\vec{v}}{Dt} + g\nabla H = 0,$$

$$(4.7) \qquad \frac{DH}{Dt} + H\,\mathrm{div}\,\vec{v} = 0,$$

so that the equations in (4.6) and (4.7) are identified with the isentropic compressible fluid flow equations (4.3) and (4.4), provided that we identify H with ρ, and the state equation relating p and ρ in (4.5) takes the particular values of $A = g/2$ and $\gamma = 2$.

We include, for future reference, the conservation form of the compressible fluid flow equations and the rotating shallow water equations. The *compressible isentropic fluid flow equations in conservation form* are

$$(4.8) \qquad \frac{\partial(\rho\vec{v})}{\partial t} + \mathrm{div}(\rho\vec{v}\otimes\vec{v} + pI) = 0,$$

$$(4.9) \qquad \frac{\partial\rho}{\partial t} + \mathrm{div}(\rho\vec{v}) = 0.$$

The *rotating shallow water equations in conservation form* are

$$(4.10) \qquad \frac{\partial(H\vec{v})}{\partial t} + \mathrm{div}\left(H\vec{v}\otimes\vec{v} + \frac{g}{2}H^2 I\right) = -(f\vec{v}^{\perp} + g\nabla h_B)H,$$

$$(4.11) \qquad \frac{\partial H}{\partial t} + \mathrm{div}(H\vec{v}) = 0,$$

where I in equation (4.8) and (4.10) is the 2×2 identity matrix. Equation (4.10) is derived by adding H times the momentum equation (4.1) to $\vec{v}$ times the conservation of mass equation (4.2) (and similarly for the momentum equation (4.8)). In the conservation form of the momentum equation we have retained the rotation and bottom topography contributions as source terms. If we assume no rotation and flat bottom topography, then the equations again reduce to a special case of the compressible flow equations (4.8) and (4.9).

Although the rotating shallow water equations are mathematically similar to the compressible flow equations, whether or not "compressibility" effects are important depends on the scales associated with the fluid motion. In gas dynamics the measure of the importance of the compressibility effects is given by the Mach number, the ratio of the representative fluid velocity U to the speed of sound $c = \sqrt{dp/d\rho}$ (see [16]). For the rotating shallow water equations, the analogue of the Mach number is given by the Froude number Fr, defined here as the ratio of the representative fluid velocity U to the gravity wave speed

$$c = \sqrt{\frac{d}{dH}\left(\frac{g}{2}H^2\right)} = \sqrt{gH}.$$

For the relevant situation concerning large-scale motions in the atmosphere, the typical fluid flow velocity is $U = 10\ \mathrm{ms}^{-1}$ and the typical value of the total depth is $H = 10$ km, so that the Froude number is Fr $= 0.03 \ll 1$. Therefore, in this situation we do not expect that the compressibility effects due to the nonlinear nature of the gravity wave velocity will be relevant in describing the behavior of the flow in such regimes.

4.2. Conservation of Potential Vorticity

For the two-dimensional rotating shallow water equations in (4.1) and (4.2), the vorticity ω is basically a scalar, $\omega = (0, 0, \omega)$, with

$$\omega = \frac{\partial v_2}{\partial x_1} - \frac{\partial v_1}{\partial x_2}.$$

To derive an equation for the evolution of the vorticity, we take the curl of equation (4.1) for the conservation of momentum, and use the calculus identities

$$\text{curl}\left(\frac{D\vec{v}}{Dt}\right) = \frac{D\omega}{Dt} + \omega \, \text{div} \, \vec{v}, \quad \text{curl}(f\vec{v}^\perp) = \text{div}(f\vec{v}) = f \, \text{div} \, \vec{v} + \frac{Df}{Dt},$$

which are straightforward to check (alternatively, the reader can use the approach developed in Chapter 2 where an evolution equation for the gradient matrix V was derived, and where the antisymmetric part of that equation produces the equation for the vorticity). By using the above two calculus identities, the resulting equation for the vorticity is given by

$$(4.12) \qquad \frac{D}{Dt}(f + \omega) = \frac{Df}{Dt} + \frac{D\omega}{Dt} = -(\omega + f) \, \text{div} \, \vec{v}.$$

We can eliminate the term involving div $\vec{v}$ in (4.12) as follows: Multiply the vorticity equation (4.12) by H and the conservation of mass equation (4.2) by $\omega + f$, and subtract them to yield

$$(4.13) \qquad H\frac{D}{Dt}(f + \omega) - (\omega + f)\frac{DH}{Dt} = 0.$$

We can then divide (4.13) by H^2 to finally get

$$(4.14) \qquad \frac{D}{Dt}\left(\frac{\omega + f}{H}\right) = 0.$$

Equation (4.14) says that the ratio of the absolute vorticity $\omega + f$ and the effective depth H is conserved along the particle trajectories of the flow. It is the analogue of Ertel's potential vorticity conservation that we derived for the Boussinesq equations in Chapter 2. Thus, the ratio $\Pi_s = \frac{\omega + f}{H}$ is called the *potential vorticity* and provides a powerful constraint on the large-scale motions of the atmosphere. For example, if ω is constant initially, then the only way that ω remains constant at a latter time is if H itself is constant. In general, the conservation of potential vorticity tells us that if H increases then ω must increase, and conversely, if H decreases, then ω must decrease provided f is constant.

4.3. Nonlinear Conservation of Energy

Besides the principles of conservation of mass and momentum upon which the rotating shallow water equations (4.1) and (4.2) are formulated, the conservation of energy is another principle of great physical and mathematical importance. Here we derive the differential and integral forms of the conservation of energy, and for simplicity we assume that the bottom topography is flat, so that $h_B(x_1, x_2) \equiv 0$ and $H = h$. We know that the kinetic energy K.E. is given by K.E. $= h\vec{v} \cdot \vec{v}/2$; to derive the equation for the total energy we need to find an appropriate potential energy so that the time derivative of the resulting total energy is a perfect divergence. A helpful observation is that the equation in (4.2) for the conservation of mass implies the general identity

$$(4.15) \qquad \frac{\partial}{\partial t}(hW) + \mathrm{div}(hW\vec{v}) = W\left(\frac{\partial h}{\partial t} + \mathrm{div}(h\vec{v})\right) + h\frac{DW}{Dt} = h\frac{DW}{Dt},$$

which is valid for any function $W(x_1, x_2, t)$. If we apply first the identity in (4.15) to the kinetic energy K.E. $= h\vec{v} \cdot \vec{v}/2$, we obtain

$$\frac{\partial}{\partial t}\left(\frac{h}{2}\vec{v} \cdot \vec{v}\right) + \mathrm{div}\left(\left(\frac{h}{2}\vec{v} \cdot \vec{v}\right)\vec{v}\right) = h\frac{D}{Dt}\left(\frac{1}{2}\vec{v} \cdot \vec{v}\right) = h\vec{v} \cdot \frac{D\vec{v}}{Dt},$$

and to evaluate the right-hand side of the expression above we utilize the conservation of momentum

$$h\vec{v} \cdot \frac{D\vec{v}}{Dt} = h\vec{v} \cdot (-f\vec{v}^{\perp} - g\nabla h) = -g\vec{v} \cdot \nabla\left(\frac{h^2}{2}\right),$$

so that we conclude that the equation for the kinetic energy is

$$(4.16) \qquad \frac{\partial}{\partial t}\left(\frac{h}{2}\vec{v} \cdot \vec{v}\right) + \mathrm{div}\left(\left(\frac{h}{2}\vec{v} \cdot \vec{v}\right)\vec{v}\right) = -g\vec{v} \cdot \nabla\left(\frac{h^2}{2}\right).$$

For the potential energy P.E. we consider P.E. $= gh^2/2$. We apply the general identity (4.15) to the potential energy and substitute the conservation of mass equation (4.2) to obtain

$$(4.17) \qquad \frac{\partial}{\partial t}\left(\frac{1}{2}gh^2\right) + \mathrm{div}\left(\frac{1}{2}gh^2\vec{v}\right) = h\frac{D}{Dt}\left(\frac{1}{2}gh\right) = -\frac{1}{2}gh^2\,\mathrm{div}\,\vec{v}.$$

Now let the total energy E be defined as the sum of the kinetic and the potential energies. Adding equations (4.16) and (4.17) for the kinetic and potential energies yields

$$\frac{\partial E}{\partial t} + \mathrm{div}(E\vec{v}) = -\mathrm{div}\left(\frac{1}{2}gh^2\vec{v}\right).$$

In summary, we have the following:

PROPOSITION 4.1 *The rotating shallow water equations satisfy the nonlinear differential equation of conservation of energy density*

$$(4.18) \qquad \frac{\partial E}{\partial t} + \mathrm{div}\left(\left(E + \frac{1}{2}gh^2\right)\vec{v}\right) = 0,$$

where the total energy E in equation (4.18) *is given by the sum of the kinetic and potential energies,* $E = (h\vec{v} \cdot \vec{v} + gh^2)/2$.

If we integrate the conservation of energy equation (4.18) over the disk $|\vec{x}| \leq R$, and assume that the energy flux density decays fast enough at infinity, for example, $(E + gh^2/2)\vec{v} = o(R^{-2})$ as $R \to \infty$, then the divergence theorem shows that the energy flux contributions on the surface $|\vec{x}| = R$ vanish in the limit of $R \to \infty$ and we obtain as a corollary the conservation of energy over the plane.

COROLLARY 4.2 *For sufficiently rapidly decaying smooth solutions of the rotating shallow water equations, the total energy is conserved, that is*

$$(4.19) \qquad \frac{d}{dt} \int_{R^2} E(\vec{x}, t)d\vec{x} = 0.$$

An interesting elementary exercise is to formulate and prove a principle of conservation of energy when the topography is nonzero.

There are other conserved quantities associated with the shallow water equations. For example, if $q = \frac{f+\omega}{H}$ is the potential vorticity, then for any function $G(q)$,

$$\frac{\partial}{\partial t}(HG(q)) + \mathrm{div}(HG(q)\vec{v}) = 0;$$

thus

$$\int HG(q) \text{ is a conserved quantity}$$

provided that this integral converges. These statements are an easy exercise for the reader to check.

4.4. Linear Theory for the Rotating Shallow Water Equations

In the study of the rotating shallow water equations a great deal of insight can be gained by investigating the associated linearized equations. In order to derive the linearized rotating shallow water equations, we assume a constant background state of the form $H = H_0$ and $\vec{v} = 0$, and we assume flat topography, so that $h_B(x_1, x_2) = 0$. Here we also assume that the rotation rate f is constant for simplicity. We consider small perturbations around the background state of the form

$$H = H_0 + \delta h, \qquad \vec{v} = \delta\vec{v},$$

where $\delta \ll 1$. Then we plug these expressions for H and $\vec{v}$ into the rotating shallow water equations (4.1) and (4.2), collect the terms of order $O(\delta)$, and discard the terms of higher order. The resulting equations are *linearized rotating shallow water equations*

$$(4.20) \qquad \frac{\partial \vec{v}}{\partial t} + f\vec{v}^{\perp} = -g\nabla h,$$

$$(4.21) \qquad \frac{\partial h}{\partial t} + H_0 \, \mathrm{div} \, \vec{v} = 0.$$

The linear equations in (4.20) and (4.21) will be analyzed in great detail below. We want to know what the solutions look like and what new effects are introduced by the rotation ($f \neq 0$). We will also make comparisons with the corresponding linearized equations for compressible gas flow when appropriate.

Case 1: Steady (Geostrophic Balance) Solutions. The steady state solutions of the linearized equations (4.20) and (4.21) must satisfy

$$(4.22) \qquad f \vec{v}^\perp = -g \nabla h \,,$$

$$(4.23) \qquad \operatorname{div} \vec{v} = 0 \,.$$

From (4.22) it follows that $\vec{v}$ satisfies

$$(4.24) \qquad \vec{v} = \frac{g}{f} \nabla^\perp h$$

where $\nabla^\perp = (-\partial/\partial x_2, \partial/\partial x_1)$ is the perpendicular gradient. The steady solution given in (4.24) represents an equilibrium state where the Coriolis force due to rotation is in balance with the pressure force due to the fluctuations in height. It is straightforward to verify that given any steady height function $h(x_1, x_2)$, the steady velocity field defined by (4.24) together with h satisfy equations (4.22) and (4.23). For the special form of $\vec{v}$ given in (4.24), the incompressibility condition in (4.23) reduces to

$$\operatorname{div} \vec{v} = \frac{g}{f} \left(\frac{\partial}{\partial x_1} \left(-\frac{\partial h}{\partial x_2} \right) + \frac{\partial}{\partial x_2} \left(\frac{\partial h}{\partial x_1} \right) \right) = 0 \,.$$

Clearly the solution $\vec{v}$ in (4.24) satisfies $\vec{v} \cdot \nabla h \equiv 0$, that is, $\vec{v}$ is tangent to the level curves of h. Therefore, the geostrophic balance solution in (4.24) is characterized by the fact that the streamlines of $\vec{v}$ coincide with the level curves (isobaths) of the height function h. In Section 4.6 we will formulate an important generalization of the geostrophic balanced flows, known as the quasi-geostrophic equations, in which h and $\vec{v}$ evolve in time while satisfying the geostrophic balance condition in equation (4.24).

Case 2: Poincaré Waves ("Inertio-Gravity Waves"). From the linearized rotating shallow water equations (4.20) and (4.21), let us try to derive a second-order wave equation in similar fashion as it is usually done for the acoustic waves in gas dynamics: If we differentiate in time the linearized equation for conservation of mass in (4.21), we get

$$\frac{\partial^2 h}{\partial t^2} = -H_0 \operatorname{div} \left(\frac{\partial \vec{v}}{\partial t} \right) = H_0 (f \operatorname{div} \vec{v}^\perp + g \Delta h) \,,$$

and utilizing the fact that $\operatorname{div} \vec{v}^\perp = \operatorname{curl} \vec{v} = \omega$, we get

$$(4.25) \qquad \frac{\partial^2 h}{\partial t^2} = h_0 f \omega + g H_0 \Delta h \,.$$

To compute an equation for ω, we linearize equation (4.14) for the potential vorticity

$$(4.26) \qquad \frac{\partial}{\partial t} \left(\frac{\omega}{H_0} \right) - \frac{f}{H_0^2} \frac{\partial h}{\partial t} = 0 \,,$$

so that if we now differentiate (4.25) once more in time and substitute the time derivative of ω by (4.26), we finally get the following:

PROPOSITION 4.3 *If h and $\vec{v}$ satisfy the linearized rotating shallow water equations (4.20) and (4.21), then $h_t = \frac{\partial h}{\partial t}$ solves the Klein-Gordon equation*

$$(4.27) \qquad \frac{\partial^2}{\partial t^2}(h_t) = g H_0 \Delta(h_t) - f^2 h_t \,.$$

We remark that this result does not contradict the geostrophic balance solutions discussed in case 1; although the height h in equation (4.24) is arbitrary, h is a steady state and therefore its time derivative $h_t \equiv 0$ solves the Klein-Gordon equation in (4.27). It is straightforward to compute harmonic plane wave solutions of the Klein-Gordon equation.

COROLLARY 4.4 $h = h_0 \exp i(\vec{k} \cdot \vec{x} - \omega(\vec{k})t)$, $\vec{k} = (k_1, k_2)$, *is a plane wave solution of the Klein-Gordon equation in (4.27) provided that $\omega(\vec{k})$ and $\vec{k}$ satisfy the dispersion relation*

$$(4.28) \qquad \omega(\vec{k}) = \pm\sqrt{g H_0 |\vec{k}|^2 + f^2} \,.$$

The dispersion relation in (4.28) shows that the Poincaré waves are dispersive in nature as described in detail in the next chapter, and that the frequency $\omega(\vec{k})$ is always greater than the rotation frequency f. In the limit of $f \to 0$, the dispersion relation (4.28) approaches the linear relation $\omega(\vec{k}) = \sqrt{g H_0}\,|\vec{k}|$; in this limit the wave is no longer dispersive, and in fact it becomes a gravity wave traveling with the phase velocity $c = \sqrt{gH}$. On the other hand, when $f \to \infty$ then $\omega(\vec{k}) \approx f$, the frequency of the rotating coordinate system, and in this case the Poincaré wave reduces to an inertia wave. Since Poincaré waves exhibit the behavior of both gravity and inertia waves, they are also known as inertia-gravity waves.

It is interesting to interpret the previous discussion in nondimensional units that suggest the length scales where rotation becomes important. Since the units of $|\vec{k}|$ are (length)$^{-1}$, we can rewrite the dispersion relation in (4.28) in the form

$$\omega(\vec{k}) = \pm\sqrt{gH}(|\vec{k}|^2 + L_R^{-2})^{1/2}$$

where the length scale is given by

$$L_R = \frac{\sqrt{gH}}{f}.$$

This length scale is called the *Rossby radius of deformation*. For large length scales, $L \gg L_R$, i.e., $|\vec{k}| \ll L_R^{-1}$, the effects of rotation dominate the dispersion relation, while for small length scales, $L \ll L_R$, i.e., $|\vec{k}| \gg L_R^{-1}$, the effects of rotation are insignificant.

Case 3: General Plane Wave Solutions for the Linearized Rotating Shallow Water Equations. Next we consider the construction of plane wave solutions for the full 3×3 system of equations in (4.20) and (4.21) for linearized rotating shallow water flow. If we denote by $\vec{u}$ the column vector $\vec{u} = {}^T(h, v_1, v_2)$, then the equations in (4.20) and (4.21) can be written in the general form

$$\vec{u}_t + A_1 \vec{u}_{x_1} + A_2 \vec{u}_{x_2} + B\vec{u} = 0 \,, \tag{4.29}$$

where A_1, A_2, and B are constant-coefficient matrices (the reader is encouraged to write them down from equations (4.20) and (4.21)). We seek plane wave solutions of (4.29) of the form

$$\vec{u} = e^{i(\vec{k}\cdot\vec{x} - \omega(\vec{k})t)} \vec{r} \,, \tag{4.30}$$

where, as usual, the solution of (4.29) is given by the real part of $\vec{u}$ in (4.30). In order to determine the constant vector $\vec{r}$ and the frequency $\omega(\vec{k})$, $\vec{k} = (k_1, k_2)$, we plug (4.30) back into (4.29) and conclude that $\vec{r}$ and $\omega(\vec{k})$ are the right eigenvector and eigenvalue of the associated eigenvalue problem

$$\mathcal{A}(\vec{k})\vec{r} = (-\omega(\vec{k})I + A_1 k_1 + A_2 k_2 - iB)\vec{r} = 0 \,, \tag{4.31}$$

where the matrix $\mathcal{A}(\vec{k})$ is given explicitly by

$$\mathcal{A}(\vec{k}) = \begin{pmatrix} -\omega(\vec{k}) & H_0 k_1 & H_0 k_2 \\ g k_1 & -\omega(\vec{k}) & if \\ g k_2 & -if & -\omega(\vec{k}) \end{pmatrix} \,. \tag{4.32}$$

The eigenvalues in (4.32) are given by the roots of the characteristic polynomial

$$0 = \det(\mathcal{A}(\vec{k})) = -\omega\big(\omega^2 - \big(gH_0|\vec{k}|^2 + f^2\big)\big) \,, \tag{4.33}$$

whose solutions are given by

$$\omega = 0, \pm\sqrt{gH_0|\vec{k}|^2 + f^2} \,,$$

corresponding to the steady waves and Poincaré waves studied earlier in cases 1 and 2. Next we construct the plane wave solutions for these wave velocities.

Case 3a: $\omega = \pm\sqrt{gH_0|\vec{k}|^2 + f^2}$ **(Poincaré Waves).** With this value of the frequency $\omega(\vec{k})$ the matrix $\mathcal{A}(\vec{k})$ in (4.32) can be reduced by rows to the equivalent form

$$\begin{pmatrix} 0 & 0 & 0 \\ -\dfrac{\omega k_1 + if k_2}{H_0|\vec{k}|^2} & 1 & 0 \\ -\dfrac{\omega k_2 - if k_1}{H_0|\vec{k}|^2} & 0 & 1 \end{pmatrix} \,, \tag{4.34}$$

and v_1 and v_2 are immediately solved in terms of h,

$$v_1 = \frac{\omega k_1 + if k_2}{H_0|\vec{k}|^2} h \,, \qquad v_2 = \frac{\omega k_2 - if k_1}{H_0|\vec{k}|^2} h \,. \tag{4.35}$$

Now, if we let h be the real part of $h_0 e^{i(\vec{k}\cdot\vec{x} - \omega(\vec{k})t)}$, then utililzing (4.35) we obtain the plane wave solution explicitly as follows: The height h is given by

$$h = h_0 \cos(\vec{k}\cdot\vec{x} - \omega(\vec{k})t) \,, \tag{4.36}$$

and the velocity $\vec{v}$ has an orthogonal decomposition in terms of the wavenumber vector $\vec{k} = (k_1, k_2)$ and the orthogonal wavenumber vector $\vec{k}^\perp = (-k_2, k_1)$,

$$(4.37) \qquad \vec{v} = v_\| \frac{\vec{k}}{|\vec{k}|} + v_\perp \frac{\vec{k}^\perp}{|\vec{k}|}$$

where the parallel and perpendicular components $v_\|$ and $v_\perp$ are given by

$$(4.38) \qquad v_\| = \frac{h_0 \omega(\vec{k})}{H_0 |\vec{k}|} \cos(\vec{k} \cdot \vec{x} - \omega(\vec{k}) t),$$

$$(4.39) \qquad v_\perp = \frac{h_0 f}{H_0 |\vec{k}|} \sin(\vec{k} \cdot \vec{x} - \omega(\vec{k}) t).$$

Equations (4.36), (4.37), (4.38), and (4.39) describe the plane wave solution when the frequency is

$$\omega = \pm\sqrt{g H_0 |\vec{k}|^2 + f^2}.$$

The parallel component $v_\|$ is associated with pressure waves, whereas the orthogonal component $v_\perp$ is associated with shearing motion. It is interesting to observe that the rotation is responsible for the appearance of the shear component $v_\perp$ in equation (4.39). Also notice that $v_\perp$ has a phase shift of $\pi/2$ relative to both $v_\|$ and h. From equations (4.37), (4.38), and (4.39) it also follows that the magnitude v of the velocity is given by

$$(4.40) \qquad v = \sqrt{v_\|^2 + v_\perp^2},$$

where $v_\|$ and $v_\perp$ lie on the ellipse

$$(4.41) \qquad \frac{v_\|^2}{\omega(\vec{k})^2} + \frac{v_\perp^2}{f^2} = \left(\frac{h_0}{H_0 |\vec{k}|}\right)^2.$$

Case 3b: $\omega = 0$ (Steady Waves). For the value of the frequency $\omega = 0$, the matrix $\mathcal{A}(\vec{k})$ is reduced by rows to the equivalent form

$$(4.42) \qquad \begin{pmatrix} 0 & 0 & 0 \\ -\frac{igk_1}{f} & 0 & 1 \\ \frac{igk_2}{f} & 1 & 0 \end{pmatrix},$$

and from equation (4.42) the components v_1 and v_2 of the velocity are given in terms of the height h by

$$(4.43) \qquad v_1 = -\frac{igk_2}{f} h, \qquad v_2 = \frac{igk_1}{f} h.$$

Therefore, if h is given by the real part of $f h_0 e^{i\vec{k} \cdot \vec{x}}$, then the plane wave solution corresponding to $\omega = 0$ is given by

$$(4.44) \qquad h = f h_0 \cos(\vec{k} \cdot \vec{x}),$$

$$(4.45) \qquad \vec{v} = -g h_0 |\vec{k}| \sin(\vec{k} \cdot \vec{x}) \frac{\vec{k}^\perp}{|\vec{k}|}.$$

The plane wave solution in equations (4.44) and (4.45) corresponds to pure shear motion, where $\vec{v}$ moves perpendicular to the direction $\vec{k}$ of propagation and with a phase shift of $\pi/2$ relative to the oscillations in the height h. The reader can also verify that the plane wave solution in equations (4.44) and (4.45) is a special case of the geostrophic balance solution previously discussed in case 1 (4.24).

Before concluding the discussion on the plane wave solutions of the rotating shallow water equations, it is appropriate to compare them with the corresponding plane wave solutions for the linearized equations of compressible fluid flow. We know that when we discard the effects of rotation and the topography ($f = 0$ and $h_B = 0$), the rotating shallow water equations become a particular case of the compressible fluid flow equations where we identify h and ρ, and the equation of state has the particular form $p = gh^2/2$. The linear theory for the compressible fluid flow equations is well known and can be done in entirely analogous and simpler fashion as the linear theory presented in this section for the rotating shallow water equations. Here we just summarize the results of the theory.

Linearized Compressible Fluid Flow Equations. The linearized equations associated with the compressible fluid flow equations (4.3) and (4.4) around the ground state $\rho = \rho_0$ and $\vec{v} = 0$ are given by

$$(4.46) \qquad \frac{\partial \rho}{\partial t} + \rho_0 \operatorname{div} \vec{v} = 0,$$

$$(4.47) \qquad \frac{\partial \vec{v}}{\partial t} + \frac{c_0^2}{\rho_0} \nabla \rho = 0,$$

where $c_0 = \sqrt{A\gamma \rho_0^{\gamma-1}}$ represents the speed of sound at the ground state. With $\vec{u} = {}^{\mathrm{T}}(\rho, v_1, v_2)$, the plane wave solutions $\vec{u}$ are given in (4.30), where ω and $\vec{r}$ solve the eigenvalue problem in (4.31). The associated matrix $\mathcal{A}(\vec{k})$ now has the form

$$(4.48) \qquad \mathcal{A}(\vec{k}) = \begin{pmatrix} -\omega(\vec{k}) & \rho_0 k_1 & \rho_0 k_2 \\ \frac{c_0^2}{\rho_0} k_1 & -\omega(\vec{k}) & 0 \\ \frac{c_0^2}{\rho_0} k_2 & 0 & -\omega(\vec{k}) \end{pmatrix}.$$

Acoustic Waves: $\omega(\vec{k}) = \pm c_0 |\vec{k}|$. The eigenvalues of $\mathcal{A}(\vec{k})$ are $\omega = 0$ and $\omega = \pm c_0 |\vec{k}|$, and the corresponding plane wave solutions are acoustic waves: $\omega(\vec{k}) = \pm c_0 |\vec{k}|$. In this case the plane wave solution has the form

$$(4.49) \qquad \rho = \tilde{\rho}_0 \cos(\vec{k} \cdot \vec{x} - \omega(\vec{k})t),$$

$$(4.50) \qquad \vec{v} = \pm \frac{\tilde{\rho}_0 c_0}{\rho_0} \cos(\vec{k} \cdot \vec{x} - \omega(\vec{k})t) \frac{\vec{k}}{|\vec{k}|}.$$

Vorticity Waves: $\omega(\vec{k}) = \mathbf{0}.$ In this case the plane wave solution has the form

$$(4.51) \qquad\qquad \rho = 0\,,$$

$$(4.52) \qquad\qquad \vec{v} = |\vec{v}|\cos(\vec{k}\cdot\vec{x})\frac{\vec{k}^{\perp}}{|\vec{k}|}\,.$$

As expected, when we set $f = 0$ to suppress the rotation effects, there is a complete correspondence between the linearized rotating shallow water equations (4.20) and (4.21) and the linearized compressible fluid flow equations (4.46) and (4.47). The matrix $\mathcal{A}(\vec{k})$ in (4.32) corresponds to (4.48) once we identify the sound speed c_0 with the gravity wave speed $\sqrt{gH_0}$. In particular, notice that when $f = 0$ there are no dispersive effects. The acoustic plane waves in (4.49) and (4.50) correspond to the Poincaré waves in (4.36), (4.37), (4.38), and (4.39). It is important to observe that because $f = 0$ the shear wave component $v_{\perp}$ vanishes and only the compressional wave component $v_{\|}$ remains. Finally, the vorticity plane wave in (4.51) and (4.52) corresponds to the steady state, geostrophic balance, plane wave solution in (4.44) and (4.45). Looking back at the geostrophic balance conditions in (4.22) and (4.23), we see that when $f = 0$ a steady solution cannot support spatial variations in h and that the incompressibility condition forces the periodic velocity field to move in the direction perpendicular to $\vec{k}$.

4.5. Nondimensional Form of the Rotating Shallow Water Equations

In order to study further the rotating shallow water equations, one has to take into account the magnitudes of the different physical parameters and the scales of the problem under consideration, where the relative importance of the different physical mechanisms is then measured by ratios of the corresponding scales. In this section we recast the rotating shallow water equations in nondimensional form, and in the next section we use the nondimensional equations to derive the quasi-geostrophic approximation, an important distinguished limit equation that applies in the limit of high rotation, and can be placed at an intermediate level between the steady state geostrophic approximation considered in Section 4.4 (4.24) and the full rotating shallow water equations.

Next we proceed to write the rotating shallow water equations (4.1) and (4.2) in nondimensional form. We denote by U the typical velocity scale for the fluid flow and by L the typical horizontal length scale. With these choices of velocity and length scales, the associated time scale is given by $t_U = U/L$, and typically it represents the eddy turnover time. For the vertical length scale we consider two scales. The first scale, H_0, represents the mean height of the fluid, so that H is then given by

$$(4.53) \qquad\qquad H = H_0 + h - h_B\,.$$

The second scale, N_0, represents the mean size of the height perturbation h. With this choice of scales we introduce the nondimensional variables $\vec{x}'$, t', $\vec{v}'$, and h' by

$$(4.54) \qquad\qquad \vec{x}' = \frac{\vec{x}}{L}\,, \quad t' = \frac{t}{t_U}\,, \quad \vec{v}' = \frac{\vec{v}}{U}\,, \quad h' = \frac{h}{N_0}\,.$$

We also have three important nondimensional parameters associated with the scaling in equation (4.54):

$$(4.55) \qquad \text{Ro} = \frac{U}{Lf}, \quad \text{Fr} = \frac{U}{\sqrt{gH_0}}, \quad \Theta = \frac{N_0}{H_0}.$$

We recall the Ro is the Rossby number and represents the ratio of the rotation time $t_R = 1/f$ divided by the fluid flow time scale $t_U = L/U$. The nondimensional number Fr is the Froude number and represents the ratio of the fluid flow velocity U divided by the gravity wave speed $\sqrt{gH_0}$. Finally, the nondimensional parameter Θ represents the ratio of the characteristic height variation N_0 to the mean height H_0. The derivation of the nondimensional equations is straightforward. We introduce the new primed variables given in (4.54) into the rotating shallow water equations (4.1) and (4.2), and simplify the resulting algebra with the help of the nondimensional parameters given in (4.55) and (4.53). After dropping the primes we end up with the following:

Nondimensional Rotating Shallow Water Equations.

$$(4.56) \qquad \frac{D\vec{v}}{Dt} + \text{Ro}^{-1}\vec{v}^{\perp} + \text{Fr}^{-2}\Theta\nabla h = 0,$$

$$(4.57) \qquad \frac{Dh}{Dt} - \Theta^{-1}\vec{v}\cdot\nabla\left(\frac{h_B}{H_0}\right) + \Theta^{-1}\left(1 + \Theta h - \frac{h_B}{H_0}\right)\text{div }\vec{v} = 0.$$

For the derivation of the quasi-geostrophic equation in the next section it is also convenient to have the nondimensional version of the conservation of potential vorticity in (4.14)

$$(4.58) \qquad \frac{D}{Dt}\left(\frac{\text{Ro }\omega + 1}{1 + \Theta h - \frac{h_B}{H_0}}\right) = 0.$$

Note that the topography, h_B, in (4.57) and (4.58) is also nondimensionalized by H_0. In the next section we will derive the quasi-geostrophic equations as a distinguished asymptotic limit of equations (4.56), (4.57), and (4.58), in which both the Rossby and Froude numbers are assumed to be small. This situation is realized in mesoscale motions of the atmosphere and oceans in the midlatitudes. For example, in the atmosphere typical scales of motion include wind speeds $U = 10 \text{ ms}^{-1}$, horizontal length scale $L = 1000$ km, vertical length scale $H_0 = 10$ km, and rotation frequency $f = 10^{-4} \text{ s}^{-1}$ for the midlatitude. With these scales the Rossby number is $\text{Ro} = U/Lf = 0.1$ and the Froude number is $\text{Fr} = U/\sqrt{gH_0} = 0.03$. Similarly, on mesoscales in the ocean, the current velocity is $U = 1 \text{ ms}^{-1}$, the horizontal length scale is $L = 100$ km, the vertical depth scale is $H_0 = 10^2$ m (the active part of the upper ocean), and again the rotation frequency for the midlatitude is $f = 10^{-4} \text{ s}^{-1}$. Therefore, for these oceanic current motions the Rossby number $\text{Ro} = 0.1$ and the Froude number $\text{Fr} = 0.03$. Therefore, we conclude that both the Rossby and Froude numbers are small for the atmospheric and oceanic motions under consideration, and that the quasi-geostrophic formulation to be derived in the next section is appropriate as a leading-order approximation.

4.6. Derivation of the Quasi-Geostrophic Equations

In the late 1940s Charney developed the quasi-geostrophic equations. These equations are at a level of complexity that is intermediate between the steady geostrophic balance equations and the full rotating shallow water equations, and they allow for the equilibrium of rotation and pressure forces and at the same time for the time evolution of the flow. From the mathematical standpoint, the quasi-geostrophic equations are a distinguished asymptotic limit of the rotating shallow water equations where the nondimensional parameters Ro, Fr, and Θ are linked as small parameters. More precisely, the derivation of the quasi-geostrophic equations from the rotating shallow water equation in (4.56), (4.57), and (4.58) is based on the following set of assumptions:

Assumption 1: The Rossby number is assumed to be a small parameter, so that

$$(4.59) \qquad \mathrm{Ro} = \epsilon \quad \text{where } \epsilon \ll 1 .$$

Assumption 2: The pressure forces from the gradients of height balance exactly with the effects of rotation (geostrophic balance), that is,

$$(4.60) \qquad \Theta (\mathrm{Fr})^{-2} = (\mathrm{Ro})^{-1} .$$

If we substitute Ro, Fr, and Θ for their definitions in (4.55), we see that (4.60) is equivalent to the requirement that

$$(4.61) \qquad N_0 = \frac{fUL}{g} .$$

Assumption 3: As discussed after (4.28), define the *Rossby deformation radius*, L_R, as the length scale beyond which rotation is more important than gravity waves,

$$(4.62) \qquad L_R = \frac{\sqrt{gH_0}}{f} ;$$

then we assume that the length scale L is comparable to the Rossby deformation radius, that is,

$$(4.63) \qquad \left(\frac{L}{L_R} \right)^2 = F = O(1) .$$

(We recall that the Rossby length L_R appears in the plane wave analysis of the Poincaré waves, where the velocity magnitude v is given by equation (4.40), with the components $v_\parallel$ and $v_\perp$ lying in the ellipse given by (4.41). Then we observe that the ratio of the axes of the ellipse in (4.41) is $\omega/f = \sqrt{L_R^2 |\vec{k}|^2 + 1}$.) Utilizing (4.55), it is clear that F in (4.63) can be rewritten as

$$F = \left(\frac{L}{L_R} \right)^2 = \frac{L^2 f^2}{gH_0} = \left(\frac{Lf}{U} \right)^2 \frac{U^2}{gH_0} = (\mathrm{Ro})^{-2}(\mathrm{Fr})^2 ,$$

so that $\text{Fr}^2 = F\text{Ro}^2$, and combining this with assumption 1 we conclude that

$$(4.64) \qquad \text{Fr} = F^{1/2}\text{Ro} = F^{1/2}\epsilon \,.$$

In addition, if we combine (4.60) with (4.61), we also conclude that

$$(4.65) \qquad \Theta = (\text{Fr})^2(\text{Ro})^{-1} = F\text{Ro} = F\epsilon \,.$$

Assumption 4: Assume that the scale for the topography is of the same order of magnitude as the scale of the height perturbations; that is,

$$(4.66) \qquad h_B = N_0 \bar{h}_B \,.$$

In summary, scaling assumptions 1 through 4 imply that we are considering the following:

Distinguished Asymptotic Limit for Rotating Shallow Water Equations.

$$(4.67) \qquad \begin{array}{c} \text{Ro} = \epsilon \ll 1 \,, \quad \text{Fr} = F^{1/2}\epsilon \quad \text{where } F = O(1) \,, \\ \Theta = F\epsilon \,, \qquad h_B = N_0 \bar{h}_B \,. \end{array}$$

With the scaling assumptions summarized in equation (4.67), the nondimensional rotating shallow water equations in (4.56) and (4.57), and the potential vorticity equation in (4.58) take the following form:

Distinguished Scaling for Rotating Shallow Water Equations.

$$(4.68) \qquad \frac{D\vec{v}}{Dt} + \epsilon^{-1}\vec{v}^{\perp} = -\epsilon^{-1}\nabla h \,,$$

$$(4.69) \qquad F(1 + \epsilon F h - \epsilon F\tilde{h}_B)^{-1}\left(\frac{Dh}{Dt} - \vec{v} \cdot \nabla\tilde{h}_B\right) + \epsilon^{-1}\operatorname{div}\vec{v} = 0 \,,$$

$$(4.70) \qquad \frac{D}{Dt}\left(\frac{1 + \epsilon\omega}{1 + \epsilon F h - \epsilon F\tilde{h}_B}\right) = 0 \,,$$

where $\omega = \operatorname{curl}\vec{v}$. The nondimensional equations for rotating shallow water in (4.68), (4.69), and (4.70) are a singular limit on the time scale of interest because some of the coefficients for the equations become singular as $\epsilon \to 0$.

Formal Derivation of Quasi-Geostrophic Equations. For the derivation of the quasi-geostrophic equations we assume that the velocity $\vec{v}$ and the height h have an asymptotic expansion of the form

$$(4.71) \qquad \begin{array}{c} \vec{v}(\vec{x}, t) = \vec{v}_0(\vec{x}, t) + \epsilon\vec{v}_1(\vec{x}, t) + \epsilon^2\vec{v}_2(\vec{x}, t) + \cdots \,, \\ h(\vec{x}, t) = h_0(\vec{x}, t) + \epsilon h_1(\vec{x}, t) + \epsilon^2 h_2(\vec{x}, t) + \cdots \,, \end{array}$$

and then we introduce this expansion into the equations (4.68), (4.69), and (4.70), and solve for the powers of ϵ. For the leading power of ϵ^{-1} in the momentum equation (4.68), we get

$$\vec{v}_0^{\perp} = -\nabla h_0 \,,$$

which says that, to leading order, the fluid is in geostrophic balance, with the pressure gradient from h_0 in equilibrium with the rotational effect given by $\vec{v}_0^{\perp}$. Solving for $\vec{v}_0$ gives

(4.72) $$\vec{v}_0 = \nabla^{\perp} h_0 \,.$$

On the other hand, the leading power ϵ^{-1} of the height equation (4.69) yields

(4.73) $$\text{div } \vec{v}_0 = 0 \,,$$

but the incompressibility condition in (4.73) is automatically satisfied by $\vec{v}_0$ given by the formula in (4.72). In order to derive an equation for the leading term h_0 of the height function h in (4.71), we consider the potential vorticity equation in (4.70). We recall that the vorticity has the expansion

(4.74) $$\omega(\vec{x}, t) = \omega_0(\vec{x}, t) + \epsilon \omega_1(\vec{x}, t) + \epsilon^2 \omega_2(\vec{x}, t) + \cdots \,,$$

where the leading term ω_0 of the vorticity is related to the leading term $\vec{v}_0$ of the velocity by

(4.75) $$\omega_0 = \text{curl } \vec{v}_0 \,.$$

The leading power ϵ in the potential vorticity equation (4.70) yields

(4.76) $$\left(\frac{\partial}{\partial t} + \vec{v}_0 \cdot \nabla \right) (\omega_0 - F h_0 + F \tilde{h}_B) = 0 \,.$$

In addition, combining equations (4.72) and (4.75) we have

(4.77) $$\omega_0 = \text{curl } \vec{v}_0 = \text{curl } \nabla^{\perp} h_0 = \frac{\partial}{\partial x_1} \left(\frac{\partial h_0}{\partial x_1} \right) - \frac{\partial}{\partial x_2} \left(-\frac{\partial h_0}{\partial x_2} \right) = \Delta h_0 \,.$$

Finally, collecting the results in equations (4.72), (4.76), and (4.77), we obtain the *quasi-geostrophic equations*

(4.78) $$\vec{v}_0 = \nabla^{\perp} h_0 \,,$$

(4.79) $$\Delta h_0 = \omega_0 \,,$$

(4.80) $$\left(\frac{\partial}{\partial t} + \vec{v}_0 \cdot \nabla \right) (\omega_0 - F h_0 + F \tilde{h}_B) = 0 \,.$$

As we mentioned earlier, in the quasi-geostrophic equations, there is geostrophic balance of the rotation and pressure gradients in equation (4.78), and to leading order the vorticity, which is given by the Laplacian of the height in equation (4.79), evolves in time according to equation (4.80). These equations have been successful in describing major features of large-scale motions of the atmosphere and the oceans in the midlatitudes in a qualitative fashion.

Remarkable mathematical and statistical properties of these equations are discussed extensively in the book by Majda and Wang [26]; also see chapter 3 of Pedlosky's book [29]. Before concluding this section, we remark that the quasi-geostrophic equations (4.78), (4.79), and (4.80) are formally similar to the vorticity-stream formulation of the two-dimensional incompressible Euler equations (see, for example, [**2, 19**])

(4.81) $$\vec{v} = \nabla^{\perp}\psi \,,$$

(4.82) $$\Delta\psi = \omega \,,$$

(4.83) $$\frac{D\omega}{Dt} = 0 \,.$$

However, we point out that the quasi-geostrophic equations in (4.78), (4.79), and (4.80) include important additional effects due to both the nonzero contribution from $F \neq 0$ and the topography (represented by $\tilde{h}_B$), which are not present in the two-dimensional vorticity-stream formulation in equations (4.81), (4.82), and (4.83), and that can change considerably the character of the solutions of the quasi-geostrophic equations due to the effects of dispersion, which we discuss in subsequent chapters. Also note that the same equations were derived in Chapter 1 by considering special depth-averaged solutions of the Boussinesq equations involving y-variation of the Coriolis force. To see this, we set $Fh_0 \equiv 0$ and $F\tilde{h}_\beta \equiv \beta y$ in (4.80) to recover the general barotropic equations derived in Section 1.2. Thus, we see that formally linear variations in the Coriolis force are equivalent to a constant topographic slope in the quasi-geostrophic limit. This dual interpretation is often utilized by oceanographers. One can include the variations of Coriolis force systematically in the formal derivation presented above; see chapter 3 of Pedlosky's book. We omitted this feature here to gain simplicity in exposition.

4.7. The Quasi-Geostrophic Equations as a Singular PDE Limit

We recall that we derived the quasi-geostrophic equations as a distinguished asymptotic limit of the rotating shallow water equations as the Rossby number $\text{Ro} = \epsilon$ goes to 0, and where the Froude number Fr, the ratio of height variation to mean height Θ, and the bottom topography h_B satisfied the *scaling assumptions*

(4.84)
$$\text{Ro} = \epsilon \ll 1, \quad \text{Fr} = F^{1/2}\epsilon \quad \text{where } F = O(1),$$
$$\Theta = F\epsilon, \quad \frac{h_B}{H_0} = \epsilon\tilde{h}_B \,.$$

With these scaling assumptions the rotating shallow water equations together with the potential vorticity equation take the form

CONSERVATION OF MOMENTUM:

(4.85) $$\frac{D\vec{v}}{Dt} + \epsilon^{-1}\vec{v}^{\perp} = -\epsilon^{-1}\nabla h \,,$$

HEIGHT EQUATION:

(4.86) $$(1 + \epsilon Fh - \epsilon\tilde{h}_B)^{-1}\left(\frac{D}{Dt}(Fh) - \vec{v}\cdot\nabla\tilde{h}_B\right) + \epsilon^{-1}\,\text{div}\,\vec{v} = 0 \,,$$

CONSERVATION OF POTENTIAL VORTICITY:

(4.87) $$\frac{D}{Dt}\left(\frac{1 + \epsilon\omega}{1 + \epsilon Fh - \epsilon\tilde{h}_B}\right) = 0 \,.$$

The quasi-geostrophic equations were formally derived by introducing asymptotic expansions for the velocity $\vec{v}$, the height h, and the vorticity ω,

(4.88)

$$\vec{v}(\vec{x}, t) = \vec{v}_0(\vec{x}, t) + \epsilon \vec{v}_1(\vec{x}, t) + O(\epsilon^2),$$
$$h(\vec{x}, t) = h_0(\vec{x}, t) + \epsilon h_1(\vec{x}, t) + O(\epsilon^2),$$
$$\omega(\vec{x}, t) = \omega_0(\vec{x}, t) + \epsilon \omega_1(\vec{x}, t) + O(\epsilon^2).$$

Then we plugged the asymptotic expansion in (4.88) into the rotating shallow water equations (4.85) and (4.86) and the potential vorticity equation (4.87). Collecting the leading term of order ϵ^{-1} in equations (4.85) and (4.86) and the leading term of order ϵ in the potential vorticity equation (4.87) yields the beautiful system of equations first formulated by Charney and known as the *quasi-geostrophic equations*

$$(4.89) \qquad \qquad \vec{v}_0 = \nabla^{\perp} h_0,$$

$$(4.90) \qquad \qquad \Delta h_0 = \omega_0,$$

$$(4.91) \qquad \left(\frac{\partial}{\partial t} + \vec{v}_0 \cdot \nabla\right)(\omega_0 - F h_0 + \tilde{h}_B) = 0.$$

In quasi-geostrophic equations (4.89), (4.90), and (4.91), the fast gravity waves moving at speeds of the order ϵ^{-1} have been filtered out in the asymptotic limiting process. Instead, equation (4.89) shows that the velocity $\vec{v}_0$ and the height h_0 are in geostrophic balance, with h evolving in time according to equation (4.91), and the vorticity ω given by equation (4.90).

In this section we want to investigate rigorously how the rotating shallow water equations approximate the quasi-geostrophic equations. Here we will build the necessary mathematical machinery to prove that the solution of equations (4.85), (4.86), and (4.87) converges to the solution of equations (4.89), (4.90), and (4.91) as $\epsilon \to 0$. The proof will be constructive, and in the process we will prove the existence of solutions of the quasi-geostrophic equations. We point out that the asymptotic limit of the rotating shallow water equations to the quasi-geostrophic equations is a singular limit because some of the coefficients in equations (4.85), (4.86), and (4.87) become infinite as $\epsilon \to 0$. However, there is a theoretical framework called *balanced symmetric hyperbolic systems*, which was designed to handle singular limits for systems of equations similar to the rotating shallow water equations in (4.85), (4.86), and (4.87). This theory was originally developed by Klainerman and Majda [14, 15], where the theory was applied to a variety of physical systems including the compressible fluid flow equations and the magnetohydrodynamic flow equations. A simplified version of the theory is also presented in the book by Majda [16, chap. 2].

We point out that one key feature in this theory is that in the balanced symmetric hyperbolic systems the potentially dangerous terms of order ϵ^{-1} appear in constant-coefficient symmetric matrices, making it possible to control the solution in the singular limit. In our case, this feature will be manifest in the derivation of the energy estimate for the rotating shallow water equations below, where we will see that the contributions of the fast terms of order ϵ^{-1} cancel out, thus yielding the key uniform estimates upon which the proof of convergence rests.

In order to simplify the technicalities involved in the proof, we will consider below a slightly modified version of the rotating shallow water equations in (4.85), (4.86), and (4.87), where the variable coefficient $C = (1 + \epsilon F h - \epsilon \tilde{h}_B)^{-1}$ in equation (4.86) has been replaced by the leading constant value 1. The resulting modified equations still have the quasi-geostrophic equations as their asymptotic limit, but they will be technically simpler to handle because we won't need to estimate the time commutator terms that would have been present had we kept the time-dependent coefficient C.

4.8. The Model Rotating Shallow Water Equations

The model equations for rotating shallow water that we are going to study are motivated from equations (4.85) and (4.86) by keeping the leading term of the time coefficient $(1 + \epsilon F h - \epsilon \tilde{h}_B)^{-1}$ in the height equation (4.85), yielding

CONSERVATION OF MOMENTUM:

$$(4.92) \qquad \frac{D\vec{v}}{Dt} + \epsilon^{-1}\vec{v}^{\perp} = -\epsilon^{-1}\nabla h\,,$$

HEIGHT EQUATION:

$$(4.93) \qquad \frac{D}{Dt}(Fh) - \vec{v} \cdot \nabla \tilde{h}_B + \epsilon^{-1} \operatorname{div} \vec{v} = 0\,.$$

To derive the potential vorticity equation we take the curl of the momentum equation (4.92) and utilize the identities

$$\operatorname{curl}\left(\frac{D\vec{v}}{Dt}\right) = \frac{D\omega}{Dt} + \omega \operatorname{div} \vec{v} \quad \text{and} \quad \operatorname{curl}\vec{v}^{\perp} = \operatorname{div}\vec{v}$$

so that $\omega = \operatorname{curl}\vec{v}$ satisfies the equation

$$(4.94) \qquad \frac{D\omega}{Dt} = -(\omega + \epsilon^{-1})\operatorname{div}\vec{v}\,.$$

The term $\epsilon^{-1}\operatorname{div}\vec{v}$ is evaluated with the help of the height equation (4.93) yielding

$$(4.95) \qquad \frac{D\omega}{Dt} = -\omega \operatorname{div}\vec{v} + \frac{D}{Dt}(Fh - \tilde{h}_B)\,.$$

In equation (4.95) we used the fact that the topography term $\tilde{h}_B$ is time independent, so that $D\tilde{h}_B/Dt = \vec{v} \cdot \nabla \tilde{h}_B$.

Collecting the material derivative terms in the left side of equation (4.95), we get the

POTENTIAL VORTICITY EQUATION:

$$(4.96) \qquad \frac{D}{Dt}(\omega - Fh + \tilde{h}_B) = -\omega \operatorname{div}\vec{v},$$

and collecting equations (4.92), (4.93), and (4.96) gives the

MODEL ROTATING SHALLOW WATER EQUATIONS:

$$(4.97) \qquad \frac{D\vec{v}}{Dt} + \epsilon^{-1}\vec{v}^{\perp} = -\epsilon^{-1}\nabla h \,,$$

$$(4.98) \qquad \frac{D}{Dt}(Fh) - \vec{v} \cdot \nabla \tilde{h}_B + \epsilon^{-1}\,\mathrm{div}\,\vec{v} = 0 \,,$$

$$(4.99) \qquad \frac{D}{Dt}(\omega - Fh + \tilde{h}_B) = -\omega\,\mathrm{div}\,\vec{v} \,.$$

In the potential vorticity equation (4.99) for the model, we notice that the potential vorticity q, given by $q = \omega - Fh + \tilde{h}_B$, represents the order ϵ term in the expansion of the potential vorticity $\Pi = (1 + \epsilon\omega)/(1 + \epsilon Fh - \epsilon\tilde{h}_B)$ for the rotating shallow water equations. Although equation (4.99) shows that the potential vorticity q is not a conserved quantity, i.e., transported with the flow, we can derive an equation for a conserved "nonlinear potential vorticity" as follows: If $G_\epsilon(\omega)$ is any function of the vorticity ω, then equation (4.94) implies that

$$(4.100) \qquad \frac{DG_\epsilon(\omega)}{Dt} = -G'_\epsilon(\omega)(\omega + \epsilon^{-1})\,\mathrm{div}\,\vec{v} = \epsilon^{-1}\,\mathrm{div}\,\vec{v}((1 + \epsilon\omega)G'_\epsilon(\omega)) \,.$$

If we now let $G_\epsilon(\omega)$ be the function $G_\epsilon(\omega) = (\ln(1 + \epsilon\omega) - 1)/\epsilon$, which clearly satisfies $G'_\epsilon(\omega) = (1 + \epsilon\omega)^{-1}$, then equation (4.100) simplifies to

$$(4.101) \qquad \frac{DG_\epsilon(\omega)}{Dt} = -\epsilon^{-1}\,\mathrm{div}\,\vec{v} \,,$$

and evaluating the term $\epsilon^{-1}\,\mathrm{div}\,\vec{v}$ with the help of the height equation (4.93) in the same fashion as we did earlier for equation (4.94), we finally obtain the conservation equation

$$(4.102) \qquad \frac{D}{Dt}(G_\epsilon(\omega) - Fh + \tilde{h}_B) = 0 \,,$$

where $G_\epsilon(\omega) = (\ln(1 + \epsilon\omega) - 1)/\epsilon$. In particular, expanding the nonlinear potential vorticity $G_\epsilon(\omega)$ around $\epsilon = 0$ shows that $G_\epsilon(\omega) - Fh + \tilde{h}_B = \omega - Fh + \tilde{h}_B + O(\epsilon)$, so that the nonlinear potential vorticity $G_\epsilon(\omega) - Fh + \tilde{h}_B$ agrees with the potential vorticity $q = \omega - Fh + \tilde{h}_B$ to order ϵ.

Next we verify that the asymptotic limit of the model rotating shallow water equations (4.97), (4.98), and (4.99) as $\epsilon \to 0$ are the quasi-geostrophic equations (4.89), (4.90), and (4.91). If we introduce the asymptotic expansions in (4.88) for $\vec{v}, h$, and ω into the model rotating shallow water equations (4.97), (4.98), and (4.99), then to the leading-order ϵ^{-1}, the momentum equation (4.97) yields the geostrophic balance equation

$$(4.103) \qquad \vec{v}_0 = \nabla^{\perp}h_0 \,.$$

Also, to leading-order ϵ^{-1}, the height equation (4.98) yields the incompressibility constraint $\mathrm{div}\,\vec{v}_0 = 0$, which is automatically satisfied by the formula for $\vec{v}_0$ given in equation (4.103).

Finally, with the leading-order vorticity ω_0 and height h_0 related through equation (4.103) by

(4.104) $$\omega_0 = \Delta h_0 \,,$$

then to leading-order $O(1)$ in the potential vorticity equation (4.99), we have

(4.105) $$\frac{D}{Dt}(\omega_0 - Fh_0 + \tilde{h}_B) = -\omega \operatorname{div} \vec{v}_0 = 0 \,,$$

and we conclude that asymptotically, in the limit of $\epsilon \to 0$, the model rotating shallow water equations satisfy the quasi-geostrophic equations (4.103), (4.104), and (4.105).

The rest of this chapter will be devoted to proving rigorously that the solution of the model rotating shallow water equations in (4.97), (4.98), and (4.99) converge to the solution of the quasi-geostrophic equations (4.89), (4.90), and (4.91) as $\epsilon \to 0$. The technical tools needed for the proof involve the derivation of "good" uniform energy estimates, *independent* of ϵ, for the solutions of the model rotating shallow water equations, and the analytical machinery of Sobolev spaces, which are the natural function spaces to consider when there are energy estimates available for the solutions. These technical tools are discussed in the next section.

4.9. Preliminary Mathematical Considerations

One key tool in the proof of convergence of the model rotating shallow water equations (4.97), (4.98), and (4.99) to the quasi-geostrophic equations (4.89), (4.90), and (4.91) consists in the derivation of uniform energy estimates, *independent of ϵ*, for the solution of (4.97), (4.98), and (4.99) and its spatial derivatives. For that purpose we start with the derivation of the basic energy identity for the *model rotating shallow water equations with forcing terms*

(4.106) $$\frac{D\vec{v}}{Dt} + \epsilon^{-1}\vec{v}^{\perp} = -\epsilon^{-1}\nabla h + \vec{\mathcal{F}}_v \,,$$

(4.107) $$\frac{D}{Dt}(Fh) - \vec{v} \cdot \nabla\tilde{h}_B + \epsilon^{-1}\operatorname{div} \vec{v} = \mathcal{F}_h \,,$$

and denote by $\vec{u}$ the vector $\vec{u} = (h, \vec{v})$. As we will see later, the reason for considering the forcing terms in equations (4.106) and (4.107) is that in the derivation of energy estimates for the higher-order spatial derivatives of the solution of the model rotating shallow water equations, these higher-order derivatives will satisfy equations like (4.106) and (4.107) where the forcing terms are going to be given by space derivative commutator terms.

We start with the derivation of the differential form of the conservation of energy. Define the energy density $\mathcal{E}$ by

(4.108) $$\mathcal{E} = \frac{1}{2}\vec{v} \cdot \vec{v} + \frac{1}{2}Fh^2 \,;$$

then the conservation of energy for the model rotating shallow water equations in differential form is as follows:

LEMMA 4.5 (Differential Form of the Conservation of Energy) *If $\vec{u} = (\vec{v}, h)$ satisfy equations (4.97) and (4.98), then the energy density $\mathcal{E}$ satisfies the differential equation*

$$(4.109) \quad \frac{\partial \mathcal{E}}{\partial t} = -\operatorname{div}(\vec{v}\mathcal{E}) + (\operatorname{div}\vec{v})\mathcal{E} - \epsilon^{-1}\operatorname{div}(h\vec{v}) + h\vec{v} \cdot \nabla \tilde{h}_B + \vec{v} \cdot \vec{\mathcal{F}}_v + h\mathcal{F}_h .$$

PROOF: Differentiating the energy density $\mathcal{E}$ and utilizing the model shallow water equations (4.97) and (4.98) to evaluate the time derivatives yields

$$
\begin{aligned}
(4.110) \quad \frac{\partial \mathcal{E}}{\partial t} &= \vec{v} \cdot \frac{\partial \vec{v}}{\partial t} + h\frac{\partial(Fh)}{\partial t} \\
&= \vec{v} \cdot (-\vec{v} \cdot \nabla \vec{v} - \epsilon^{-1}\vec{v}^{\perp} - \epsilon^{-1}\nabla h + \vec{\mathcal{F}}_v) \\
&\quad + h(-\vec{v} \cdot \nabla(Fh) + \vec{v} \cdot \nabla \tilde{h}_B - \epsilon^{-1}\operatorname{div}\vec{v} + \mathcal{F}_h) \\
&= -\vec{v} \cdot \nabla \mathcal{E} - \epsilon^{-1}\operatorname{div}(h\vec{v}) + h\vec{v} \cdot \nabla \tilde{h}_B + \vec{v}\vec{\mathcal{F}}_v + h\mathcal{F}_h ,
\end{aligned}
$$

and since $\vec{v} \cdot \nabla \mathcal{E} = \operatorname{div}(\mathcal{E}\vec{v}) - \mathcal{E}\operatorname{div}\vec{v}$, then equation (4.110) reduces to equation (4.109). This proves the lemma. $\qquad\square$

We observe that in equation (4.109) there are potentially dangerous terms of order ϵ^{-1}, which are in fact the manifestation of the presence of fast-moving gravity waves. However, the ϵ^{-1} term in equation (4.109) is a perfect divergence and upon integration it will disappear. This key fact implies that in the energy estimates for the model shallow water equations the fast gravity waves effectively are filtered out. This crucial fact is shared by the balanced symmetric hyperbolic systems studied by Klainerman and Majda in the papers previously cited.

Next we derive the integral form of the energy identity that will be utilized in the proof of the convergence theorem in Section 4.11. However, in order to simplify the proof of the convergence theorem, it is convenient to avoid the consideration of boundary term contributions. This is done by assuming that the flow is periodic in space, say of period 1. In other words, from now on we assume that the spatial domain is the 2-dimensional torus $\mathbb{T}^2$.

Let us define the (total) energy $E(t)$ as the integral of the energy density $\mathcal{E}$

$$(4.111) \qquad E(t) = \int_{\mathbb{T}^2} \left(\frac{1}{2}\vec{v} \cdot \vec{v} + \frac{1}{2}Fh^2 \right) dx .$$

Then conservation of energy for the model rotating shallow water equations takes the following form:

LEMMA 4.6 (Integral Form of the Conservation of Energy) *If $\vec{u} = (\vec{v}, h)$ satisfy equations (4.97) and (4.98), then the total energy $E(t)$ satisfies the differential equation*

$$(4.112) \quad \frac{dE}{dt} = \int_{\mathbb{T}^2} (\operatorname{div}\vec{v})\mathcal{E}\,dx + \int_{\mathbb{T}^2} h\vec{v} \cdot \nabla \tilde{h}_B\,dx + \int_{\mathbb{T}^2} \vec{v} \cdot \vec{\mathcal{F}}_v\,dx + \int_{\mathbb{T}^2} h\mathcal{F}_n\,dx .$$

PROOF: The conservation of energy in equation (4.112) follows directly by integrating over the torus $\mathbb{T}^2$ the differential conservation of energy in equation

(4.109) and noticing that the individual contributions from the integrals of $\mathrm{div}(\mathcal{E}\vec{v})$ and $\epsilon^{-1}\,\mathrm{div}(h\vec{v})$ in equation (4.109) are zero. $\qquad\square$

With the help of Lemma 4.6 we can derive the basic energy estimate needed for the proof of the convergence theorem in Section 4.11.

PROPOSITION 4.7 (Energy Estimate) *If* $\vec{u} = (\vec{v}, h)$ *satisfy equations* (4.97) *and* (4.98), *then the total energy* $E(t)$ *satisfies the differential inequality*

$$(4.113) \qquad \frac{dE}{dt} \leq |\mathrm{div}\,\vec{v}|_{L^\infty} E(t) + C\|\vec{\mathcal{F}}_v\|_0 (E(t))^{1/2} + C|\nabla \tilde{h}_B|_{L^\infty} E(t),$$

where $\|\cdot\|_0$ *is the* L^2 *norm, and* $|\cdot|_{L^\infty}$ *is the* L^∞ *norm.* $\vec{\mathcal{F}}$ *is the vector* $\vec{\mathcal{F}} = (\vec{\mathcal{F}}_v, \mathcal{F}_h)$, *and* C *is a constant independent of* ϵ.

PROOF: To derive the inequality in (4.113), we estimate each one of the integral expressions in the right side of the energy identity in (4.112) with the help of the Hölder and Cauchy-Schwarz inequalities, and the fact that the L^2 norm $\|\vec{u}\|_0$ is equivalent to the energy norm $(E(t))^{1/2}$. More specifically, let us estimate each term in the right side of equation (4.112). The first term is estimated with the Hölder inequality and yields

$$(4.114) \qquad \left| \int_{\mathbb{T}^2} (\mathrm{div}\,\vec{v})\mathcal{E}\,dx \right| \leq |\mathrm{div}\,\vec{v}|_{L^\infty} \int_{\mathbb{T}^2} \mathcal{E}\,dx \leq |\mathrm{div}\,\vec{v}|_{L^\infty} E(t).$$

To estimate the second term we combine the Hölder and Cauchy-Schwarz inequality

$$(4.115) \qquad \begin{aligned} \left| \int_{\mathbb{T}^2} h\vec{v} \cdot \nabla \tilde{h}_B \, dx \right| &\leq |\nabla \tilde{h}_B|_{L^\infty} \int_{\mathbb{T}^2} |h\vec{v}|\,dx \\ &\leq |\nabla \tilde{h}_B|_{L^\infty} \|h\|_0 \|\vec{v}\|_0 \leq C|\nabla \tilde{h}_B|_{L^\infty} E(t), \end{aligned}$$

where the last inequality utilized the fact that the L^2 norm is equivalent to the energy norm $(E(t))^{1/2}$. Finally, the estimates of the remaining two terms are a straightforward application of the Cauchy-Schwarz inequality and the equivalence of the L^2 and energy norms. This proves the energy estimate in (4.113). $\qquad\square$

The proof of the convergence theorem in Section 4.11 involves energy estimates not only of the solution $\vec{u}$ but also of higher-order spatial derivatives of $\vec{u}$. The appropriate function space must satisfy the requirement that the functions and their spatial derivatives (up to a given order) are square-integrable, that is, we have to consider Sobolev spaces H^s. For this reason the second technical tool involves the Sobolev spaces H^s and some of its properties, including the sharp calculus inequalities for the products of functions in H^s and a compactness criterion for functions $\vec{u}(t)$ with values in H^s. A good reference on the basic facts about Sobolev spaces is the book by Folland [10].

DEFINITION 4.8 The Sobolev space $H^s(\mathbb{T}^N)$ of periodic functions on the N-dimensional torus $\mathbb{T}^N$ is defined as the set of functions $\vec{u}$ in $L^2(\mathbb{T}^N)$ so that its distribution derivatives $D^\alpha \vec{u}$ are in $L^2(\mathbb{T}^N)$ for all $|\alpha| \leq s$. The Sobolev norm of order s is defined by

$$(4.116) \qquad \|\vec{u}\|_s^2 = \sum_{|\alpha| \leq s} \int_{\mathbb{T}^N} D^\alpha \vec{u} \cdot D^\alpha \vec{u} \, dx.$$

As usual, the short-hand multi-index notation D^α for the higher-order derivatives has been utilized: If $\alpha = (\alpha_1, \alpha_2, \ldots, \alpha_N)$, then D^α is defined by

$$D^\alpha = \frac{\partial}{\partial x_1^{\alpha_1}} \frac{\partial}{\partial x_2^{\alpha_2}} \cdots \frac{\partial}{\partial x_N^{\alpha_N}}.$$

One attractive feature of the Sobolev spaces is that if $s > 0$ is large enough, then any function in H^s has r classical derivatives for some number $r < s$. More precisely, we have the following:

PROPOSITION 4.9 (Sobolev's Lemma) *If $s > \frac{N}{2} + k$, then any function $\vec{u}$ in $H^s(\mathbb{T}^N)$ is also in $C^k(\mathbb{T}^N)$, and in addition*

$$(4.117) \qquad \max_{\vec{x} \in \mathbb{T}^N} \sum_{|\alpha| \leq k} |D^\alpha \vec{u}(\vec{x})| \leq C_s \|\vec{u}\|_s,$$

where C_s depends only on s and $\mathbb{T}^N$.

In order to estimate the nonlinear products of functions in H^s, we will need the following sharp calculus inequalities:

PROPOSITION 4.10 (Calculus Inequalities) *If f and g are functions in $H^s(\mathbb{T}^N)$ and $s > \frac{N}{2}$, then for $|\alpha| \leq s$ the following estimates hold:*

$$(4.118) \qquad \|D^\alpha(fg)\|_0 \leq C_s \{|f|_{L^\infty} \|g\|_s + |g|_{L^\infty} \|f\|_s\},$$

$$(4.119) \qquad \|D^\alpha(fg) - f D^\alpha g\|_0 < C_s \{|\nabla f|_{L^\infty} \|g\|_{s-1} + |g|_{L^\infty} \|f\|_s\}.$$

The proof of the sharp calculus inequalities in (4.118) and (4.119) can be found in the 1981 paper of Klainerman and Majda previously mentioned [**14, 15**], or in Majda and Bertozzi [**19**]. It is interesting to remark that, in fact, the inequalities in (4.118) and (4.119) were first developed in the aforementioned paper of Klainerman and Majda to deal specifically with the problem of convergence of compressible to incompressible flows. In particular, the estimate in (4.119) will be crucially needed in the proof of the convergence theorem below in order to estimate commutator terms involving the higher-order derivatives D^α and the convective operator $\vec{v} \cdot \nabla$.

The last technical tool we need is a compactness criterion for sequences of functions $\{\vec{u}^\epsilon(t)\}$ in $C([0, T], H^s(\mathbb{T}^N))$, which generalizes the well-known classical compactness criterion of Arzela-Ascoli for sequences of functions in $C([0, T], \mathbb{R}^N)$.

PROPOSITION 4.11 (Lions-Aubin Compactness Lemma) *Assume that $\{\vec{u}^\epsilon\}$ is a sequence of functions in $C([0, T], H^s(\mathbb{T}^N))$, where $s > 0$, such that*

(i) $\{\vec{u}^\epsilon\}$ is uniformly bounded in $C([0, T], H(\mathbb{T}^N))$, that is,

(4.120)
$$\max_{0 \le t \le T} \|\vec{u}^\epsilon\|_s \le M_0 \,.$$

(ii) The sequence of time derivatives $\{\vec{u}_t^\epsilon\}$ is uniformly bounded in $C([0, T], H^q(\mathbb{T}^N))$, for some q with $0 \le q < s$, that is,

(4.121)
$$\max_{0 \le t \le T} \|\vec{u}_t^\epsilon\|_q \le M_1 \,.$$

Then there is a subsequence of $\{\vec{u}^\epsilon\}$ that converges in $C([0, T], H^r(\mathbb{T}^N))$ for $0 \le r < s$.

A proof of the Lions-Aubin lemma can be found in Temam [36, p. 271]. In the proof of the convergence theorem we will derive uniform bounds for the time derivative of the solution in the L^2 norm so that $q = 0$, and then take advantage of the Lions-Aubin compactness lemma to extract a convergence subsequence.

4.10. Rigorous Convergence of the Model Rotating Shallow Water Equations to the Quasi-Geostrophic Equations

With the necessary mathematical machinery developed in the previous section, we are now in the position to state the main convergence theorem of the model rotating shallow water equations (4.97), (4.98), and (4.99) to the quasi-geostrophic equations (4.89), (4.90), and (4.91).

THEOREM 4.12 *Consider the model rotating shallow water equation for $\vec{u}^\epsilon = (h^\epsilon, \vec{v}^\epsilon)$ and with flat bottom topography $\tilde{h}_B \equiv 0$*

(4.122)
$$\frac{D\vec{v}^\epsilon}{Dt} + \epsilon^{-1}\vec{v}^{\epsilon\perp} = -\epsilon^{-1}\nabla h^\epsilon \,,$$

(4.123)
$$\frac{D}{Dt}(Fh^\epsilon) + \epsilon^{-1}\operatorname{div}\vec{v}^\epsilon = 0 \,,$$

(4.124)
$$\frac{D}{Dt}(\omega^\epsilon - Fh^\epsilon) = -\omega^\epsilon \operatorname{div}\vec{v}^\epsilon \,,$$

with initial data given by

(4.125) $\quad \vec{v}^\epsilon(\vec{x}, t)\big|_{t=0} = \vec{v}_0(\vec{x}) + \epsilon\vec{v}_1(\vec{x}) \,, \quad h^\epsilon(\vec{x}, t)\big|_{t=0} = h_0(\vec{x}) + \epsilon h_1(\vec{x}) \,,$

with $\vec{v}_0, \vec{v}_1, h_0,$ and h_1 in $H^s(\mathbb{T}^2)$. Then we have the following:

(i) *If $s \ge 3$, then the solutions $\vec{u}^\epsilon$ of equations (4.122), (4.123), and (4.124) exist on a common time interval $[0, T]$ independent of ϵ and satisfy the uniform H^s estimate*

(4.126)
$$\max_{0 \le t \le T} \|\vec{u}^\epsilon(t)\|_s \le C \,,$$

where C is independent of ϵ.

(ii) *If the initial data $\vec{u}_0$ is in geostrophic balance and has zero average*

(4.127)
$$\vec{v}_0 = \nabla^\perp h_0 \,,$$

(4.128)
$$\langle \vec{v}_0 \rangle = 0 \,,$$

then the time derivative $\vec{u}_t^\epsilon$ of the solution $\vec{u}^\epsilon$ satisfies the uniform L^2 estimate

(4.129)
$$\max_{0 \le t \le T} \|\vec{u}_t^\epsilon(t)\|_0 \le C\,,$$

where C is independent of ϵ. Moreover, when the initial data satisfies the approximate geostrophic balance in equations (4.127) and (4.128), then there is a function $\vec{u}^0(\vec{x}, t) = (\vec{v}^0(\vec{x}, t), h^0(\vec{x}, t))$ in $C([0, T], H^s(\mathbb{T}^N))$ so that

(4.130)
$$\max_{0 \le t \le T} \left\{ \left|\vec{u}^\epsilon(t) - \vec{u}^0(t)\right|_{L^\infty} + \left|\nabla\vec{u}^\epsilon(t) - \nabla\vec{u}^0(t)\right|_{L^\infty} \right\} \to 0 \quad as\ \epsilon \to 0\,,$$

and $\vec{u}^0$ solves the quasi-geostrophic equations

(4.131)
$$\vec{v}^0 = \nabla^\perp h^0\,,$$

(4.132)
$$\Delta h^0 = \omega^0\,,$$

(4.133)
$$\left(\frac{\partial}{\partial t} + \vec{v}^0 \cdot \nabla\right)(\omega^0 - Fh^0) = 0\,.$$

The details of the proof of the theorem will be given in the next section. Here we want to make some remarks about the theorem and the strategy of the proof. The theorem states that if the initial data is smooth enough ($s \le 3$), then the solutions exist and are uniformly bounded for a fixed time interval independent of ϵ. However, in order to have uniform estimates for the time derivative of the solution, equation (4.127) tells us that we are restricted to consider initial data that is in *geostrophic balance* within terms of order $O(\epsilon)$. The zero average condition for $\vec{v}^0$ in equation (4.128) is simply a consistency condition on the initial data obtained by integrating equation (4.127) in space and remembering that the solutions are periodic. Finally, the assumption of no topography, $\tilde{h}_B \equiv 0$, is not a crucial one, and is made for simplicity in the exposition.

The proof of the theorem follows in several steps:

Step 1: *Derivation of nonlinear energy estimates for high-order spatial derivatives.* These estimates are uniform in ϵ and only require that the initial data be smooth enough. To derive the estimates we combine the basic energy estimate in Proposition 4.7 with the sharp calculus inequalities for commutators in Proposition 4.10.

Step 2: *Derivation of nonlinear energy estimates for the time derivative of the solution in a low Sobolev norm.* In contrast with step 1, a uniform estimate for the time derivative of $\vec{u}^\epsilon$ can be derived only if the initial data is in geostrophic balance as expressed by equations (4.127) and (4.128). If these balance conditions are satisfied initially, then one can show that $\vec{u}_t^\epsilon$ is initially bounded in L^2, and then an energy estimate for $\vec{u}_t^\epsilon$ shows that it is uniformly bounded on the time interval $[0, T]$.

Step 3: *Passage to the limit by a compactness argument.* Here we utilize the uniform estimates for the solution $\vec{u}^\epsilon$ in H^s and for its time derivative

$\vec{u}_t^\epsilon$ and L^2, together with the Lions-Aubin compactness lemma to obtain a convergence to a function $\vec{u}^0$.

Step 4: *Identification of the limit problem.* In this final step we show that the limit $\vec{u}^0$ satisfies the quasi-geostrophic equations.

4.11. Proof of the Convergence Theorem

Next we proceed to prove the theorem following the steps outlined in the previous section.

Step 1. *Derivation of nonlinear energy estimates for high-order spatial derivatives.*

In order to accomplish this, we must first find out what the equation satisfied by $D^\alpha \vec{u}^\epsilon$ is. This is done by first applying D^α to the momentum and height equations in (4.122) and (4.123). Then we can rearrange the resulting equation to look like equations (4.122) and (4.123) applied to the derivative $D^\alpha \vec{u}^\epsilon$ by adding and subtracting the term $D(D^\alpha \vec{u}^\epsilon)/Dt$. The resulting equations for $D^\alpha \vec{u}^\epsilon$ are the model rotating shallow water equations with forcing,

$$(4.134) \qquad \frac{D\vec{v}^\epsilon}{Dt} + \epsilon^{-1} \vec{v}^{\epsilon\perp} = -\epsilon^{-1}\nabla h^\epsilon + \vec{\mathcal{F}}_v^\alpha \,,$$

$$(4.135) \qquad \frac{D}{Dt}(Fh^\epsilon) + \epsilon^{-1}\,\mathrm{div}\,\vec{v}^\epsilon = \mathcal{F}_h^\alpha \,,$$

where the forcing function $\vec{\mathcal{F}}^\alpha = (\vec{\mathcal{F}}_v^\alpha, \mathcal{F}_h^\alpha)$ is made of spatial commutator terms

$$(4.136) \qquad \vec{\mathcal{F}}_v^\alpha = \vec{v}^\epsilon \cdot \nabla(D^\alpha \vec{v}^\epsilon) - D^\alpha(\vec{v}^\epsilon \cdot \nabla \vec{v}^\epsilon)\,,$$

$$(4.137) \qquad \mathcal{F}_h^\alpha = \vec{v}^\epsilon \cdot \nabla(FD^\alpha h^\epsilon) - D^\alpha(\vec{v}^\epsilon \cdot \nabla Fh^\epsilon)\,.$$

We remark that there are no time commutator terms in (4.136) and (4.137) because in this model the coefficient of the time derivative terms was set to be a constant.

Now we can apply Proposition 4.7 to equations (4.135) and (4.136) for $D^\alpha \vec{u}^\epsilon$ and conclude that $E_\alpha(t)$ defined by $E_\alpha(t) = E(D^\alpha \vec{u}^\epsilon)$ satisfies the energy inequality

$$(4.138) \qquad \frac{dE_\alpha}{dt} \leq |\mathrm{div}\,\vec{v}^\epsilon|_{L^\infty} E_\alpha(t) + C\|\vec{\mathcal{F}}_v^\alpha\|_0 (E_\alpha(t))^{1/2}\,.$$

To estimate the forcing term $\vec{\mathcal{F}}^\alpha$, we exploit the calculus inequality (4.119) in Proposition 4.10. For example, to estimate $\mathcal{F}_h^\alpha$ given by equation (4.137), we set $f = \vec{v}^\epsilon$ and $g = \nabla(Fh^\epsilon)$ in the calculus inequality (4.119) and conclude that

$$(4.139) \qquad \|\mathcal{F}_h^\alpha\|_0 \leq C_s\{|\nabla \vec{v}^\epsilon|_{L^\infty} \|h^\epsilon\|_s + |\nabla h^\epsilon|_{L^\infty} \|\vec{v}^\epsilon\|_s\}\,,$$

and collecting these results for $|\alpha| \leq s$ we get

$$(4.140) \qquad \sum_{|\alpha|\leq s} \|\mathcal{F}_h^\alpha\|_0 \leq C_s |\nabla \vec{u}^\epsilon|_{L^\infty} \sum_{|\alpha|\leq s} E_\alpha(\vec{u}^\epsilon(t))^{1/2}\,.$$

Combining the estimate in (4.140) with the energy inequality in (4.138), we get the key estimate

$$(4.141) \qquad \frac{d}{dt}\left(\sum_{|\alpha|\le s}\|\mathcal{F}_h^\alpha\|_0\right) \le C_s|\nabla\vec{u}^\epsilon|_{L^\infty}\left(\sum_{|\alpha|\le s}E_\alpha(\vec{u}^\epsilon(t))^{1/2}\right),$$

where the constant C_s does not depend on ϵ. Since the norms $(\sum_{|\alpha|\le s}E_\alpha(t))^{1/2}$ and $\|\vec{u}^\epsilon\|_s$ are equivalent, we also conclude that the Sobolev norm $\|\vec{u}^\epsilon\|_s$ satisfies

$$(4.142) \qquad \frac{d}{dt}\|\vec{u}^\epsilon(t)\|_s \le C_s|\nabla\vec{u}^\epsilon|_{L^\infty}\|\vec{u}^\epsilon(t)\|_s\,.$$

Since the estimate in (4.142) is independent of ϵ, we can use it to show that the solutions $\vec{u}^\epsilon$ of equations (4.122) and (4.123) have a common interval of existence: Since $s > \frac{N}{2}+1$, Sobolev's lemma tells us that $|\vec{u}^\epsilon|_{L^\infty} \le C_s\|\vec{u}^\epsilon\|_s$. Therefore the inequality in (4.142) reduces to

$$(4.143) \qquad \frac{d}{dt}\|\vec{u}^\epsilon(t)\|_s \le C_s\|\vec{u}^\epsilon(t)\|_s^2\,.$$

This nonlinear ODE has a fixed interval of existence $[0, T]$ uniformly in ϵ (notice that the initial data in (4.125) is uniformly bounded in ϵ), and in particular $\vec{u}^\epsilon$ satisfies the estimate in equation (4.126). Now we can combine the local existence theorem for symmetric hyperbolic systems with the a priori estimate (uniform in ϵ) that we just derived from (4.143) to continue the solution $\vec{u}^\epsilon$ of (4.122) and (4.123) to the entire interval $[0, T]$. For the application of the continuation principle to solutions of symmetric hyperbolic systems and the convergence of compressible to incompressible flows, the reader is referred to Majda [16, chap. 2].

Step 2. *Derivation of nonlinear energy estimates for the time derivative of the solution in a low Sobolev norm.*

First of all, we must show that initially $\vec{u}_t^\epsilon$ is uniformly bounded in L^2. For this set it is crucial to require that the initial data be in geostrophic balance to order $O(\epsilon)$, as in equation (4.127). The geostrophic constraint for the initial data in (4.127) implies that

$$(4.144) \qquad \vec{v}_0^\perp = \nabla h_0\,,$$

and also that

$$(4.145) \qquad \mathrm{div}\,\vec{v}_0 = 0\,.$$

Combining equations (4.144) and (4.145) for the initial data with equations (4.122) and (4.123) for the solution $\vec{u}^\epsilon$, we conclude that initially the time derivative $\vec{u}_t^\epsilon$ satisfies

$$(4.146) \qquad \begin{aligned}\vec{v}_t^\epsilon(\vec{x},t)\big|_{t=0} &= -(\vec{v}_0+\epsilon\vec{v}_1)\cdot\nabla(\vec{v}_0+\epsilon\vec{v}_1)\,,\\ h_t^\epsilon(\vec{x},t)\big|_{t=0} &= -(\vec{v}_0+\epsilon\vec{v}_1)\cdot\nabla(h_0+\epsilon h_1)\,,\end{aligned}$$

and from equation (4.146) it follows that

$$(4.147) \qquad \|\vec{u}_t^\epsilon(0)\|_0 \le C\,,$$

where C is independent of ϵ.

Next we must derive a uniform L^2 estimate for the time derivative $\vec{u}_t^\epsilon$. For this purpose we mimic the strategy developed in step 1 and derive the evolution equation satisfied by $\vec{u}_t^\epsilon$. Differentiating the momentum and height equations (4.122) and (4.123) with respect to time yields

$$(4.148) \qquad \frac{D}{Dt}(\vec{v}_t^\epsilon) + \epsilon^{-1}\vec{v}_t^{\epsilon\perp} = -\epsilon^{-1}\nabla h_t^\epsilon - \vec{v}_t^\epsilon \cdot \nabla \vec{v}^\epsilon ,$$

$$(4.149) \qquad \frac{D}{Dt}(h_t^\epsilon) + \epsilon^{-1}\operatorname{div}\vec{v}_t^\epsilon = -\vec{v}_t^\epsilon \cdot \nabla \vec{v}^\epsilon .$$

Now we apply the basic energy estimate in Proposition 4.7 to equations (4.148) and (4.149) and conclude that

$$(4.150) \qquad \frac{d}{dt}E(\vec{u}_t^\epsilon(t)) \leq |\operatorname{div}\vec{v}^\epsilon|_{L^\infty}E(\vec{u}_t^\epsilon(t)) + |\nabla\vec{u}^\epsilon|_{L^\infty}E(\vec{u}_t^\epsilon(t)) ,$$

but from equation (4.126) we know that $|\operatorname{div}\vec{v}^\epsilon|_{L^\infty}$ and $|\nabla\vec{v}^\epsilon|_{L^\infty}$ are uniformly bounded in the interval $[0, T]$, so that (4.150) reduces to

$$(4.151) \qquad \frac{d}{dt}E(\vec{u}_t^\epsilon(t)) \leq CE(\vec{u}_t^\epsilon(t)) ,$$

and integrating the differential inequality (4.151), and recalling that initially $\vec{u}_t^\epsilon$ in (4.149) is uniformly bounded, we conclude that

$$(4.152) \qquad \max_{0\leq t\leq T}\|\vec{u}_t^\epsilon(t)\|_s \leq C ,$$

and this proves the uniform estimate for the time derivative $\vec{u}_t^\epsilon$ in (4.129).

Step 3. *Passage to the limit by a compactness argument.*

In the two previous steps we established uniform estimates (4.126) and (4.127) for the solution $\vec{u}^\epsilon$ of equations (4.122) and (4.123), and for the time derivative $\vec{u}_t^\epsilon$ of the solution. These two estimates imply that the sequence $\{\vec{u}^\epsilon\}$ is bounded in $C([0, T], H^s(\mathbb{T}^N))$ and that the sequence of time derivative $\{\vec{u}_t^\epsilon\}$ is bounded in $C([0, T], H^0(\mathbb{T}^N))$. By the Lions-Aubin compactness lemma we know that there is a subsequence that converges to $\vec{u}^0$ in $C([0, T], H^r(\mathbb{T}^N))$ for $0 \leq r < s$. In particular, since $s > \frac{N}{2} + 1$, we can pick r so that $\frac{N}{2} + 1 < r < s$ and utilize Sobolev's lemma to conclude that

$$(4.153) \quad \max_{0\leq t\leq T}\left\{\left|\vec{u}^\epsilon(t) - \vec{u}^0(t)\right|_{L^\infty} + \left|\nabla\vec{u}^\epsilon(t) - \nabla\vec{u}^0(t)\right|_{L^\infty}\right\} \leq$$

$$\max_{0\leq t\leq T}\left\|\vec{u}^\epsilon(t) - \vec{u}^0(t)\right\|_r \to 0 \quad \text{as } \epsilon \to 0 .$$

This proves the convergence statement in (4.130). We remark that although we proved (4.153) only for a subsequence, it is easy to prove by the energy principle that the smooth solution of (4.122) and (4.123) is unique. Therefore all the subsequences converge to the same limit $\vec{u}^0$, and we eventually can conclude that the original sequence $\{\vec{u}^\epsilon\}$ must converge to $\vec{u}^0$.

Step 4. *Identification of the limit problem.*

In the last step of the proof we want to show that the limit $\vec{u}^0$ satisfies the quasi-geostrophic equations (4.131), (4.132), and (4.133). From the momentum equation (4.122) and the uniform stability estimates in (4.126) and (4.129), it follows that

$$(4.154) \qquad \max_{0 \leq t \leq T} \left\| \vec{v}^{\epsilon\perp} - \nabla h^\epsilon \right\|_0 = \epsilon \left\| \frac{D\vec{v}^\epsilon}{Dt} \right\|_0 \leq C\epsilon ,$$

so that the left side of the inequality in (4.154) goes to 0 as $\epsilon \to 0$. Combining this result with the convergence statement in (4.130), we conclude that $\vec{u}^0$ satisfies

$$(4.155) \qquad \vec{v}^0 = \nabla^\perp h^0 ,$$

which establishes the geostrophic constraint (4.131) for the limit $\vec{u}^0$. In particular, $\vec{u}^0$ automatically satisfies the divergence-free constraint

$$(4.156) \qquad \operatorname{div} \vec{v}^0 = 0 ,$$

and the vorticity $\omega^0 = \operatorname{curl} \vec{v}^0$ and the height h^0 are related by

$$(4.157) \qquad \Delta h^0 = \omega^0 .$$

To derive the vorticity equation (4.133) for the limit, we cannot take directly the strong limit in equation (4.124) because we don't know that the time derivative $\vec{u}^\epsilon_t$ converges to $\vec{u}^0_t$ as $\epsilon \to 0$. Instead, we show that $\vec{u}^0$ satisfies the vorticity equation (4.133) in the distribution sense. First, $\vec{u}^\epsilon$ is a classical solution of (4.124), so it also satisfies it in the distribution sense. If $\phi(\vec{x}, t)$ is a test function in $C_0^\infty(\mathbb{T}^2 \times [0, T])$, then the distribution formulation of equation (4.124) is

$$(4.158) \qquad \int_0^\infty \int_{\mathbb{T}^2} \left(\phi_t + \operatorname{div}\left(\phi\vec{v}^\epsilon \right) \right)\left(\omega^\epsilon - Fh^\epsilon \right) dx\, dt = \int_0^\infty \int_{\mathbb{T}^2} \phi\omega^\epsilon \operatorname{div} \vec{v}^\epsilon \, dx\, dt ,$$

and combining the convergence result in (4.130) with Lebesgue's dominated convergence theorem, we can pass to the limit inside the integrals in equation (4.158) and conclude that the limit function $\vec{u}^0$ satisfies

$$(4.159) \qquad \int_0^\infty \int_{\mathbb{T}^2} \left(\phi_t + \operatorname{div}\left(\phi\vec{v}^0 \right) \right)\left(\omega^0 - Fh^0 \right) dx\, dt = \int_0^\infty \int_{\mathbb{T}^2} \phi\omega^0 \operatorname{div} \vec{v}^0 \, dx\, dt .$$

Equation (4.159) shows that $\vec{u}^0$ satisfies

$$(4.160) \qquad \left(\frac{\partial}{\partial t} + \vec{v}^0 \cdot \nabla \right)(\omega^0 - Fh^0) = -\omega^0 \operatorname{div} \vec{v}^0$$

in the distribution sense.

Finally, since we know that $\vec{v}^0$ satisfies the incompressibility condition in equation (4.156), then equation (4.160) reduces to

$$(4.161) \qquad \left(\frac{\partial}{\partial t} + \vec{v}^0 \cdot \nabla \right)(\omega^0 - Fh^0) = 0 ,$$

and this establishes the vorticity equation (4.133) for the limit function $\vec{u}^0$ in the distribution sense and concludes the proof of the theorem.

CHAPTER 5

Linear and Weakly Nonlinear Theory of Dispersive Waves with Geophysical Examples

5.1. Linear Wave Midlatitude Planetary Equations

We begin our discussion by studying linear waves for the quasi-geostrophic equations from Section 1.2 and derived in the last chapter. These equations are also called the *midlatitude planetary equations*,

$$\frac{\partial q}{\partial t} + \psi^{\perp} \cdot \nabla q = 0, \quad q = \Delta\psi - F\psi + \beta y + h_b.$$

An alternate form for the advective term results from noting that

$$\vec{v} \cdot \nabla q = \nabla^{\perp}\psi \cdot \nabla q = \det(\nabla\psi, \nabla q) = J(\psi, q),$$

so the above equations are the following:

$$(5.1) \qquad \frac{\partial q}{\partial t} + J(\psi, q) = 0, \quad q = \Delta\psi - F_0\psi + \beta y + h_b.$$

Next we consider the quasi-geostrophic equations as an example of how to do linear wave theory.

The following two-step process tells us how to obtain knowledge about solutions through linear waves:

Step 1. Find nonlinear states.

We look for time-independent solutions $\partial q/\partial t = 0$ to the midlatitude quasi-geostrophic equations (5.8). We will consider the simplest case $h_b = 0$, and look for solutions to

$$J(\psi, q) = \det(\nabla\psi, \nabla q) = 0.$$

One family of solutions is as follows: Pick any function of latitude $\overline{\psi}(y)$. Then $\overline{\psi}$ generates a potential vorticity and velocity field

$$(5.2) \qquad \bar{q} = \frac{\partial^2 \overline{\psi}}{\partial y^2} - F_0\overline{\psi} + \beta y, \quad \overline{\vec{v}} = \begin{pmatrix} -\frac{\partial \overline{\psi}}{\partial y} \\ 0 \end{pmatrix} = \begin{pmatrix} \overline{V} \\ 0 \end{pmatrix}.$$

This generates an exact solution because $\bar{q}$ and $\overline{\psi}$ are functions of y only, so $\nabla\bar{q}$ and $\nabla\overline{\psi}$ are collinear.

These states are an example of a shear flow, with flows in the x-direction but varying in the y-direction. In fact, weather data averaged over long time periods of say, 30 days, looks like these meridional shear flow states. It is perturbations and oscillations about these states that govern the weather.

Step 2. Linearize about these states.

Assume that you have a family of solutions q^ϵ and ψ^ϵ such that

(1) $q^\epsilon\big|_{\epsilon=0} = \bar{q}$, $\psi^\epsilon\big|_{\epsilon=0} = \overline{\psi}$,
(2) q^ϵ and ψ^ϵ are smooth functions of ϵ, and
(3) q^ϵ and ψ^ϵ provide solutions to the full nonlinear equations (5.8) for all ϵ.

Then linearization means to Taylor-expand the equations in ϵ about the point $\epsilon = 0$ and keep the linear terms. Excuse an abuse of notation, and let

$$(5.3) \qquad q = \frac{\partial q^\epsilon}{\partial \epsilon}\bigg|_{\epsilon=0}, \qquad \psi = \frac{\partial \psi^\epsilon}{\partial \epsilon}\bigg|_{\epsilon=0}.$$

The linear equations solved by q and ψ are *linearized midlatitude QG equations*

$$(5.4) \qquad (1)\ q = \Delta\psi - F\psi, \qquad (2)\ \frac{\partial q}{\partial t} + \overline{V}\frac{\partial q}{\partial x} + \frac{d\bar{q}}{dy}\frac{\partial \psi}{\partial x} = 0.$$

Derivation: Result (1) was trivial as the linearization of a linear equation is the same equation. Result (2) begins with the equation of motion for q^ϵ

$$\frac{\partial q^\epsilon}{\partial t} + J(q^\epsilon, \psi^\epsilon) = 0.$$

Take the derivative with respect to ϵ,

$$\frac{\partial}{\partial t}\frac{\partial q^\epsilon}{\partial \epsilon} + \frac{d}{d\epsilon}\det(\nabla q^\epsilon, \nabla\psi^\epsilon) = 0,$$

and evaluate at $\epsilon = 0$,

$$\frac{\partial}{\partial t}\frac{\partial q^\epsilon}{\partial \epsilon}\bigg|_{\epsilon=0} + \det\left(\nabla\frac{\partial q^\epsilon}{\partial \epsilon}\bigg|_{\epsilon=0}, \nabla\overline{\psi}\right) + \det\left(\nabla\bar{q}, \nabla\frac{\partial\psi^\epsilon}{\partial \epsilon}\bigg|_{\epsilon=0}\right) = 0.$$

Since $\bar{q}$ and $\overline{\psi}$ are functions of y only,

$$\frac{\partial q}{\partial t} - \frac{d\overline{\psi}}{dy}\frac{\partial q}{\partial x} + \frac{d\bar{q}}{dy}\frac{\partial\psi}{\partial x} = 0 \quad \text{and} \quad \overline{V} = -\frac{d\overline{\psi}}{dy}.$$

EXERCISE 5.1. Compute the linearization of midlatitude quasi-geostrophic equations in the case that includes topography.

For the special choice $\overline{\psi}(y) = -\overline{V}y$ with $\overline{V}$ constant, the linearized equations become

$$(5.5) \qquad \frac{\partial q}{\partial t} + \overline{V}\frac{\partial q}{\partial x} + (-F\overline{V} + \beta)\frac{\partial\psi}{\partial y} = 0, \qquad q = \Delta\psi - F\psi.$$

In this case, we rewrite the above equations

$$(5.6) \qquad q = \Delta\psi - F\psi, \qquad \frac{\partial q}{\partial t} + V\frac{\partial q}{\partial x} + \tilde{\beta}\frac{\partial\psi}{\partial x} = 0, \qquad \tilde{\beta} = FV + B.$$

In the "effective beta" $\tilde{\beta}$, the mean wind adds to or competes with the effect from the curvature of the earth created by β when $F \neq 0$.

We will generally think in terms of the Northern Hemisphere, and set longi-tude-latitude coordinates (x, y) with increasing x being eastward, decreasing x

westward, increasing y to the north, and decreasing y to the south. The term "westerlies" means winds that come from the west and point to the east. The earth's rotation leads to a dominance of westerlies in both hemispheres, and a positive value of u would reflect a mean wind blowing to the east.

Plane Wave Ansatz. The simplest way to do wave analysis on this constant-coefficient linear equation (5.6) is simply to insert a plane wave ansatz into the equations of motion. The plane wave ansatz is

$$q(\vec{x}, t) = A e^{-\iota\omega(\vec{k})t} e^{\iota(k_1 x + k_2 y)} \quad \text{or} \quad \tilde{A} e^{-i\omega(\vec{k})t} e^{i(k_1 x + k_2 y)}.$$

We will interchange the notation $(x, y) \leftrightarrow (x_1, x_2)$ and perhaps also $(k_1, k_2) \leftrightarrow (k, l)$. k_1 is the longitudinal wavenumber, and k_2 is the meridional or latitudinal wavenumber.

Each operation in the linear equations acts on the plane wave as a multiplication. The derivative in time acts as multiplication by the factor $-\iota\omega$, while derivatives in spatial variables bring down factors of the appropriate component of the wavenumber vector. By guessing a similar plane wave form for the stream function, we find the relation between the two variables,

$$\psi = -\frac{q}{|\vec{k}|^2 + F},$$

while in the evolution equation, the plane wave itself drops out, leaving a relation between the frequency ω and the wavenumber vector $\vec{k}$, called the *dispersion relation*. We have already seen other examples of such a dispersion relation for internal gravity waves in Section 2.4 and inertio-gravity waves in rotating shallow water in Section 4.4.

Dispersion Relation: Midlatitude Planetary Waves.

(5.7) $$\omega(\vec{k}) = k_1 \left(V - \frac{\tilde{\beta}}{|\vec{k}|^2 + F} \right).$$

If the frequency ω is real, as it is here, then the given forms will in fact be pure wave solutions. The wave points in the direction given by the wavenumber vector $\vec{k}$ and moves with the *phase velocity*

$$\vec{c}_p = \begin{pmatrix} \frac{\omega(\vec{k})}{k_1} \\ \frac{\omega(\vec{k})}{k_2} \end{pmatrix}.$$

The following naive approach to the problem is presented as a pitfall to be avoided: In the absence of a mean wind ($V = 0$), the dispersion relation $\omega_\beta = -\beta k_1/(|\vec{k}|^2 + F)$ applies, while without the beta effect ($\beta = 0$) the dispersion relation is $\omega_0 = V k_1$. As a first guess, the law of Galilean invariance would suggest that the effect of the mean wind could be treated as a Doppler shift and added in through $\omega = \omega_\beta + \omega_0$. This approach gives the wrong answer for $F \neq 0$, however, as the correct dispersion relation contains the effective beta $\tilde{\beta}$ rather than the original β. This example suggests that something subtle is happening. Note that these waves are retrograde and move slower than the mean wind.

The group velocity is given by the gradient of the frequency in the wave-number vector ($\vec{c}_g = \nabla_k \omega$) and tells the velocity of energy propagation by a group of waves. This fact, which will be demonstrated later in the chapter, points up the importance of group velocity, which can have a totally different direction from phase velocity.

For the case $V = 0$, the dispersion relation is $\omega = -\beta k_1/(|\vec{k}|^2 + F)$. Since $\beta > 0$, we see that the x-component of phase velocity (ω/k_1) is always negative, i.e., propagating to the west.

The group velocity is given by

$$(5.8) \qquad \frac{d\omega}{dk_1} = \beta \frac{k_1^2 - k_2^2 - F}{(|\vec{k}|^2 + F)^2}, \qquad \frac{d\omega}{dk_2} = \beta \frac{2k_1 k_2}{(|\vec{k}|^2 + F)^2}.$$

Most of our weather systems come from the west. These systems are not mono-chromatic, but contain many wavelengths. When we speak of a weather system's movement, often we are not interested in the progress of the individual wave crests, but rather the net flux of a packet of waves. When there is no mean flow, we see that despite the fact that the x-component of phase velocity always points west, the energy from a significant number of these waves propagates to the east, a remark-able fact that says this rough linearized model has some agreement with what is observed. It is seen by checking the sign of the x-component of group velocity that

- for short-wavelength waves $k_1^2 > k_2^2 + F$, the group velocity points east-ward
- for long-wavelength waves $k_1^2 < k_2^2 + F$, the group velocity points west-ward.

Our intuition from nondispersive systems such as acoustics, where phase ve-locity and group velocity are equal, goes out the window. The dispersive property of geophysical flows is related to their ability to travel long distances with slow dissipation. We show this below.

We leave it as an amusing exercise for the reader to apply the same considera-tions with similar conclusions for an eastward mean wind, $V > 0$.

EXERCISE 5.2 (Plane Waves). Consider the plane waves

$$\psi_{\text{total}} = \text{Re}(A e^{-\iota \omega(\vec{k})t} e^{\iota(k_1 x + k_2 y)}) - vy, \qquad q_{\text{total}} = \Delta \psi_{\text{total}} - F \psi_{\text{total}} + \beta y,$$

with $\omega(\vec{k})$ given by the dispersion relation (5.7).

(1) Show q_{total} and ψ_{total} form an exact solution of the nonlinear midlatitude planetary equations, that is, $J(q_{\text{total}}, \psi_{\text{total}}) = 0$.
(2) Linearize about these states.
(3) Graph streamlines ($\psi_{\text{total}} = $ constant) for flow at a fixed time t.
(4) Find conditions for which this solution is steady or time independent.

These solutions are called *Rossby waves in a mean current*.

5.2. Dispersive Waves: General Properties

Here we will develop a general context for discussing dispersive waves, first for single equations like the midlatitude planetary equations, and then generalized to systems such as the rapidly rotating shallow water equation or internal gravity waves. Our primary tool of wave analysis will be the Fourier transform.

DEFINITION 5.1 The *Fourier transform* for functions in $\mathbb{R}^d$ is defined as

$$\hat{f}(\vec{k}) = \frac{1}{(2\pi)^d} \int\limits_{\mathbb{R}^d} e^{-\iota \vec{k}\cdot\vec{x}} f(\vec{x}) d\vec{x}, \quad f(\vec{x}) = \int\limits_{\mathbb{R}^d} e^{\iota \vec{k}\cdot\vec{x}} \hat{f}(\vec{k}) d\vec{k},$$

where $\vec{x} = (x_1, x_2, \ldots, x_d)$ and $\vec{k} = (k_1, k_2, \ldots, k_d)$ in d dimensions.

Fourier analysis treats a function u as a linear combination of plane wave basis vectors. In the linearized midlatitude planetary equations just discussed, we saw how assuming a plane waveform led to a relation between the frequency and wavenumber vector. Fourier analysis allows us to do the same thing in a more general context, separating a linear wave equation into individual wavenumber components. A generic linear scalar equation for amplitude at a given wavenumber is

$$(5.9) \qquad \frac{\partial A}{\partial t} + \iota \omega(\vec{k}) A = 0,$$

whose solution is simply rotation in the complex plane at the frequency ω. Take A to be the Fourier transform of function $u(\vec{x})$: $A = \hat{u}(\vec{k})$, and define an operator on the derivative via the inverse Fourier transform

$$(5.10) \qquad W\left(\frac{\partial}{\partial x}\right) u \equiv \int\limits_{\mathbb{R}^d} e^{\iota \vec{k}\cdot\vec{x}} \iota \omega(\vec{k}) \hat{u}(\vec{k}) d\vec{k}$$

to obtain the x-space equation of motion in a form that points up the linear wave structure

$$(5.11) \qquad \frac{\partial u(\vec{x}, t)}{\partial t} + W\left(\frac{\partial}{\partial x}\right) u(\vec{x}, t) = 0.$$

This equation is a general linear scalar wave equation with plane wave solutions $A e^{-\iota \omega(\vec{k})t} e^{\iota \vec{k}\cdot\vec{x}}$. The linearized planetary equations provide a beautiful example, and the reader is invited to test the tool of Fourier transform on it and rederive the dispersion relation. But we have not yet said what makes a system dispersive.

DEFINITION 5.2 An equation $\frac{\partial u}{\partial t} + W\left(\frac{\partial}{\partial x}\right) u = 0$ with the operator W defined in (5.10) is *dispersive* if the determinant of the Hessian matrix of frequency is nonzero, i.e.,

$$\det\left(\frac{\partial^2 \omega(\vec{k})}{\partial k_i \partial k_j}\right) \neq 0, \quad \text{except for a set of measure zero in } \vec{k}.$$

While planetary waves are generally dispersive, a strong easterly mean wind ($v < 0$) can lead to an effective beta $\tilde{\beta} = Fv + \beta$ that is zero. In this case, the dispersion relation reduces to $\omega(\vec{k}) = vk_1$, and the determinant of its Hessian matrix is zero. Indeed, the waves in this circumstance look like acoustic waves in the longitudinal direction. The moral is that the dispersive properties of the waves depend on the state you linearize about.

Dispersive Waves for Systems. We limit ourselves in this context to linear, first-order systems with constant coefficients. An important example of a system to keep in mind is the linearized rotating shallow water equation discussed in Chapter 4. A general system of this type, obtained by linearization, is

$$(5.12) \qquad \frac{\partial \vec{u}}{\partial t} + \sum_{j=1}^{d} A_j \frac{\partial \vec{u}}{\partial x_j} + B\vec{u} = 0$$

where d is the number of space dimensions and $\vec{x} \in \mathcal{R}^d$. There are m equations in the m components of $\vec{u}(\vec{x}, t)$, and A_j and B are $m \times m$ matrices. By inserting $\vec{u} = Ae^{-\iota\omega(\vec{k})t}e^{\iota\vec{k}\cdot\vec{x}}\vec{R}$ into equation (5.12), we obtain an *eigenvalue problem* for $\vec{R}$,

$$(5.13) \qquad \left[-\omega_p(\vec{k})I + \sum_{j=1}^{d} A_j k_j - \iota B\right]\vec{R}_p(\vec{k}) = 0, \quad i = \sqrt{-1},$$

where I is the $m \times m$ identity matrix. There will be m eigenfrequencies ω_p and eigenvectors $\vec{R}_p$ labeled for $p = 1, 2, \ldots, m$. For pure wave motion, the complete set of frequencies must be real. For the pth mode to be dispersive, the condition given in the above definition applies. In $\mathbb{R}^2$, we need

$$\det\left(\frac{\partial^2 \omega_p(\vec{k})}{\partial k_1 \partial k_2}\right) \neq 0$$

except on a set of measure zero.

EXAMPLE 5.1 (Rotating Shallow Water Equations). This provides an example of the systems context described above. When that system is linearized about a steady state with a static fluid $\vec{v} = 0$ at a constant height $h = H_0$ and flat topography $h_b(\vec{x}) = 0$, we find

$$(5.14) \qquad \frac{\partial \vec{v}}{\partial t} + f\vec{v}^\perp = -g\nabla h, \qquad \frac{\partial h}{\partial t} + H_0 \operatorname{div} \vec{v} = 0.$$

Recall from Chapter 4 that this 3×3 system of equations for the state vector $\vec{u} = \binom{\vec{v}}{h}$ is of the form (5.12) with the matrices

$$A_1 = \begin{pmatrix} 0 & 0 & g \\ 0 & 0 & 0 \\ H_0 & 0 & 0 \end{pmatrix}, \quad A_1 = \begin{pmatrix} 0 & 0 & 0 \\ 0 & 0 & g \\ 0 & H_0 & 0 \end{pmatrix}, \quad B = \begin{pmatrix} 0 & -f & 0 \\ f & 0 & 0 \\ 0 & 0 & 0 \end{pmatrix}.$$

Recall from Chapter 4 that there are three real eigenvalues, but only two of these are dispersive.

Eigenvalues of Linearized Rotating Shallow Water Equations.

    (A) $\omega_0 = 0$ for all $\vec{k}$ steady waves,

    (B) $\omega_\pm(\vec{k}) = \pm\sqrt{f^2 + gH_0|\vec{k}|^2}$ inertio-gravity or Poincaré waves.

Waves of type (A) are steady state, so $\partial\vec{v}/\partial t = 0$ and $\partial h/\partial t = 0$, and the system is in a geostrophic balance with the Coriolis force balancing the pressure gradient:

$$(5.15) \qquad f\vec{v}^\perp = -g\nabla h, \quad \operatorname{div} \vec{v} = 0.$$

Type (B) waves, one of whose dispersion relations is convex and the other concave, are easily seen to be dispersive, with group velocities that are smaller than the phase velocities. In the absence of rotation ($f = 0$), the waves would be acoustic rather than dispersive. This can be seen just by looking at the dispersion relation, but is also a corollary to the forthcoming claim that there is no dispersion when the constant matrix B is zero.

EXAMPLE 5.2 (Linearized Stratified Flow Equations). The stably stratified Boussinesq equations from Chapter 1 are not in the form (5.12) because the divergence-free equation is not a time evolution equation like the others. As discussed subsequently in Section 5.7, this feature can be overcome, however, by eliminating the pressure, and the following eigenfrequencies are found:

EIGENVALUE OF LINEARIZED STRATIFIED FLOW EQUATIONS:

$$\text{(A)} \quad \omega_0 = 0 \qquad\qquad \text{vortical modes},$$

$$\text{(B)} \quad \omega_\pm(\vec{k}) = \pm N \frac{|k_N|}{|k|} \quad \text{internal gravity waves}.$$

The vortical waves are an important feature of Boussinesq equations in three dimensions, but the internal gravity waves have interesting properties. Group velocities are directed perpendicular to phase velocities, as described in Chapter 2. The fundamental frequency of oscillation for gravity waves is the familiar *buoyancy frequency* from Chapters 2 and 3,

$$N = \left(-\frac{g}{\rho_b} \frac{d\bar{\rho}}{dz} \right)^{1/2} .$$

More physical properties of dispersive waves in the three examples are discussed in Gill's book [**11**, chaps. 6, 8, 12]. Other notable examples of dispersive wave systems with the structure in (5.13) are the infinite number of 3×3 symmetric systems arising for linear equatorial waves in Section 9.2. In the linear first-order system, dispersion results from the interaction of derivative terms along with the constant term B, as the following claim suggests.

CLAIM. If B vanishes in (5.12), no wave is dispersive.

PROOF: With $B \equiv 0$, the eigenvalue problem is

$$\left(-\omega_p(\vec{k})I + \sum_{j=1}^{d} A_j k_j \right) \vec{R}_p = 0 .$$

From this we see that $\omega(\vec{k})$ scales with $\vec{k}$, that is,

$$\omega_p(\lambda\vec{k}) = \lambda\omega_p(\vec{k}) .$$

We know that when we take the derivative of a function that is homogeneous of degree 1, we obtain a function homogeneous of degree 0. Thus,

$$(\nabla_k \omega_p)(\lambda\vec{k}) = (\nabla_k \omega_p)\vec{k} .$$

We have found that the gradient of frequency is constant along radial directions, so the component of its gradient aligned with the radial direction is zero. For any radial unit vector $\vec{e}_{\vec{k}} = \vec{k}/|\vec{k}|$,

$$(\vec{e}_{\vec{k}} \cdot \nabla_k) \nabla_k \omega_p = 0 \,.$$

This is the same as saying that the Hessian matrix of ω has a null vector $\vec{e}_{\vec{k}}$,

$$\left(\frac{\partial^2 \omega_p}{\partial k_i \partial k_j} \right) \vec{e}_{\vec{k}} = 0 \,,$$

and the fact that the Hessian matrix has a zero eigenvalue immediately tells us that its determinant is zero

$$\det \left(\frac{\partial^2 \omega_p}{\partial k_i \partial k_j} \right) = 0$$

and so the waves are not dispersive, and the claim is proven. $\square$

5.3. Interpretation of Group Velocity

It was remarked earlier that the group velocity is the significant representative of the velocity for a group of waves as it indicates the speed and direction of the propagation of energy by the waves. This interpretation will now be explained. We return to our generic scalar wave equation

$$\frac{\partial u(\vec{x})}{\partial t} + W\left(\frac{\partial}{\partial x} \right) u(\vec{x}) = 0 \,,$$

and consider the simple plane wave solution $u(\vec{x}) = e^{-\iota \omega(\vec{k}_0)t + \iota \vec{k}_0 \cdot \vec{x}}$. This pure sinusoidal solution with wavelength $2\pi/|\vec{k}_0|$ has a Fourier spectrum that consists of a single delta function centered at wavenumber $\vec{k}_0$. Since a structure like this is rarely seen in the real world, it makes sense to consider a finite approximation to the delta function spectrum by considering a wave packet, a bundle of wavenumber vectors that has a frequency distribution that looks like a smoothed delta function near $\vec{k}_0$.

Wave Packet Initial Data.

$$u\big|_{t=0} = e^{\iota \vec{k}_0 \cdot \vec{x}} A_0(\epsilon \vec{x}) \,.$$

Here, ϵ is a small parameter ($\epsilon \ll 1$) and A_0 is a smooth, rapidly decaying function. Written in physical space, this initial data looks like the original sinusoid of wavelength $2\pi/|\vec{k}_0|$ but with an amplitude that is now modulated on a much longer length scale, longer by a factor $1/\epsilon$. In fact, we will be more interested in the large-scale envelope than in every small-scale wiggle produced by the phase $\vec{k}_0$.

We need to show that this initial data given in physical space has the described wave bundle structure in Fourier space. We Fourier-transform the wave packet initial data and obtain

$$\hat{u}(\vec{k})\big|_{t=0} = (2\pi)^{-d} \int e^{-\iota \vec{k} \cdot \vec{x}} e^{\iota \vec{k}_0 \cdot \vec{x}} A_0(\epsilon \vec{x}) d\vec{x} = \widehat{A_0(\epsilon \vec{x})}(\vec{k} - \vec{k}_0) \,.$$

To evaluate this, we change variables inside the integral to the new scale $\vec{x}' = \epsilon\vec{x}$ and find

$$\widehat{A_0(\epsilon\vec{x})}(\vec{k} - \vec{k}_0) = \epsilon^{-d}\int e^{-\iota\frac{\vec{k}-\vec{k}_0\cdot\vec{x}'}{\epsilon}}A_0(\vec{x}')d\vec{x}' = \epsilon^{-d}\widehat{A_0}\left(\frac{\vec{k} - \vec{k}_0}{\epsilon}\right).$$

We can now see that this is a smoothed delta function centered at $\vec{k}_0$ for appropriate functions A_0 such as a Gaussian. As ϵ gets smaller, the amplitude becomes larger and the support of the function gets clustered near $\vec{k}_0$.

With this wave packet initial data clearly understood, we now inquire into its time evolution under the wave equation. The Fourier amplitude evolves according to $\hat{u}(\vec{k}, t) = \hat{u}(\vec{k}, 0)e^{-\iota\omega(\vec{k})t}$. Thus the physical variable is given by

$$u(\vec{x}, t) = \int_{\mathbb{R}^d} e^{\iota\vec{k}\cdot\vec{x}}e^{-\iota\omega(\vec{k})t}\epsilon^{-d}\widehat{A_0}\left(\frac{\vec{k} - \vec{k}_0}{\epsilon}\right)d\vec{k}.$$

Let $\tilde{\vec{k}} = (\vec{k} - \vec{k}_0)/\epsilon$; that is, rescale the wavenumbers so that the spectrum looks like it has width 1 rather than width ϵ,

$$u(\vec{x}, t) = \int_{\mathbb{R}^d} e^{\iota(\epsilon\tilde{\vec{k}}+\vec{k}_0)\cdot\vec{x}}e^{-\iota\omega(\epsilon\tilde{\vec{k}}+\vec{k}_0)t}\widehat{A_0}(\tilde{\vec{k}})d\tilde{\vec{k}} = e^{-\iota\omega(\vec{k}_0)t+\iota\vec{k}_0\cdot\vec{x}}A_\epsilon(\epsilon\vec{x}, \epsilon t),$$

where A_ϵ is the envelope that modulates on the long time and space scales

$$A_\epsilon(\epsilon\vec{x}, \epsilon t) = \int_{\mathbb{R}^d} e^{\iota\epsilon\vec{x}\cdot\tilde{\vec{k}}}e^{-\iota(\omega(\vec{k}_0+\epsilon\tilde{\vec{k}})-\omega(\vec{k}_0))t}\widehat{A_0}(\tilde{\vec{k}})d\tilde{\vec{k}}.$$

In the rescaled space and time variable $\vec{x}' = \epsilon\vec{x}$ and $t' = \epsilon t$,

$$A_\epsilon(\vec{x}', t') = \int_{\mathbb{R}^d} e^{\iota\vec{x}'\cdot\tilde{\vec{k}}}e^{-\iota\frac{\omega(\vec{k}_0+\epsilon\tilde{\vec{k}})-\omega(\vec{k}_0)}{\epsilon}t'}\widehat{A_0}(\tilde{\vec{k}})d\tilde{\vec{k}}.$$

We now wish to consider the limit $\epsilon \to 0$. The Lebesgue dominated convergence theorem tells us that we can interchange the limit with the integral,

$$(5.16) \qquad \lim_{\epsilon\to 0} A_\epsilon(\vec{x}', t') = \int_{\mathbb{R}^d} e^{\iota\vec{x}'\cdot\tilde{\vec{k}}}e^{-\iota\nabla_k\omega|_{\vec{k}_0}\cdot\tilde{\vec{k}}t'}A_0(\tilde{\vec{k}})d\tilde{\vec{k}} = A_0(\vec{x}' - \nabla_k\omega|_{\vec{k}_0}t').$$

We conclude that the slowly varying large-wavelength amplitude travels at the group velocity $\nabla_k\omega|_{\vec{k}_0}$.

Energetics. The wave equation $\frac{\partial u}{\partial t} + W(\frac{\partial}{\partial x})u = 0$ conserves energy,

$$(5.17) \qquad\qquad E = \frac{1}{2}\int |\vec{u}|^2\, d\vec{x}, \qquad \frac{\partial}{\partial t}E = 0.$$

To see this, we compute

$$\frac{\partial}{\partial t}E = \frac{1}{2}\int (u_t\bar{u} + u\bar{u}_t)d\vec{x} = \mathrm{Re}\int u_t\bar{u}\, d\vec{x} = -\mathrm{Re}\int Wu\bar{u}\, d\vec{x}$$

and use a standard fact from Fourier analysis, the *isometry property*,

$$\int f(\vec{x})\bar{g}(\vec{x})d\vec{x} = \int \hat{f}(\vec{k})\bar{\hat{g}}(\vec{k})d\vec{k}$$

to obtain

$$\frac{\partial}{\partial t}E = -\operatorname{Re}\int \widehat{Wu}(\vec{k})\bar{\hat{u}}(\vec{k})d\vec{k} = -\operatorname{Re}\int \iota\omega(\vec{k})\hat{u}(\vec{k})\bar{\hat{u}}(\vec{k})d\vec{k}$$

$$= -\operatorname{Re}\int \iota\omega(\vec{k})|u(\vec{k})|^2\,d\vec{k} = 0$$

since the integrand is pure imaginary.

It remains to observe that for the wave packet initial data, the solution $u(\vec{x}, t) = e^{\iota\vec{k}_0\cdot\vec{x}}e^{-\iota\omega(\vec{k}_0)t}A_\epsilon$ has energy

$$(5.18) \qquad\qquad E(t) = \int |A_\epsilon(\epsilon\vec{x}, \epsilon t)|^2 d\vec{x}\,.$$

The phase information has disappeared in (5.18), and we interpret $|A|^2$ as the energy density. In physical problems, we care about where the energy propagates, and we have found above in (5.16) that it travels with the wave packet envelope at the group velocity. These considerations generalize to systems as discussed below.

Recall the linear scalar wave equation

$$(5.19) \qquad\qquad \frac{\partial u(\vec{x})}{\partial t} + W\left(\frac{\partial}{\partial x}\right)u(\vec{x}) = 0$$

where the operator $W(\frac{\partial}{\partial x})$ is defined by

$$(5.20) \qquad\qquad W\left(\frac{\partial}{\partial x}\right)u \equiv \int_{\mathbb{R}^d} e^{\iota\vec{k}\cdot\vec{x}}\iota\omega(\vec{k})\hat{u}(\vec{k})d\vec{k}\,.$$

With the wave packet initial data

$$u\Big|_{t=0} = e^{\iota\vec{k}_0\cdot\vec{x}}A_0(\epsilon\vec{x})$$

the solution to (5.19) is given by

$$u^\epsilon(\vec{x}, t) = A_0(\epsilon\vec{x} - \nabla_k\omega(\vec{k}_0)\epsilon t)e^{\iota(\vec{k}_0\cdot\vec{x}_0 - \omega(\vec{k}_0)t)}\,.$$

In this slowly modulated wave train, energy travels with the large-scale amplitude envelope, which solves the partial differential equation

$$\frac{\partial A}{\partial t} + \nabla_k\omega(\vec{k}_0)\cdot\nabla_x A = 0\,, \quad A\Big|_{t=0} = A_0\,.$$

The velocity that governs the propagation of energy is the group velocity $\vec{c}_g = \nabla_k\omega(\vec{k})$.

Recall also the first-order linear wave equation for systems,

$$(5.21) \qquad\qquad \frac{\partial\vec{u}}{\partial t} + \sum_{j=1}^{d} A_j\frac{\partial\vec{u}}{\partial x_j} + B\vec{u} = 0\,.$$

The plane wave ansatz $\vec{u} = Ae^{-\iota\omega(\vec{k})t + \iota\vec{k}\cdot\vec{x}}\vec{R}$ leads to the eigenvalue problem

$$\left[-\iota\omega_p(\vec{k})I + \iota\sum_{j=1}^{d}A_jk_j + B\right]\vec{R}_p(\vec{k}) = 0.$$

We will ask that the eigenfrequencies $\omega_p(\vec{k})$ for $p = 1, 2, \ldots, m$ be real and distinct so that there will be a complete set of right eigenvectors $\vec{R}_p(\vec{k})$ and left eigenvectors $\vec{L}_p(\vec{k})$ determined by

$$\vec{L}_p(\vec{k})\left[-\iota\omega_p(\vec{k})I + \iota\sum_{j=1}^{d}A_jk_j + B\right] = 0$$

and also that the eigenvectors be normalized so that the matrix with right eigenvectors as columns and the matrix with left eigenvectors as rows are inverses,

(5.22) $$L_q(\vec{k}) \cdot R_p(\vec{k}) = \delta_{qp}.$$

EXERCISE 5.3. Show that the solution to (5.21) with wave packet initial data

$$\vec{u}_0(\vec{x}) = \vec{A}_0(\epsilon\vec{x})e^{\iota\vec{k}_0\cdot\vec{x}}$$

in the limit $\epsilon \to 0$ is given by

(5.23) $$\vec{u}^\epsilon(\vec{x}, t) \sim \sum_{p=1}^{m}A_0^p(\epsilon\vec{x} - \nabla_k\omega_p(\vec{k}_0)\epsilon t)\vec{R}_p(\vec{k}_0)e^{\iota(\vec{k}_0\cdot\vec{x} - \omega_p(\vec{k}_0)t)} + o(1)$$

where A_0^p is the inverse Fourier transform of $\widehat{A_0^p}$, the projection of the Fourier transform of the initial wave amplitude onto the pth left eigenvector

$$\widehat{A_0^p}(\vec{k}) = \vec{L}_p(\vec{k}) \cdot \widehat{A_0}(k).$$

We see that general initial data splits into component modes of the system, each moving with its own phase speed and group velocity.

EXERCISE 5.4. Compute for the midlatitude quasi-geostrophic in the presence of a mean wind the determinant of the Hessian matrix for frequency and show that the waves are in fact dispersive,

$$\omega(\vec{k}) = vk_1 - \frac{\tilde{\beta}k_1}{|\vec{k}|^2 + F}, \quad \det\left(\frac{\partial^2\omega}{\partial k_1 \partial k_2}\right) \neq 0.$$

The stratified Boussinesq equations are both nonlocal like the planetary equations and a big system like the rotating shallow water equations. The reader is invited to consider how to combine the strategies used on these other models to analyze the waves in stratified flow. This will be treated later in Section 5.7.

All three AOS models are dispersive, but for different reasons. The Poincaré waves have one convex and one concave frequency as a function of wavenumber magnitude. The planetary waves are neither concave nor convex, but rather have a point where $\partial_{k_1}\omega = 0$. The internal gravity wave dispersion relation is highly anisotropic as discussed in Section 2.4.

5.4. Distant Propagation from a Localized Source

In the next few sections, we will temporarily be leaving our AOS models to concentrate on general methods for linear dispersive waves. Two sections ago we defined the term *dispersive*, but we haven't shown what it really means. To that end, we consider the evolution of an initial distribution under a linear wave equation and focus on the behavior at large distance and long time from the source. The scalar case will be done in class; generalization to systems merely involves synthesizing some linear algebra in a fashion as sketched above.

Consider $\frac{\partial u}{\partial t} + Wu = 0$ with the operator $W(\frac{\partial}{\partial x})$ defined by (5.20) where the dispersion relation $\omega(\vec{k})$ satisfies

$$\det\left(\frac{\partial^2 \omega}{\partial k_i \partial k_j}\right) \neq 0 \,,$$

and let the initial condition $u(\vec{x}, 0) = u_0(\vec{x})$ have a Fourier transform with compact support, i.e., $\hat{u}(\vec{k}) = 0$ for $|\vec{k}| > M$. This condition is stronger than we need, but is chosen to avoid the complications of discussing numerous subcases. As we have seen, the solution is given via inverse Fourier transform:

$$u(\vec{x}, t) = \int\limits_{\mathbb{R}^d} e^{-\iota\omega(\vec{k})t + \iota\vec{x}\cdot\vec{k}} \hat{u}_0(\vec{k}) d\vec{k} \,.$$

We will examine the velocities present in this solution in the regime of long time by taking the limit $t \to \infty$ with the ratio $\vec{c} = \vec{x}/t$ fixed,

$$(5.24) \qquad u(\vec{c}t, t) = \int\limits_{\mathbb{R}^d} e^{\iota(\vec{c}\cdot\vec{k} - \omega(\vec{k}))t} \hat{u}_0(\vec{k}) d\vec{k} \,.$$

We ask, At what velocities $\vec{c}$ will energy be carried by the waves at long times?

Consider the following warmup example.

NONDISPERSIVE $1 - d$ EQUATION:

$$(5.25) \qquad \frac{\partial u}{\partial t} + c_0 \frac{\partial u}{\partial x} = 0 \,, \quad u|_{t=0} = u_0(x) \,,$$

$$u(x, t) = \int e^{\iota(x - c_0 t)k} \hat{u}_0(k) dk \,,$$

$$(5.26) \qquad u(ct, t) = \int e^{\iota(c - c_0)tk} \hat{u}_0(\vec{k}) d\vec{k} \,.$$

These first steps have followed those taken in the general case. We investigate what are the significant values of the parameter $c = x/t$ in the limit $t \to \infty$. We know from the solution, $u(x, t) = u_0(x - c_0 t)$, that information is carried along the characteristic at speed c_0. That nothing is carried at other speeds is the subject of the following claim. The essence is that the exponential in (5.26), since the gradient of phases is never small when $c \neq c_0$, oscillates rapidly, inducing cancellations that strongly diminish the amplitude of the integral. This is the first principle of stationary phase.

CLAIM. If $c \neq c_0$, then $\lim_{t\to\infty} u(ct, t) = O(t^{-L})$ for any large L.

PROOF: Observe that for $c \neq c_0$,

$$e^{\iota t k (c - c_0)} = \frac{1}{t^L \iota^L (c - c_0)^L} \left(\frac{d^L}{dk^L} e^{\iota t k (c - c_0)} \right).$$

Then with the insignificant constant $\tilde{c} = \iota^{-L} (c - c_0)^{-L}$,

$$u(x, t) = t^{-L} \tilde{c} \int \frac{d^L}{dk^L} e^{\iota t k (c - c_0)} \hat{u}(k) dt.$$

By repeated integration by parts, assuming $\hat{u}(k)$ is sufficiently smooth and dropping the boundary terms due to the compact support of $\hat{u}(k)$, we find

$$u(ct, t) = t^{-L} \tilde{c} (-1)^L \int e^{\iota t k (c - c_0)} \frac{d^L \hat{u}(k)}{dk^L} dk.$$

The integral is bounded, so the long-time behavior is t^{-L}, as claimed. $\qquad \square$

Principle of Stationary Phase. The first part of the principle of stationary phase was contained in the above claim—that nothing propagates when the phase has nonzero gradient. The second part concerns the propagation that occurs at those points of stationary phase. In the linear dispersion of the warmup, the energy propagated at only a single speed. But we will see that in the dispersive regime where second derivatives of phase are not zero, completely different behavior occurs. We will work in the general context of a d-dimensional integral $\int A(\vec{x}, \vec{c}) \exp(\iota \Psi(\vec{x}, \vec{c}) / \epsilon)$ with a large factor $1/\epsilon$ in the phase.

LEMMA 5.3 *Provided*

    (i) *$\epsilon \to 0$ is a small parameter, $A(\vec{x}, \vec{c})$ has compact support in $\vec{x}$, and $A(\vec{x}, \vec{c})$ and $\Psi(\vec{x}, \vec{c})$ are smooth functions of the parameter $\vec{c}$,*

    (ii) *$\nabla_x \Psi(\vec{x}_j, \vec{c}) = 0$ for $j = 1, 2, \ldots, q$, and*

    (iii) *$\det(\nabla_x^2 \Psi(\vec{x}_j, \vec{c})) \neq 0$,*

the following precise asymptotic expansion applies:

$$(5.27) \quad \int_{\mathbb{R}^d} A(\vec{x}, \vec{c}) e^{\frac{\iota \Psi(\vec{x}, \vec{c})}{\epsilon}} d\vec{x} \sim$$

$$\sum_{j=1}^{q} (2\pi \epsilon)^{d/2} A(\vec{x}_j, \vec{c}) \left| \det \nabla_x^2 \Psi(\vec{x}_j, \vec{c}) \right|^{-1/2} e^{\frac{\iota \Psi(\vec{x}_j, \vec{c})}{\epsilon}} e^{\frac{\iota \sigma_j \pi}{4}} + o(\epsilon^{d/2}),$$

where σ_j is the signature of the Hessian matrix $\nabla_x^2 \Psi(\vec{x}_j, \vec{c})$, that is, the number of positive eigenvalues minus the number of negative eigenvalues.

The reader is directed to Duistermaat's 1973 Courant lecture notes [5] for a readable and rigorous proof. The ideas behind (5.27) are simple. Where the phase is not stationary, a change of variables allows the application of the ideas in the warmup problem to show that the contribution here is asymptotically small. At the isolated points where the phase is stationary, the phase locally looks like a quadratic,

$$\Psi(\vec{x}) = \Psi(\vec{x}_j) + \frac{1}{2} \nabla_x^2 \Psi(\vec{x} - \vec{x}_j, \vec{x} - \vec{x}_j),$$

and we need to compute a Gaussian integral explicitly.

Application to Long-Time Propagation. Our solution for the evolution of a localized source (5.24) in the limit of long times is an example of an integral of the type discussed above with a large multiplier in the phase. With the associations

$$\vec{x} \leftrightarrow \vec{k}, \quad A(\vec{x}, \vec{c}) \leftrightarrow \hat{u}_0(\vec{k}), \quad \Psi(\vec{x}, \vec{c}) \leftrightarrow -\omega(\vec{k}) + \vec{c} \cdot \vec{k}, \quad t \leftrightarrow \frac{1}{\epsilon},$$

the points of stationary phase $\nabla \Psi = 0$ are given by

$$\text{(A)} \quad \nabla_k \omega(\vec{k}_j) = \vec{c} \quad \text{for } j = 1, 2 \ldots, q.$$

The condition on $\nabla^2 \Psi = -\nabla_k^2 \omega(\vec{k})$ is

$$\text{(B)} \quad \det(\nabla_k^2 \omega(\vec{k}_j)) \neq 0,$$

which is our same dispersive condition. Finally, $\hat{u}_0(\vec{k})$ has compact support, so we satisfy all the conditions of the lemma and apply it to find the asymptotic limit for $t \to \infty$,

$$\text{(5.28)} \quad u(\vec{c}t, t) \sim t^{-d/2} \sum_{j=1}^{q} e^{it(-\omega(\vec{k}_j) + \vec{c} \cdot \vec{k}_j)} e^{i\sigma_j \pi/4} \hat{u}(\vec{k}_j(\vec{c})) \left| \det \nabla_k^2 \omega(\vec{k}_j) \right|^{-1/2}.$$

We have found that information will only propagate at velocity $\vec{c}$ if it is the group velocity of some wave vector. Several (q) wave vectors may have the same group velocity, but they will be finite in number and isolated. If $\vec{c}$ is allowed to vary, say, on the surface of a ball, generally every direction will be a group velocity and will carry energy. The conclusion is that, unlike the warmup problem where only one velocity was involved, the waves spread out and travel in an open set of many different directions. These directions are associated with the group velocity from (A). From (B) and the implicit function theorem, it is easy to see that the vectors $\vec{c}$ satisfying (A) belong to an open set. Thus the term *dispersive*.

Since energy is conserved while the wave spreads in many directions, the amplitude of the wave must decrease, and a simple physicist's argument sketched below also allows us to predict the decay rate. In one dimension with $\frac{d\omega}{dk} = c$ and $\frac{d^2\omega}{dk^2} > 0$, let

$$-\alpha \leq \frac{d\omega}{dk} \leq \beta$$

so c takes values between $-\alpha$ and β. The length of the interval on which u is supported grows linearly with t. The energy is proportional to the integral of the square of the wave amplitude, so that for the energy to be conserved, $\int u^2 \, dx = $ const; thus, the wave amplitude must be proportional to the inverse square root of $t : u \sim t^{-1/2}$. Similarly, in higher-dimension d, the wave amplitude decays as $u \sim t^{-d/2}$. The result in (5.28) confirms this intuition.

EXERCISE 5.5. Apply the method of stationary phase to the constant-coefficient linear first-order system (5.21) where all of the wave fields are dispersive to find the asymptotic form for the solution at large t and fixed $\vec{c} = \vec{x}/t$.

5.5. WKB Methods for Linear Dispersive Waves

In nonlinear problems, we will no longer have the Fourier transform to save the day, and we should look for methods that will continue to serve us in the nonlinear case. We consider the linear first-order system again, but now with coefficients that vary slowly with space and time. A natural application of this is rapidly rotating shallow water equations where the beta plane expansion is not used, but the rotation parameter is left as a function of latitude, and you linearize about any geostrophic state. Consider the system

$$\frac{\partial \vec{u}}{\partial t} + \sum_{j=1}^{d} A_j(\epsilon \vec{x}, \epsilon t) \frac{\partial \vec{u}}{\partial x_j} + B(\epsilon \vec{x}, \epsilon t)\vec{u} = 0,$$

$$\vec{u}\big|_{t=0} = \vec{\mathcal{A}}_0(\epsilon \vec{x}) e^{\frac{\iota \phi_0(\epsilon \vec{x})}{\epsilon}}, \quad \epsilon \ll 1.$$

Notice that the common situation for WKB is at work here. The wavelength of the disturbance u, which is $O(1)$, is much smaller than the scale of variation of the medium parameters, which is $O(1/\epsilon)$. The initial data is also varying on the large scale, but the gradient of the initial phase is $O(1)$. If you take $\phi_0 = \vec{x} \cdot \vec{k}_0$, $A_j =$ const, and $B =$ const, then you get the wave packet problem of Exercise 5.3. We can get an expectation for what the solutions will look like in the limit $\epsilon \ll 1$ by recalling the solution to Exercise 5.3, equation (5.23).

The first step is to rescale into the large-scale variable $\vec{x}' = \epsilon \vec{x}, t' = \epsilon t$, although we immediately rename the variables by removing primes,

$$(5.29) \qquad \frac{\partial \vec{u}}{\partial t} + \sum_{j=1}^{d} A_j(\vec{x}, t) \frac{\partial \vec{u}}{\partial x_j} + \frac{1}{\epsilon} B(\vec{x}, t)\vec{u} = 0, \quad \vec{u}\big|_{t=0} = \vec{\mathcal{A}}_0(\vec{x}) e^{\frac{\iota \phi_0(\vec{x})}{\epsilon}}.$$

In this equation, everything varies on large scales and long times except the phase of the initial data, which is highly oscillatory. Also, the magnitude of the constant term has become large. Our basis vectors will be obtained from the following eigenvalue problem:

Eigenvalue Problem. Assume that

$$(5.30) \qquad \left[-\iota \omega I + \iota \sum_{j=1}^{d} A_j(\vec{x}, t)\vec{k}_j + B(\vec{x}, t) \right] \vec{R} = 0$$

has a full set of eigenvalues $\omega_p(\vec{k}, \vec{x}, t)$ and right eigenvectors $\vec{R}_p(\vec{k}, \vec{x}, t)$ for $p = 1, 2, \ldots, m$ at every location $\vec{x}$ and time t. Left eigenvectors are defined by $\vec{L}_p(\vec{k}, \vec{x}, t) \cdot \vec{R}_q(\vec{k}, \vec{x}, t) = \delta pq$.

We can expand the arbitrary initial data of (5.29) in terms of the left eigenvectors at time $t = 0$ with $\vec{k}(\vec{x}) = \nabla_x \phi_0(\vec{x})$,

$$\mathcal{A}(\vec{x}) = \sum_{p=1}^{m} \mathcal{A}_0^p(\vec{x}) \vec{R}_p(\nabla \phi_0, \vec{x}, t = 0), \quad \mathcal{A}_0^p(\vec{x}) = \vec{L}_p(\nabla \phi_0, \vec{x}, t = 0) \cdot \vec{\mathcal{A}}_0(\vec{x}).$$

However, for our purposes, since the problem is linear, we consider initial data that is nonzero in only one of the modes,

$$(5.31) \qquad \vec{u}\big|_{t=0} = \vec{R}_p(\nabla\phi_0, \vec{x}, 0)\mathcal{A}_0^p(\vec{x})e^{\frac{\iota\phi_0}{\epsilon}}.$$

We make an informed guess as to the structure of the final solution. To leading order we expect that the wave would remain in the same channel # p, so we take that for our leading-order term, and allow evolution into the other channels at the higher order in ϵ. The amplitudes will vary on the slow-time and large-space scales, while the phase will have the fast small scales built in through a $\frac{1}{\epsilon}$ factor. Thus, our ansatz for the asymptotic solution to (5.29) with initial data (5.31) in the limit $\epsilon \to 0$ is

$$(5.32) \qquad \vec{u}(\vec{x}, t) = \vec{u}^{(0)}(\vec{x}, t)e^{\frac{\iota\phi_p(\vec{x},t)}{\epsilon}} + \epsilon\sum_{q=1}^{m} \vec{u}_q^{(1)}(\vec{x}, t)e^{\frac{\iota\phi_q(\vec{x},t)}{\epsilon}} + O(\epsilon^2).$$

We build the solution formally by inserting our ansatz into the equation of motion and matching powers of ϵ. The leading-order terms are of size $O(\epsilon^{-1})$,

$$(5.33) \qquad \left(\iota\partial_t\phi_p I + \iota\sum_{j=1}^{d} A_j(\vec{x}, t)\partial_{x_j}\phi_p + B(\vec{x}, t)\right)\vec{u}^{(0)}(\vec{x}, t)e^{\frac{\iota\phi_p}{\epsilon}} = 0.$$

Thus we see our eigenvalue problem (5.30) emerging as the asymptotically dominant term. We immediately know two things: that the eigenfrequencies will be the solutions already found, and that the leading-order wave amplitude will be proportional to the appropriate eigenvector already found; that is,

(A) $\quad\dfrac{\partial\phi_p(\vec{x}, t)}{\partial t} + \omega_P(\nabla_x\phi_p, \vec{x}, t) = 0 \quad$ (eikonal equation),

$\qquad \phi_p(\vec{x}, t = 0) = \phi_0(\vec{x})$,

(B) $\quad\vec{u}^{(0)}(\vec{x}, t) = \mathcal{A}_0^p(\vec{x}, t)\vec{R}_p(\nabla\phi_p, \vec{x}, t)$, $\quad \mathcal{A}_0^p(\vec{x}, t = 0) = \mathcal{A}_0^p(\vec{x})$.

The solution matches the initial conditions at $t = 0$. At this stage in the argument $A^p(\vec{x}, t)$ is arbitrary except at $t = 0$. Obviously, a channel other than # p could not have done this. The asymptotic match at the leading order has provided an equation for the fast phase. We now proceed to the next order, $O(\epsilon^0)$. In our ansatz (5.32), we included all modes in the order ϵ term, but let's try to get by with only mode # p and see what happens:

$$\vec{u}(\vec{x}, t) = \left(\vec{u}^{(0)}(\vec{x}, t) + \epsilon\vec{u}_p^{(1)}(\vec{x}, t)\right)e^{\frac{\iota\phi_p(\vec{x},t)}{\epsilon}} + O(\epsilon^2).$$

Then in the asymptotic expansion, the order $O(\epsilon^0)$ equation is

$$(5.34) \quad \left(\iota\partial_t\phi_p I + \iota\sum_{j=1}^{d} A_j(\vec{x}, t)\partial_{x_j}\phi_p + B(\vec{x}, t)\right)\vec{u}_p^{(1)}(\vec{x}, t)e^{\frac{\iota\phi_p(\vec{x},t)}{\epsilon}}$$

$$+ \left(\partial_t\vec{u}^{(0)}(\vec{x}, t) + \sum_{j=1}^{d} A_j(\vec{x}, t)\partial_{x_j}\vec{u}^{(0)}(\vec{x}, t)\right)e^{\frac{\iota\phi_p(\vec{x},t)}{\epsilon}} = 0.$$

Canceling the phase, we can write this as

$$(5.35) \qquad M\vec{u}_p^{(1)} = \vec{F}$$

where

$$M = \iota \partial_t \phi_p I + \iota \sum_{j=1}^{d} A_j(\vec{x}, t) \partial_{x_j} \phi_p + B(\vec{x}, t)$$

is the matrix we met in (5.33), which said $M\vec{u}^{(0)} = 0$. This shows that a solution to (5.35) is not unique, as a term proportional to $\vec{u}^{(0)}$ could be added to $\vec{u}_p^{(1)}$ and we'd still have a solution. The Fredholm alternative of linear algebra tells us that equation (5.35) can be solved if and only if F is perpendicular to the left eigenvector.

Fredholm Alternative.

$$\vec{L}_p \cdot \vec{F} = 0.$$

This solvability condition at higher order gives a scalar evolution equation for the evolution of the amplitude from the previous order. This equation is called the *transport equation*:

$$(5.36) \qquad \begin{aligned} \vec{L}_p(\nabla \phi_p, \vec{x}, t) \bigg(& \partial_t \left(\mathcal{A}_0^p(\vec{x}, t) \vec{R}_p(\nabla \phi_p, \vec{x}, t) \right) \\ & + \sum_{j=1}^{d} A_j(\vec{x}, t) \partial_{x_j} \left(\mathcal{A}_0^p(\vec{x}, t) \vec{R}_j(\nabla \phi_p, \vec{x}) \right) \bigg) = 0, \end{aligned}$$

$$\mathcal{A}_0^p(\vec{x}, t = 0) = \mathcal{A}_0^p(\vec{x}).$$

When the transport equation is solved, we have the leading-term wave amplitude $\mathcal{A}_0^p(\vec{x}, t)$ for all times. To continue, notice that $\vec{u}_p^{(1)}$ is not proportional to $\vec{R}_p$, since when hit by the matrix this vector does not give zero. Therefore $\vec{u}_p^{(1)}$ must contain pieces of the other basis vectors. Let

$$\vec{u}_p^{(1)} = A_p^{(1)} \vec{R}_p(\nabla \phi_p, \vec{x}, t) + \sum_{q \neq p} A_p^{(1)} \vec{R}_q(\nabla \phi_p, \vec{x}, t)$$

where the coefficients $A_q^{(1)}$ are given by

$$A_q^{(1)} = \vec{L}_q(\nabla \phi_p, \vec{x}, t) \cdot \vec{u}_p^{(1)} = \vec{L}_q \cdot M^{-1} \vec{F}.$$

As mentioned above, the value of $A_p^{(1)}$ is arbitrary, but has initial condition $A_p^{(1)}\big|_{t=0} = 0$ so as not to interfere with the already matched initial conditions, even at order ϵ. On the other hand, we cannot match the initial conditions with nonzero components for the other basis vectors $A_q^{(1)} \neq 0$ and cancel the effect of nonzero $A_q^{(1)}$ at the initial time $t = 0$. This is where the assumption of only one mode at the $O(\epsilon)$ term breaks down, and we need the additional terms in (5.32) in order to fix this. These new terms $\vec{u}_q^{(1)}(\vec{x}, t)$ will lead to eikonal and transport equations in the qth channel. This process can be continued to arbitrary order. For example, to solve the equations to the order $O(1)$ that we have displayed above, repeating the argument in (5.33) for the qth channel yields the eikonal equation for ϕ_q,

$$\frac{\partial \phi_q}{\partial t} + \omega_q(\nabla_x \phi_q, \vec{x}, t) = 0, \quad \phi_q(\vec{x}, t = 0) = \phi_q(x).$$

To recap, at leading order, the solution stays in the channel it started in, with a fast phase and slow amplitude modulation according to the eikonal and transport equations, respectively. At next order in epsilon, the other channels become excited with their own equations for phase and amplitude.

To summarize these results, we recall the first-order linear wave equation for systems with coefficients that are varying slowly in time and space compared to the wavelength

$$\frac{\partial \vec{u}}{\partial t} + \sum_{j=1}^{d} A_j(\epsilon \vec{x}, \epsilon t) \frac{\partial \vec{u}}{\partial x_j} + B(\epsilon \vec{x}, \epsilon t)\vec{u} = 0$$

$$\vec{u}\big|_{t=0} = \vec{\mathcal{A}}_0(\epsilon \vec{x})\vec{R}_p(\nabla_{\epsilon x}\phi_0, \epsilon \vec{x}, \epsilon t = 0)e^{\frac{\iota \phi_0(\epsilon \vec{x})}{\epsilon}}$$

where the $\vec{R}_p$ is one of the right eigenvectors of the eigenvalue problem

$$(5.37) \qquad \left[-\iota \omega I + \iota \sum_{j=1}^{d} A_j(\vec{x}, t)k_j(\vec{x}, t) + B(\vec{x}, t) \right]\vec{R} = 0.$$

The problem is to describe the solution in the limit $\epsilon \ll 1$. It should be stressed that there is no loss of generality from using only a single mode in the initial condition. General initial data are expanded in component modes, and the WKB procedure is carried out for each mode. Rescaling of space and time variables $\epsilon \vec{x} \to \vec{x}$ and $\epsilon t \to t$,

$$\frac{\partial \vec{u}}{\partial t} + \sum_{j=1}^{d} A_j(\vec{x}, t) \frac{\partial \vec{u}}{\partial x_j} + \epsilon^{-1} B(\vec{x}, t)\vec{u} = 0$$

(5.38)

$$\vec{u}\big|_{t=0} = \vec{\mathcal{A}}_0(\vec{x})\vec{R}_p(\nabla_x \phi_0, \vec{x}, t = 0)e^{\frac{\iota \phi_0(\vec{x})}{\epsilon}}.$$

The asymptotic solution to (5.38) is given by

$$\vec{u}(\vec{x}, t) = \mathcal{A}_0^p(\vec{x}, t)\vec{R}_p(\nabla_x \phi_p, \vec{x}, t)e^{\frac{\iota \phi_p(\vec{x},t)}{\epsilon}}$$

(5.39)

$$+ \epsilon \sum_{q=1}^{m} \mathcal{A}_1^p(\vec{x}, t)\vec{R}_q(\nabla_x \phi_p, \vec{x}, t)e^{\frac{\iota \phi_q(\vec{x},t)}{\epsilon}} + O(\epsilon^2).$$

This leads to remarkably simplified PDEs for the scalars phase (nonlinear first order) and amplitude (linear first order) of the asymptotically leading term.

(A) *Eikonal equation* for the phase:

$$\frac{\partial \phi_p(\vec{x}, t)}{\partial t} + \omega_p(\nabla_x \phi_p, \vec{x}, t) = 0, \quad \phi_p(\vec{x}, t)\big|_{t=0} = \phi_0(\vec{x}),$$

(B) *Transport equation* for the amplitude:

$$\vec{L}_p(\nabla\phi_p, \vec{x}, t)\Bigg(\partial_t\big(\mathcal{A}_0^p(\vec{x}, t)\vec{R}_p(\nabla\phi_p, \vec{x}, t)\big)$$

(5.40)

$$+ \sum_{j=1}^{d} A_j(\vec{x}, t)\partial_{x_j}\big(\mathcal{A}_0^p(\vec{x}, t)\vec{R}_p(\nabla\phi_p, \vec{x}, t)\big)\Bigg) = 0,$$

$$\mathcal{A}_0^p(\vec{x}, t)\big|_{t=0} = \mathcal{A}_0^p(\vec{x}).$$

The transport equation can be written

(5.41) $\qquad \dfrac{\partial\mathcal{A}_0^p}{\partial t} + \sum_{j=1}^{d} c_j \dfrac{\partial\mathcal{A}_0^p}{\partial x_j} + b\mathcal{A}_0^p = 0 \quad$ where $c_j = \vec{L}_p A_j \vec{R}_p$.

The term b contains derivatives on the eigenvectors. From our previous experience, we expect group velocity to play a role here, and the following remarkable claim tells us that the leading-order amplitude moves with the group velocity associated with the phase ϕ_p. So energy is propagated along curves that are directions of differentiation following group velocities.

CLAIM. $c_j = \dfrac{\partial\omega_p}{\partial k_j}\Big|_{(\nabla\phi_p, \vec{x}, t)}$.

PROOF: (1) Take the eigenvalue problem (5.37) with the pth mode solution in place,

$$\Bigg(-\omega_p(\vec{k}, \vec{x}, t)I + \iota \sum_{j=1}^{d} A_j(\vec{x}, t)k_j + B(\vec{x}, t)\Bigg)\vec{R}_p(\vec{k}, \vec{x}, t) = 0,$$

and differentiate with respect to k_j,

$$\Bigg(-\frac{\partial\omega_p}{\partial k_j}I + \iota A_j\Bigg)\vec{R}_p + \underbrace{\Bigg(-\iota\omega_p I + \iota \sum_{j'=1}^{d} A_{j'}k_{j'} + B\Bigg)}_{M}\frac{\partial\vec{R}_j}{\partial k_j} = 0.$$

(2) Apply the dot product from the left with the left eigenvector $\vec{L}_p(\vec{k}, \vec{x}, t)$

$$-\iota\frac{\partial\omega_p}{\partial k_j} + \iota\vec{L}_p A_j \vec{R}_p + \vec{L}_p M \frac{\partial\vec{R}_p}{\partial k_j} = 0.$$

(3) Since $\vec{L}_p$ is a left eigenvector, $\vec{L}_p M = 0$, and

$$\frac{\partial\omega_p}{\partial k_j} = \vec{L}_p A_j \vec{R}_p \quad \text{so that } c_j = \frac{\partial\omega_p}{\partial k_j}.$$

Thus, the characteristic direction for the transport equation is defined by the group velocity.

□

The above asymptotic expansion is readily justified rigorously provided the solution of the eikonal equation remains smooth by using energy estimates. This is an exercise for the interested reader.

Linear WKB for Midlatitude Planetary Waves. The rotating shallow water equation exactly fits the framework just described when the rotation rate is variable, and we will discuss equatorial waves later in the course. In this section, we consider the midlatitude quasi-geostrophic equations. These are not a first-order system, but their scalar nature means only one phase speed and no concern for basis eigenvectors. The stratified equations provided a good exercise for doing linear WKB in the full generality of a nonlocal system and the results of nonlinear WKB are presented in Section 5.7.

Recall the linearization of the midlatitude planetary equations about the zonal mean state $\bar{q}(y) = \overline{\psi}_{yy} - F\overline{\psi} + \beta y$ in the absence of topography. With $\bar{v} = -\overline{\psi}_h$,

$$\frac{\partial q}{\partial t} + \bar{v}\frac{\partial q}{\partial x} + \frac{d\bar{q}}{dy}\frac{\partial \psi}{\partial x} = 0, \quad q = \Delta\psi - F\psi + \beta y.$$

Suppose the initial condition is given for potential vorticity

$$q(\vec{x}, t)\big|_{t=0} = \tilde{A}(\epsilon x, \epsilon y)e^{\frac{\iota\phi_0(\epsilon x, \epsilon y)}{\epsilon}}.$$

To set up our WKB situation we want the coefficients to be functions of the slow variable ϵy, but they should be of magnitude $O(1)$ to leading order. For this to happen, we take $\overline{\psi}_\epsilon = \epsilon^{-1}\overline{\psi}(\epsilon y)$ so that

(5.42) $\bar{v}(\epsilon y) = \overline{\psi}'$, $\quad \bar{q}_y(\epsilon y) = \epsilon^2\bar{v}'' + F\bar{v} + \beta$.

One should ask whether ϵ is in fact small in physical situations. Is there scale separation between the disturbance wavelength and the zonal background scale? The answer is sometimes yes and sometimes no, and sometimes one comes up with an ϵ of one-quarter or one-half. But experience has shown that WKB can provide qualitative information even when series terms do not decay rapidly. We are also, of course, building our WKB technique.

Now that the coefficients are slowly varying, we transform the equations into the long-time and large-scale variables $t \to \epsilon t$ and $\vec{x} \to \epsilon\vec{x}$. With $\tilde{\beta} = F\bar{v} + \beta$,

(5.43)   (A) $\dfrac{\partial q}{\partial t} + \bar{v}\dfrac{\partial q}{\partial x} + \tilde{\beta}\dfrac{\partial \psi}{\partial x} + \epsilon^2\bar{v}''\dfrac{\partial \psi}{\partial x} = 0$,   (B) $q = \epsilon^2\Delta\psi - F\psi$.

The initial condition is

(5.44) $q(\vec{x}, t)\big|_{t=0} = \tilde{A}(x, y)e^{\frac{\iota\phi_0(x, y)}{\epsilon}}.$

We proceed to find a formal solution as ϵ asymptotes to 0. For students interested in rigor, it is not difficult to use energy estimates to bound errors and turn these formal solutions into convergent series, but that will not be done here. As in the systems case, our ansatz will have a rapidly turning phase and an amplitude expressed as a series starting at $O(1)$.

(5.45) $\psi(x, y, t) = (A_0(x, y, t) + \epsilon A_1(x, y, t))e^{\frac{\iota\phi(x,y,t)}{\epsilon}} + O(\epsilon^2).$

For computing q, we use

$$\epsilon^2\Delta(Ae^{\frac{\iota\phi}{\epsilon}}) = \left(-|\nabla\phi|^2 + \epsilon(2\iota\nabla\phi \cdot \nabla A + \iota A\Delta\phi) + \epsilon^2\Delta A\right)e^{\frac{\iota\phi}{\epsilon}}.$$

Then we get q up to first order using (5.43B),

$$q(x, y, t) = (B_0(x, y, t) + \epsilon B_1(x, y, t))e^{\frac{\iota\phi(x,y,t)}{\epsilon}} + O(\epsilon^2)$$

where

(5.46)
$$B_0 = -(|\nabla\phi|^2 + F)A_0\,,$$
$$B_1 = -(|\nabla\phi|^2 + F)A_1 + 2\iota\nabla\phi \cdot \nabla A_0 + \iota A_0 \Delta\phi\,.$$

Now that we have these relationships between the series terms for the two variables, we insert q and ψ into the wave equation (5.43A) and match powers of ϵ. The lowest terms are $O(\epsilon^{-1})$

$$\iota\frac{\partial\phi}{\partial t}B_0 e^{\frac{\iota\phi}{\epsilon}} + \bar{v}\iota\phi_x B_0 e^{\frac{\iota\phi}{\epsilon}} + \tilde{\beta}\iota\phi_x A_0 e^{\frac{\iota\phi}{\epsilon}} = 0\,.$$

Substituting A_0 according to (5.46) gives

$$\left(\iota\frac{\partial\phi}{\partial t} + \bar{v}\iota\phi_x - \frac{\tilde{\beta}}{|\nabla\phi|^2 + F}\iota\phi_x\right)B_0 e^{\frac{\iota\phi}{\epsilon}} = 0\,.$$

After cancellations, we get an equation for the phase, written with the initial condition

(5.47)
$$\frac{\partial\phi}{\partial t} + \bar{v}(y)\phi_x - \tilde{\beta}(y)\frac{\phi_x}{|\nabla\phi|^2 + F} = 0\,, \quad \phi(x, y, t)\big|_{t=0} = \phi_0(x, y)\,.$$

This is our eikonal equation, once we put it into the format (5.40A).

EIKONAL EQUATION:

(5.48)
$$\phi_t + \omega(\nabla\phi, y) = 0\,, \quad \omega(\vec{k}, y) = \bar{v}(y)k_1 - \frac{\tilde{\beta}(y)k_1}{|\vec{k}|^2 + F}\,.$$

Thus, the phase velocity is given by the local dispersion relation as found in the constant-coefficient case of Section 5.1. We proceed to the next order $O(1)$,

$$\left(\iota\phi_1 B_1 + \bar{v}\iota\phi_x B_1 + \tilde{\beta}\iota\phi_x A_1 + \frac{\partial B_0}{\partial t} + \bar{v}\frac{\partial B_0}{\partial x} + \tilde{\beta}\frac{\partial A_0}{\partial x}\right)e^{\iota\phi/\epsilon} = 0\,.$$

Canceling the common phase and substituting for A_1,

$$= \left(\iota\phi_t + \bar{v}\iota\phi_x - \frac{\tilde{\beta}\iota\phi_x}{|\nabla\phi|^2 + F}\right)B_1 + \tilde{\beta}\iota\phi_x\left(2\iota\frac{\nabla\phi \cdot \nabla A_0}{|\nabla\phi|^2 + F} + \iota\frac{A_0\Delta\phi}{|\nabla\phi|^2 + F}\right)$$
$$+ \frac{\partial B_0}{\partial t} + \bar{v}\frac{\partial B_0}{\partial x} + \tilde{\beta}\frac{\partial A_0}{\partial x} = 0\,.$$

Using the phase equation (5.47), the first term multiplying B_1 drops out. This step corresponds to the Fredholm alternative of the systems case. Further substituting for A_0, we find an equation for B_0 alone.

TRANSPORT EQUATION:

(5.49)
$$\frac{\partial B_0}{\partial t} + \bar{v}\frac{\partial B_0}{\partial x} - \tilde{\beta}\frac{\partial}{\partial x}\left(\frac{B_0}{|\nabla\phi|^2 + F}\right)$$
$$+ \tilde{\beta}\phi_x\left(\frac{2\nabla\phi \cdot \nabla((|\nabla\phi|^2 + F)^{-1}B_0)}{|\nabla\phi|^2 + F} + \frac{B_0\Delta\phi}{(|\nabla\phi|^2 + F)^2}\right) = 0,$$

$$B_0|_{t=0} = \tilde{A}.$$

The equation was solved for B_0 because the initial condition was given for the potential vorticity. If the initial condition had been given for the stream function, it would have made sense to solve for A_0. As in the first-order system case, we expect to see the group velocity appearing in (5.49) to advect the wave amplitude. This is indeed the case. It is a simple matter to show that equation (5.49) can be written

$$\frac{\partial B_0}{\partial t} + c_1\frac{\partial B_0}{\partial x} + c_2\frac{\partial B_0}{\partial y} + \alpha B_0 = 0$$

where

$$c_1 = \bar{v} - \frac{\tilde{\beta}}{|\nabla\phi|^2 + F} + 2\tilde{\beta}\frac{(\phi_x)^2}{(|\nabla\phi|^2 + F)^2}, \qquad c_2 = \frac{2\tilde{\beta}\phi_x\phi_y}{(|\nabla\phi|^2 + F)^2},$$

so $\vec{c} = \nabla_k\omega(\vec{k}, y)|_{\vec{k}=\nabla\phi}$ is the group velocity associated with planetary waves. This last result is given as Exercise 5.6.

Initial Conditions and Higher Orders. The initial condition was all placed in the leading-order term B_0, so the higher-order terms are all initially zero: $B_1|_{t=0} = 0$. One may ask if B_1 will remain zero at later times. To answer this question, we must go one higher order in ϵ to obtain a transport equation for B_1. Let

$$\psi = (A_0 + \epsilon A_1 + \epsilon^2 A_2)e^{\iota\phi/\epsilon}, \qquad q = (B_0 + \epsilon B_1 + \epsilon^2 B_2)e^{\iota\phi/\epsilon}.$$

The terms are related by

$$B_0 = cA_0, \qquad B_1 = cA_1 - f_1(A_0), \qquad B_2 = cA_2 - f_1(A_1) - F_2(A_0),$$

where

$$c = -(|\nabla\phi|^2 + F), \qquad -f_1(A) = 2\iota\nabla\phi \cdot \nabla A + \iota A\Delta\phi, \qquad -f_2(A) = \Delta A.$$

The $O(\epsilon^{-1})$ terms give the equation for the phase. The $O(1)$ terms give the transport equation in (5.49) for the leading term

$$\frac{\partial B_0}{\partial t} + \bar{v}(B_0)_x + \tilde{\beta}\left(\frac{B_0}{c}\right)_x + \tilde{\beta}\frac{\iota\phi_x}{c}f_1\left(\frac{B_0}{c}\right) = 0.$$

And the $O(\epsilon)$ terms give a transport equation for the secondary term. Note that the $\bar{v}''$ term of $\bar{q}_y$ enters at this order:

$$\frac{\partial B_1}{\partial t} + \bar{v}(B_1)_x + \tilde{\beta}\left(\frac{B_1}{c} + \frac{1}{c}f_1\left(\frac{B_0}{c}\right)\right)_x + \tilde{\beta}\frac{\iota\phi_x}{c}f_1\left(\frac{B_1}{c} + f_1\left(\frac{B_0}{c}\right)\right)$$
$$+ \tilde{\beta}\frac{\iota\phi_x}{c}f_2\left(\frac{B_0}{c}\right) + \bar{v}''\frac{\iota\phi_x}{c}B_0 = 0.$$

We see that there is a forcing of the B_1 term from the leading B_0 term, so that $B_1 = 0$ is not a solution.

Properties of the Eikonal Equation for Dispersive Waves. Generally, the WKB procedure leads to simplified equations for the wave amplitude and phase, but solving these equations can be very complicated. Solutions of the eikonal equation may not exist for all times. While in the constant-coefficient case the eikonal equation has plane wave solutions $\phi_p = \vec{x} \cdot \vec{k} - \omega_p(\vec{k})t$, these simplest linear wave trains often do not represent observed physical phenomena where different types of waves are generated in different places. Also, spatial variation of coefficients can lead to different phase and group speeds in different places, and waves can travel and interact in complex ways. We recall the eikonal equation

$$\frac{\partial \phi(\vec{x}, t)}{\partial t} + \omega(\nabla_x \phi, \vec{x}, t) = 0, \quad \phi(\vec{x}, t)\big|_{t=0} = \phi_0(\vec{x}).$$

We will use Hamilton-Jacobi theory to rewrite the eikonal equation in terms of characteristics that follow the progress of "markers"; see chapter 1 of Fritz John's book [**13**]. Let $\vec{X}(\vec{\alpha}, t)$ be the position of a marker that began at position $\vec{\alpha}$. Also, let

$$z(\vec{\alpha}, t) = \phi(\vec{X}(\vec{\alpha}, t), t), \quad \vec{K}(\vec{\alpha}, t) = \nabla_x \phi(\vec{X}(\vec{\alpha}, t), t),$$

$$\tau(\vec{\alpha}, t) = \frac{\partial}{\partial t} \phi(\vec{X}(\vec{\alpha}, t), t),$$

with the

CHARACTERISTIC EQUATIONS:

(5.50)

(A) $\dfrac{d}{dt}\vec{X} = \nabla_k \omega(\vec{K}(\vec{\alpha}, t), \vec{X}(\vec{\alpha}, t), t),$

(B) $\dfrac{d}{dt}\vec{K} = -\nabla_x \omega(\vec{K}(\vec{\alpha}, t), \vec{X}(\vec{\alpha}, t), t),$

(C) $\dfrac{d}{dt}z = \vec{K}(\vec{\alpha}, t) \cdot \nabla_k \omega(\vec{K}(\vec{\alpha}, t), \vec{X}(\vec{\alpha}, t), t) + \tau,$

(D) $\dfrac{d}{dt}\tau = -\dfrac{\partial}{\partial t}\omega(\vec{K}(\vec{\alpha}, t), \vec{X}(\vec{\alpha}, t), t).$

The initial conditions are also called the *strip conditions*,

(5.51)

$$\vec{X}(\vec{\alpha}, t)\big|_{t=0} = \vec{\alpha}, \quad z(\vec{\alpha}, t)\big|_{t=0} = \phi_0(\vec{\alpha}), \quad \vec{K}(\vec{\alpha}, t)\big|_{t=0} = \nabla_\alpha \phi_0(\vec{\alpha}),$$

$$\tau(\vec{\alpha}, t)\big|_{t=0} = -\omega(\nabla_\alpha \phi_0(\vec{\alpha}), \vec{\alpha}, t).$$

Part (D) is trivially integrated to give $\tau = -\omega$. Parts (A) and (B) form nonlinear coupled equations that need to be solved, then can be plugged into part (C) and integrated to get z.

Exact Solution. When the map $\vec{\alpha} \mapsto \vec{X}(\vec{\alpha}, t)$ is invertible, let $\vec{\alpha}(\vec{X}, t)$ be the inverse. Then $\phi = z(\vec{\alpha}(\vec{X}, t), t)$ gives an exact solution to the eikonal equation for times $0 \leq t \leq T_*$ provided

$$\det \frac{d\vec{X}}{d\vec{\alpha}} > 0$$

so that the map is invertible.

Note that initially the map is the identity and the Jacobian of transformation is one, so we can be certain that the map will be invertible for at least some finite time.

For the time-independent case, the equations are the Hamiltonian equations of mechanics, with the dispersion relation serving as the Hamiltonian $H = \omega(\vec{k}, \vec{x})$. Solution trajectories remain on surfaces of constant ω.

Homogeneous and Time-Independent Dispersion Relation: $\omega = \omega(\vec{k})$. For this case the characteristic equations simplify to

$$\frac{d}{dt}\vec{K}(\vec{\alpha}, t) = 0,$$

$$\frac{d}{dt}\vec{X}(\vec{\alpha}, t) = \nabla_k \omega(\vec{K}(\vec{\alpha}, t)),$$

$$\frac{d}{dt}z(\vec{\alpha}, t) = \vec{K} \cdot \nabla_k \omega(\vec{K}(\vec{\alpha}, t)) - \omega(\vec{K}(\vec{\alpha}, t)).$$

These can be integrated immediately to give

(5.52)
$$\vec{K}(\vec{\alpha}, t) = \nabla_\alpha \phi_0(\vec{\alpha}), \quad \vec{X}(\vec{\alpha}, t) = t\nabla_k \omega(\nabla_\alpha \phi_0(\vec{\alpha})) + \vec{\alpha},$$
$$z(\vec{\alpha}, t) = -t\omega(\nabla_\alpha \phi_0(\vec{\alpha})) + t\nabla_\alpha \phi_0(\vec{\alpha}) \cdot \nabla_k \omega(\nabla_\alpha \phi_0(\vec{\alpha})) + \phi_0(\vec{\alpha}).$$

Breakdown occurs when

$$\det\left(\frac{d\vec{X}}{d\vec{\alpha}}\right) = \det\left(I + t\nabla_k^2 \omega|_{\nabla\phi_0(\vec{\alpha})} \cdot \nabla_\alpha^2 \phi_0(\vec{\alpha})\right) = 0.$$

This equation gives points $(t, \vec{\alpha})$ where the map is no longer invertible. To convert this to location, follow the characteristics $\vec{X}(\vec{\alpha}, t)$ to obtain the focusing set:

DEFINITION 5.4 The *focusing set* is the set of points $(t, \vec{x})$ where $\det d\vec{X}/d\vec{\alpha} = 0$ first occurs.

These places where the graph of $\vec{X}$ becomes multi-valued are associated with caustics, local regions where the phase begins to oscillate rapidly and the assumption of WKB breaks down.

In one space dimension, let $f_0(\alpha) = \omega'(\partial\phi_0/\partial\alpha)$. Then if f_0 is monotone increasing the eikonal solution exists for all time, while if $\partial f_0/\partial\alpha < 0$ anywhere, then the solution only exists until time $T_* = -(\min_\alpha \partial f_0/\partial\alpha)^{-1}$.

Plane Wave Initial Data. The above computation is important because when plane wave initial data is considered in the homogeneous, time-independent case, the problem reduces to one dimension. Suppose

$$\phi_0 = \tilde{\phi}_0(\vec{\alpha} \cdot \vec{e}_0)$$

where $\vec{e}_0$ is an arbitrarily directed vector with $|\vec{e}_0| = 1$. By guessing that ϕ takes the form $\phi(\vec{x}, t) = \tilde{\phi}(\vec{x} \cdot \vec{e}_0)$, we obtain the reduced eikonal equation

(5.53)
$$\frac{\partial}{\partial t}\tilde{\phi}(x, t) + \omega\left(\vec{e}_0 \frac{\partial\tilde{\phi}}{\partial x}\right) = 0, \quad \tilde{\phi}\big|_{t=0} = \tilde{\phi}_0(\alpha),$$

with the one-dimensional dispersion relation

$$\omega_{\vec{e}_0}(k) = \omega(\vec{e}_0 k) \,.$$

We apply the criterion developed above for the occurrence of focusing. With $\alpha = \vec{\alpha} \cdot \vec{e}_0$, let

$$f_0(\alpha) = \omega'_{\vec{e}_0}(\tilde{\phi}'_0(\alpha)) = \vec{e} \cdot \nabla_k \omega(\vec{e}\tilde{\phi}'_0(\alpha)) \,;$$

then

$$f'_0(\alpha) = \omega''_{\vec{e}_0}(\tilde{\phi}'_0(\alpha))\tilde{\phi}''_0(\alpha) = \vec{e} \cdot \nabla_k \big(\vec{e} \cdot \nabla_k \omega(\vec{e}\tilde{\phi}'_0(\alpha))\big)\tilde{\phi}''_0(\alpha)$$

leads to the following:

PROPOSITION 5.5 *Phase focusing occurs if and only if*

(5.54) $$\tilde{\phi}''_0(\alpha)(\nabla_k^2\omega)\big|_{\vec{e}\tilde{\phi}'_0(\alpha)}\vec{e} \cdot \vec{e} < 0 \quad \text{somewhere.}$$

Application to AOS model problems. Naturally, we are interested in applying this result for plane waves in our physical problems.

Inertio-Gravity Waves: $\omega = \pm\sqrt{f^2 + gH|\vec{k}|^2}$. Consider $\omega(k) = \omega^+(\vec{e}_0 k)$. Clearly, the dispersion relation does not depend on the direction of the plane wave

$$\omega(k) = \sqrt{f^2 + gHk^2} \,.$$

It is common that a wave train will be created, perhaps by a row of mountains, with one wavelength to the left and a different one to the right, and a transition region in between. That is, suppose $\frac{\partial\phi}{\partial\alpha}$ is monotone and

$$\frac{\partial\phi}{\partial\alpha} \to \begin{cases} k_+ & \text{for } \alpha \to \infty \\ k_- & \text{for } \alpha \to -\infty \,. \end{cases}$$

These k_+ and k_- can be thought of as the local wavelength of the wave to the right and left, respectively. The group velocity ω' is monotone increasing from $k = -\infty$ to $k = +\infty$. Intuitively, we expect that if we have two waves with positive wavenumbers, and the smaller wavelength wave is located to the left of the larger wave, the small wave will travel faster to the right and overtake the large wave. On the other hand, if the positions are switched, the waves will spread apart. This loose physical explanation motivates the following:

CLAIM. If

$$k_+ > k_- \quad \text{then there is no focusing set,}$$
$$k_+ < k_- \quad \text{then there is a focusing set.}$$

PROOF: A quick calculation shows that $\omega''(k) > 0$ everywhere, so the condition $\omega''(k)\phi''(\alpha) < 0$ depends on the sign of ϕ''. For ϕ' monotone, $k_+ < k_-$ implies that ϕ'' must be negative everywhere, while $k_+ > k_-$ implies that ϕ'' is everywhere positive. $\qquad\square$

Internal Gravity Waves: $\omega = N\frac{k_1}{(k_1^2+k_2^2)^{1/2}}$. The calculation

$$\omega_{\vec{e}_0}(k) = \omega(\vec{e}k) = \pm N\frac{e_0^1}{|\vec{e}|}$$

shows that the homogeneity of the dispersion relation leads to a plane wave frequency that is constant, and hence group velocity that is zero. We know that generally, group velocity for internal gravity waves is perpendicular to the direction of propagation, so no advection occurs in that forward direction. Thus, when we send in single monochromatic plane waves, we don't see the dispersive effects. The upshot is that there is no focusing set for this problem as internal waves in a stratified flow.

EXERCISE 5.6. Find when phase focusing occurs for midlatitude planetary waves

$$\omega = vk_1 - \tilde{\beta}\frac{k_1}{|\vec{k}|^2 + F}.$$

In the past few sections we have used the linear WKB method for first-order linear systems and for a particular scalar case and discussed properties of the eikonal equation. In each of these examples, the coefficients in the equation varied on a much larger scale than that of the disturbance wavelength. A special case of this, used in the following, would be if the coefficients did not vary at all. The WKB condition is built into the initial data with a wavelength that is $O(1)$ and amplitude that varies on $O(\epsilon^{-1})$. We consider the linear scalar case with an arbitrary (dispersive) dispersion relation

(5.55) $$\frac{\partial u}{\partial t} + W\left(\frac{\partial}{\partial \vec{x}}\right)u = 0, \quad u\big|_{t=0} = \tilde{A}_0(\epsilon x)e^{\frac{\iota\phi_0(\epsilon x)}{\epsilon}},$$

where

$$W\left(\frac{\partial}{\partial \vec{x}}\right)f(x) = \int_{\mathbb{R}^d} e^{\iota\vec{x}\cdot\vec{k}}\iota\omega(\vec{k})\hat{f}(\vec{k})d\vec{k}.$$

The WKB ansatz applies,

(5.56) $$u(\vec{x},t) = \left(A_0(\epsilon\vec{x},\epsilon t) + \epsilon A_1(\epsilon\vec{x},\epsilon t) + O(\epsilon^2)\right)e^{\frac{\iota\phi(\epsilon x,\epsilon t)}{\epsilon}},$$

and leads to an eikonal equation for the phase

(5.57) $$\phi_t + \omega(\nabla\phi) = 0, \quad \phi\big|_{t=0} = \phi_0,$$

and a transport equation for the amplitude

(5.58) $$\frac{\partial A_0}{\partial t} + \nabla_k\omega(\nabla\phi)\cdot\nabla A_0 + D(\vec{x},t,\nabla\phi)A_0 = 0, \quad A_0\big|_{t=0} = \tilde{A}_0.$$

The coefficient in front of the constant term in the transport equation could be computed in a WKB derivation like those we have done, but we will use a nice shortcut to obtain D later in this section.

Recall that we wrote down the bicharacteristic ray equations for the eikonal equations. For the simple case under consideration, their solution was given by

(5.59)
$$\vec{K}(\vec{\alpha}, t) = \nabla_\alpha \phi_0(\vec{\alpha}), \quad \vec{X}(\vec{\alpha}, t) = t \nabla_k \omega(\nabla_\alpha \phi_0(\vec{\alpha})) + \vec{\alpha},$$
$$z(\vec{\alpha}, t) = -t\omega(\nabla_\alpha \phi_0(\vec{\alpha})) + t\nabla_\alpha \phi_0(\vec{\alpha}) \cdot \nabla_k \omega(\nabla_\alpha \phi_0(\vec{\alpha})) + \phi_0(\vec{\alpha}),$$

where
$$z(\vec{\alpha}, t) = \phi(\vec{X}(\vec{\alpha}, t), t), \qquad \vec{K}(\vec{\alpha}, t) = \nabla_x \phi(\vec{X}(\vec{\alpha}, t), t).$$

The next step was to recognize that this ray-tracing procedure gives us a solution to the eikonal equation provided the map $\vec{\alpha} \to \vec{X}(\vec{\alpha}, t)$ is invertible. That is, given the map $\vec{\alpha}(\vec{x}, t)$ such that $\vec{X}(\vec{\alpha}, t), t) = \vec{x}$, then $\phi(\vec{x}, t) = z(\vec{\alpha}(\vec{x}, t), t)$ solves the eikonal equation. This all happens provided

$$\det\left(\frac{d\vec{X}}{d\vec{\alpha}}\right) \neq 0.$$

For (5.59), we have

$$\frac{d\vec{X}}{d\vec{\alpha}} = I + t\nabla_k^2 \omega|_{\nabla\phi_0} \nabla_\alpha^2 \phi_0.$$

This geometric optics approach provides a solution for short times while the map from initial marker points to later points remains one-to-one. The set where this breaks down has several names, the *phase focusing set*, *phase shock set*, or *dispersive caustic set*

$$\left\{ (\vec{x}, t) : \vec{x} = \vec{X}(\vec{\alpha}, t) \text{ and } \det\left(\frac{d\vec{X}}{d\vec{\alpha}}\right) = 0 \right\}.$$

Despite the fact that it is no longer single-valued, the map $\vec{\alpha} \mapsto \vec{X}$ remains valid for all time, and we will ask what happens beyond the focusing set.

5.5.1. Local Energy Conservation and Solution of the Transport Equation.
In Section 5.1, conservation of energy was demonstrated for the wave equation (5.55)

$$\int_{\mathbb{R}^d} |u(t)|^2 \, d\vec{x} = \int_{\mathbb{R}^d} |u_0|^2 \, d\vec{x}.$$

For solutions of form (5.56), the energy equation to leading order reads

$$\int_{\mathbb{R}^d} |A_0(\vec{x}, t)|^2 \, d\vec{x} = \int_{\mathbb{R}^d} |\tilde{A}_0(\vec{x})|^2 \, d\vec{x}.$$

We wish to convert this into a local conservation principle. Consider a ball of radius R centered at $\vec{x}_0$,

$$\mathcal{B}_R(\vec{x}_0) = \{\vec{x} : |\vec{x} - \vec{x}_0| < R\}.$$

For a smooth function ρ with support in $\mathcal{B}_R(\vec{x}_0)$, consider WKB with initial data given by $\tilde{A}_0 = \rho\bar{A}_0$. We can immediately read off from the transport equation that the support of the solution A_0 at later times will be the set given by the advection of the initial ball at the group velocity evaluated at the gradient of the phase. But compare this with the characteristic equation $d\vec{X}/dt = \nabla_k \omega(\nabla\phi)$ and realize that the flow of the support region is the same as the solution of the characteristic equations. Thus,

$$(5.60) \qquad \text{supp } A_0(\vec{x}, t) \subseteq \left\{ \vec{X}(\vec{\alpha}, t) : \vec{\alpha} \in \mathcal{B}_R(\vec{x}_0) \right\} \equiv \Omega(t) \,.$$

Clearly, $\Omega(0) = \mathcal{B}_R(\vec{x}_0)$. We are tempted to choose ρ to be the characteristic function for the ball, but since this is not smooth, we use a smooth version $\rho = \rho_\delta(|\vec{x} - \vec{x}_0|)$ where

$$\rho_\delta(|\vec{y}|) = \begin{cases} 1 & \text{for } |\vec{y}| \leq R - \delta \\ 0 & \text{for } |\vec{y}| \geq R \,. \end{cases}$$

The error can be controlled by solving the linear piece so it can be shown that energy conservation applies on the advecting ball.

Local Conservation of Energy.

$$\int_{\Omega(t)} |A_0(\vec{x}, t)|^2 \, d\vec{x} = \int_{\Omega(0)} |\tilde{A}_0(\vec{x})|^2 \, d\vec{x} \,.$$

The next step uses the *transport theorem* [2, 19], which states that if

$$\int_{\Omega(t)} f(\vec{x}, t) = \int_{\Omega(0)} f_0(\vec{x})$$

holds on a large family of sets with $\Omega(t)$ given by (5.60), then f satisfies the PDE

$$\frac{\partial f}{\partial t} + \text{div}(\vec{v} f) = 0$$

where $\vec{v} = \frac{d}{dt} \vec{X}(\vec{\alpha}, t)$. Applying the transport theorem to the local conservation of energy result gives a conservation of amplitude equation

$$(5.61) \qquad \frac{\partial}{\partial t} |A_0(\vec{x}, t)|^2 + \text{div} \left(\nabla_k \omega (\nabla \phi_0(\vec{x}, t)) |A_0(\vec{x}, t)|^2 \right) = 0 \,.$$

A corollary to this key result is to evaluate the coefficient in front of the constant term in the transport equation by noting that (5.58) and (5.61) together imply

$$(5.62) \qquad D(\vec{x}, t, \nabla \phi) = \frac{1}{2} \text{div} \left(\nabla_k \omega (\nabla \phi(\vec{x}, t)) \right) \,.$$

As promised, the coefficient was obtained without hard work. The transport equation can now be solved by flowing the local energy conservation equation backwards following the characteristics to move the volume of integration onto the original ball

$$\int_{\Omega(t)} |A_0(\vec{x}, t)|^2 \, d\vec{x} = \int_{\mathcal{B}_R(\vec{\alpha}_0)} |A_0(\vec{X}(\vec{\alpha}, t), t)|^2 \det \left(\frac{d\vec{X}}{d\vec{\alpha}} \right) d\vec{\alpha} = \int_{\mathcal{B}_R(\vec{\alpha}_0)} |\tilde{A}_0(\vec{\alpha})|^2 \, d\vec{\alpha} \,.$$

Since the ball of integration is arbitrary, we can equate the integrands across the last equal sign.

Solution of Transport Equation.

$$(5.63) \qquad |A_0(\vec{X}(\vec{\alpha}, t), t)|^2 = |\tilde{A}_0(\vec{\alpha})|^2 \left(\det \left(\frac{d\vec{X}}{d\vec{\alpha}} \right) \right)^{-1} .$$

The geometric optics procedure predicts infinite amplitude on the focusing set where the Jacobian of transformation is zero. But we can't trust the prediction because the WKB assumption breaks down as we approach the caustic set. The same ideas used here work for the systems case, once you find the conserved energy.

EXERCISE 5.7. Use the exact solution (5.63) along with the transport equation (5.58) to rederive the coefficient D (5.62) and/or conversely, show that (5.58) and (5.62) imply (5.63).

Hint. $D = J_t / 2J$ and $J_t = J \operatorname{div}(d\vec{X}/dt)$ where $J = \det(d\vec{X}/d\vec{\alpha})$.

5.6. Beyond Caustics: Eikonal Equation Revisited

We return to our scalar wave equation with rapidly oscillating phase (5.55). We write the solution through Fourier analysis

$$u(\vec{x}, t) = \int_{\mathbb{R}^d} e^{\iota(\vec{x}\cdot\vec{k} - \omega(\vec{k})t)} \hat{u}_0(\vec{k}) d\vec{k}$$

$$(5.64) \qquad = (2\pi)^{-d} \int_{\mathbb{R}^d} \int_{\mathbb{R}^d} e^{\iota(\vec{x}\cdot\vec{k} - \omega(\vec{k})t) - \iota\vec{y}\cdot\vec{k}} \tilde{A}_0(\epsilon \vec{y}) e^{\frac{\iota\phi_0(\epsilon\vec{y})}{\epsilon}} d\vec{y}\, d\vec{k} .$$

Motivated by our experience at short times, we introduce the slow variables

$$\vec{x}' = \epsilon\vec{x}, \qquad t' = \epsilon t, \qquad \vec{y}' = \epsilon\vec{y}, \qquad u = u_\epsilon\left(\frac{\vec{x}'}{\epsilon}, \frac{t'}{\epsilon}\right) = \tilde{u}_\epsilon(\vec{x}', t') ,$$

and drop primes

$$\tilde{u}_\epsilon(\vec{x}, t) = (2\pi\epsilon)^{-d} \int_{\mathbb{R}^d} \int_{\mathbb{R}^d} e^{\iota\frac{1}{\epsilon}((\vec{x} - \vec{y})\cdot\vec{k} - \omega(\vec{k})t + \phi_0(\vec{y}))} \tilde{A}_0(\vec{y}) d\vec{y}\, d\vec{k} .$$

This double integral is all set up for the method of stationary phase. To make an upcoming association, we change $\vec{y} \to \vec{\alpha}$,

$$\tilde{u}_\epsilon(\vec{x}, t) = (2\pi\epsilon)^{-d} \int_{\mathbb{R}^d \times \mathbb{R}^d} e^{\iota\frac{\Psi(\vec{x}, \vec{\alpha}, \vec{k}, t)}{\epsilon}} \tilde{A}_0(\vec{\alpha}) d\vec{\alpha}\, d\vec{k} ,$$

$$(5.65) \qquad \Psi(\vec{x}, \vec{\alpha}, \vec{k}, t) = (\vec{x} - \vec{\alpha}) \cdot \vec{k} - \omega(\vec{k})t + \phi_0(\vec{\alpha}) .$$

Recall the principles of stationary phase discussed in Section 5.4. The integral will be dominated by those critical points $p_j(\vec{x}, t) = (\vec{\alpha}_j(\vec{x}, t), \vec{k}_j(\vec{x}, t))$ where

$$\nabla_\alpha \Psi \big|_{p_j} = 0 \quad \text{and} \quad \nabla_k \Psi \big|_{p_j} = 0 .$$

A quick calculation yields

$$(5.66) \qquad \vec{x} = \vec{\alpha}_j + \nabla_k \omega(\vec{k}_j)t , \qquad \vec{k}_j = \nabla_\alpha \phi_0(\vec{\alpha}) , \qquad \text{for } j = 1, 2, \ldots, l .$$

So the critical points are just those points on the bicharacteristic rays $\vec{X}(\vec{\alpha}_j, t)$ discovered in WKB analysis. However, we are no longer confining the map to be one-to-one (although that will still be true at short times), and this solution applies even beyond the focusing set.

There is one more condition for applying stationary phase:

$$\det M \neq 0$$

where

$$(5.67) \qquad M = \begin{pmatrix} d_k^2 \Psi & d_{\alpha,k}^2 \Psi \\ d_{\alpha,k}^2 \Psi & d_\alpha^2 \Psi \end{pmatrix}\bigg|_{p_j} = \begin{pmatrix} -t \, d_k^2 \omega(\nabla\phi_0(\vec{\alpha}_j)) & -I \\ -I & d_\alpha^2 \phi(\vec{\alpha}_j) \end{pmatrix}.$$

To compute the determinant, use the following block matrix identity:

$$\begin{pmatrix} F & -I \\ -I & G \end{pmatrix} = \begin{pmatrix} 0 & I \\ I & 0 \end{pmatrix} \begin{pmatrix} -I & 0 \\ F & FG-I \end{pmatrix} \begin{pmatrix} I & -G \\ 0 & I \end{pmatrix},$$

$$\det \begin{pmatrix} F & -I \\ -I & G \end{pmatrix} = \det \begin{pmatrix} 0 & I \\ I & 0 \end{pmatrix} \det(I - FG) = (-1)^d \det(I - FG),$$

where $d \times d$ is the size of the square matrices in the blocks. Then

$$(5.68) \qquad \det M = (-1)^d \det\left(I + t \, d_k^2\omega \cdot d_\alpha^2\phi\right)\big|_{p_j} = (-1)^d \det\left(\frac{d\vec{X}}{d\vec{\alpha}}\right)\bigg|_{p_j};$$

that is, stationary phase only works if we are not on the phase shock set. We can now apply the method of stationary phase asymptotic expansion. The leading power is $(2\pi\epsilon)^{N/2}$ where N is the dimension of the integral. Since we are evaluating a double integral in d dimensions, we have $N = 2d$, and the powers of $2\pi\epsilon$ cancel in front. If anyone was worried about the large inverse power of ϵ that had been sitting in front of the integral, their fears turn out to be ungrounded. Also, we will need $\Psi(\vec{x}, \vec{\alpha}_j, \vec{k}_j, t)$,

$$\Psi(\vec{x}, \vec{\alpha}_j, \vec{k}_j, t) = (\vec{x} - \vec{\alpha}_j) \cdot \nabla\phi_0(\vec{\alpha}_j) - t\omega(\nabla\phi_0(\vec{\alpha}_j)) + \phi_0(\vec{\alpha}_j)$$
$$= t\nabla_k\omega(\nabla\phi_0(\vec{\alpha}_j)) \cdot \nabla\phi_0(\vec{\alpha}_j) - t\omega(\nabla\phi_0(\vec{\alpha}_j)) + \phi_0(\vec{\alpha}_j)$$
$$= z(\vec{\alpha}_j, t) = \phi(\vec{X}(\vec{\alpha}_j, t), t)$$

from the stationary phase principle. The conclusion is that off the phase shock set,

$$(5.69) \qquad \tilde{u}_\epsilon(\vec{x}, t) \sim \sum_{j=1}^{l} A(\vec{\alpha}_j, (\vec{x}, t)) \left| \det\left(\frac{d\vec{X}}{d\vec{\alpha}}\right) \right|^{-1/2}\bigg|_{p_j} e^{\iota\sigma_j \frac{\pi}{4}} e^{\iota\frac{\phi(\vec{X}(\vec{\alpha}_j, t), t)}{\epsilon}},$$

where σ_j is the number of positive eigenvalues minus the number of negative eigenvalues of the matrix M (5.67). For short times, the map is one-to-one ($l = 1$), and we look to compute σ_1. There is no loss of generality to consider the time $t = 0$ because the signs of the eigenvalues will not change until one crosses zero, at which point we have hit the phase shock set. Let $\vec{y}$ be an eigenvector of the matrix $G = \nabla_\alpha^2\phi_0$ with eigenvalue γ : $G\vec{y} = \gamma\vec{y}$. Since G is symmetric, γ is real. Then the length $2d$ vectors $\begin{pmatrix} \vec{y} \\ \vec{y}\lambda_+ \end{pmatrix}$ and $\begin{pmatrix} \vec{y} \\ \vec{y}\lambda_- \end{pmatrix}$ are eigenvectors of the matrix

$M(t = 0) = \begin{pmatrix} 0 & -I \\ -I & G \end{pmatrix}$ with eigenvalues λ_+ and λ_-, respectively, provided λ_+ and λ_- are the roots of the equation

$$\lambda^2 - \gamma\lambda - 1 = 0, \quad \lambda_\pm = \frac{\gamma}{2}\left(1 \pm \sqrt{1 + \frac{4}{\gamma^2}}\right).$$

Observe that one of these roots will be positive and one negative. This is true for each of the d eigenvalues of G, and we conclude that the number of positive and negative eigenvalues for the matrix M is equal, so $\sigma_1 = 0$. Thus, at short times, this answer (5.69) agrees with the conclusion (5.63) of WKB.

As advertised, we can now evaluate the solution beyond the phase shock set, where many eikonal rays can contribute. The WKB form remains valid except for the phase shift resulting from the $\alpha_j \neq 0$, but we must sum the contribution of several different phases when many different rays contribute to the same location.

To evaluate the solution on the phase shock set requires a fancy change of variables that will reduce to integrals of standard type involving special functions such as Airy's equation [5]. While the amplitudes do not become infinite as suggested in the WKB solution, they do become very large, with the highest intensity occurring at the cusp.

EXERCISE 5.8. Consider

$$\frac{\partial\vec{u}}{\partial t} + \sum_{j=1}^{d} A_j \frac{\partial\vec{u}}{\partial x_j} + Bu = 0, \quad \vec{u}\big|_{t=0} = \vec{\mathcal{A}}_0(\epsilon\vec{x})e^{\iota\frac{\phi_0(\epsilon\vec{x})}{\epsilon}}.$$

Get the asymptotic form of the solution for any large-scale point not on the focusing set by decomposing into right eigenvectors and applying the method of this section. Note that each wave will have its own caustic set.

EXERCISE 5.9. Apply the above theory to the rotating shallow water equation. Note that the geostrophic waves will have no caustics, only the dispersive inertio-gravity waves.

5.7. Weakly Nonlinear WKB for Perturbations Around a Constant State

In earlier parts of this chapter we linearized equations about interesting nonlinear states. For a general nonlinear equation, we could extend the procedure to the next order and obtain quadratic nonlinearities applicable for solutions centered at zero. So it is natural to study quadratically nonlinear equations at small amplitude. In the AOS models studied in these chapters, all nonlinearities are quadratic.

We examine the first-order, constant-coefficient system with nonlinearity entering through the matrix-valued bilinear terms B_j and S and with small amplitude initial data involving a superposition of slowly varying plane waves,

$$\frac{\partial \vec{u}}{\partial t} + \sum_{j=1}^{d} A_j \frac{\partial \vec{u}}{\partial x_j} + B\vec{u} + \sum_{j=1}^{d} B_j\left(\vec{u}, \frac{\partial \vec{u}}{\partial x_j}\right) + S(\vec{u}, \vec{u}) = 0$$

(5.70)
$$\vec{u}\big|_{t=0} = \epsilon\left(\sum_{p=1}^{r} \mathcal{A}_p^0(\epsilon\vec{x}) \vec{R}_{l_p}(\vec{k}_p) e^{\iota\vec{k}_p\cdot\vec{x}} + \text{c.c.}\right).$$

As in the earlier part of this chapter, we assume that the eigenvalue problem

(5.71)
$$\left(-\iota\omega I + \sum_{j=1}^{d} \iota A_j k_p^j + B\right)\vec{R} = 0$$

has eigenvalues $\omega_l(\vec{k}_p)$ and right eigenvectors $\vec{R}_l(\vec{k}_p)$ for $1 \le l \le m$ and also left eigenvectors $\vec{L}_l(\vec{k}_p)$. The number of waves present in the initial data is given by r, and the channel or type of wave associated with the pth wave is indicated by l_p. Below we will assume $S(\vec{u}, \vec{u}) = 0$, as is the case in the example that follows.

EXAMPLE 5.3 (Rapidly Rotating Shallow Water Equations). In Chapter 4 we linearized these equations about the mean state with no velocity and constant height H_0. Now we keep the nonlinear quadratic terms. Let the total height be given by $h(\vec{x}) = H_0 + \tilde{h}(\vec{x})$. Then the perturbation vector $\vec{u} = (v_1, v_2, \tilde{h})^\mathsf{T}$ solves an equation of type (5.70) with the linear matrices in Chapter 4 and with the nonlinear terms $S(\vec{u}, \vec{u}) = 0$ and

$$B\left(\vec{u}, \frac{\partial \vec{u}}{\partial x_j}\right) = v_j \frac{\partial \vec{u}}{\partial x_j} + \begin{pmatrix} 0 \\ 0 \\ \tilde{h}\frac{\partial v_j}{\partial x_j} \end{pmatrix}.$$

EXERCISE 5.10. Carry out the same arguments as we develop in this section for the case $B_j \equiv 0$ but $S(\vec{u}, \vec{u}) \ne 0$.

We recall the linear WKB solutions in the case of constant coefficients. The phase and amplitude at each wavenumber vector evolve independently of those at any other wavenumber

$$\vec{u} = \epsilon\left(\sum_{p=1}^{r} \mathcal{A}_p(\epsilon\vec{x}, \epsilon t) \vec{R}_{l_p}(\vec{k}_p) e^{\iota(\vec{x}\cdot\vec{k}_p - \omega_{l_p}(\vec{k}_p)t)} + \text{c.c.}\right) + O(\epsilon^2).$$

The plane wave phase $\phi_p = \vec{x} \cdot \vec{k}_p - \omega_{l_p}(\vec{k}_p)t$ satisfies the eikonal equation

$$\frac{\partial\phi_p}{\partial t} + \omega_{l_p}(\nabla\phi_p) = 0, \qquad \phi_p\big|_{t=0} = \vec{x} \cdot \vec{k}_p,$$

while the amplitude is a function of the long wave variables $\vec{X} = \epsilon\vec{x}$ and $T = \epsilon t$ and is advected on these scales by the associated group velocity

$$\frac{\partial \mathcal{A}_p}{\partial T} + \nabla_k \omega_{l_p}(\vec{k}_p) \cdot \nabla_X \mathcal{A}_p = 0, \qquad \mathcal{A}_p\big|_{t=0} = \mathcal{A}_p^0(\mathcal{X}).$$

The introduction of nonlinearity will alter this independence of wavenumbers. Waves in the initial data can create new waves at order $O(\epsilon)$. Sometimes even a single wave can generate other waves. Wave packets that crossed harmlessly in the linear theory can now interact and scatter at leading order. We seek quantitative evaluation of these processes and simplified evolution equations for the small-amplitude waves.

The procedure to be used is called the *method of multiple scales*. It assumes separation of short-wave scales $\vec{x}$ and t and long-wave scales $\vec{X} = \epsilon\vec{x}$ and $T = \epsilon t$. Formally, we allow the wave amplitudes and phases to be functions of all these variables, which we treat as independent and then evaluate at the points $\vec{X} = \epsilon\vec{x}$ and $T = \epsilon t$. Thus,

$$\frac{\partial}{\partial x_j} f(\vec{x}, t, \epsilon\vec{x}, \epsilon t) = \frac{\partial f}{\partial x_j} + \epsilon \frac{\partial f}{\partial X_j}\bigg|_{\substack{\vec{X}=\epsilon\vec{x} \\ T=\epsilon t}}$$

(5.72)
$$\frac{\partial}{\partial t} f(\vec{x}, t, \epsilon\vec{x}, \epsilon t) = \frac{\partial f}{\partial t} + \epsilon \frac{\partial f}{\partial T}\bigg|_{\substack{\vec{X}=\epsilon\vec{x} \\ T=\epsilon t}}.$$

Our ansatz is a series in the asymptotically small parameter of functions of both time and space scales,

(5.73)
$$\vec{u} = \epsilon\vec{u}_1(\vec{x}, t, \vec{X}, T) + \epsilon^2\vec{u}_2(\vec{x}, t\vec{X}, T).$$

Solvability Condition. For uniform validity of the asymptotic expansion (5.73), $u_2(\vec{x}, t, \vec{X}, T)$ should grow sublinearly in (x, t); that is,

$$\lim_{|\vec{x}|+|t|\to\infty} \frac{|u_2(\vec{x}, t, \vec{X}, T)|}{|\vec{x}| + |t|} = 0.$$

This is the sometimes hidden principle in many asymptotics methods. The asymptotics achieve nothing if the second term, $\epsilon^2 u_2$, is not small compared to first term, ϵu_1. Suppose u_2 grows linearly in the small variables: $u_2 \sim |\vec{x}| + |t|$ for large $|\vec{x}| + |t|$. Then the term $\epsilon^2 u_2 = \epsilon^2(|\vec{x}| + |t|) = \epsilon(|\vec{X}| + |T|)$ grows as ϵ times a bounded contribution in $(\vec{X}, T)$, making it the same magnitude as ϵu_1. Thus, such sublinear growth conditions are actually needed to guarantee that $\epsilon\vec{u}_1$ in (5.73) is the leading-order contribution formally.

We plug our ansatz (5.73) into the equation of motion (5.70) using the rules (5.72) and match powers of ϵ. The lowest order is $O(\epsilon)$:

$$\frac{\partial\vec{u}_1}{\partial t} + \sum_{j=1}^{d} A_j \frac{\partial\vec{u}_1}{\partial x_j} + B\vec{u}_1 = 0, \quad \vec{u}_1\big|_{t=0} = \sum_{p=1}^{r} \mathcal{A}_p^0(\vec{X})\vec{R}_{l_p}(\vec{k}_p)e^{i\vec{k}_p\cdot\vec{x}} + \text{c.c.}$$

(Here and below "c.c." denotes the complex conjugate of the previous term.) We integrate this using the eigenvalue problem (5.71). The leading-order behavior has each mode evolving independently on the fast scales t and $\vec{x}$. We will find, however, that extra modes can be created on the slow time scale T, and must be included in the solution

(5.74)
$$\vec{u}_1(\vec{x}, t, \vec{X}, T) = \sum_{p=1}^{\tilde{r}} \mathcal{A}_p(\vec{X}, T)e^{i(\vec{k}_p\cdot\vec{x}-\omega_{l_p}(\vec{k})t)} \vec{R}_{l_p}(\vec{k}_p) + \text{c.c.}$$

where

$$(5.75) \qquad \mathcal{A}_p(\vec{X}, T)\big|_{T=0} = \begin{cases} \mathcal{A}_p^0(\vec{X}) & \text{for } 1 \le p \le r \\ 0 & \text{for } r < p \le \tilde{r}. \end{cases}$$

Below, we show how these modes can be generated. Equating coefficients at order $O(\epsilon^2)$ yields

$$(5.76) \quad -\left(\frac{\partial \vec{u}_2}{\partial t} + \sum_{j=1}^{d} A_j \frac{\partial \vec{u}_2}{\partial x_j} + B\vec{u}_2 \right) = \frac{\partial \vec{u}_1}{\partial T} + \sum_{j=1}^{d} A_j \frac{\partial \vec{u}_1}{\partial X_j} + \sum_{j=1}^{d} B_j\left(\vec{u}_1, \frac{\partial u_1}{\partial x_j} \right).$$

Plugging in $\vec{u}_1$ from (5.74), the nonlinear terms will involve sums of products of different phases, which can lead to wave creation. The right-hand side of (5.76) looks like a forcing term for the left-hand side. We need to discover when the growth of $\vec{u}_2$ in the fast variables is sublinear, and for this we turn to a simpler framework.

Auxiliary Problem: Linear System with Forcing. Let $\vec{u}$ solve

$$(5.77) \qquad \frac{\partial \vec{u}}{\partial t} + \sum_{j=1}^{d} A_j \frac{\partial \vec{u}}{\partial x_j} + B\vec{u} = \vec{F} e^{\iota(\vec{k}\cdot\vec{x} - ct)}, \quad \vec{u}\big|_{t=0} = 0.$$

We look for the necessary and sufficient conditions that $\vec{u}$ be sublinear in t. We build the solution in the obvious way by expanding in modes and integrating the linear part via the eigenfrequency. Thus, we assume

$$\vec{u}(\vec{x}, t) = \sum_{s=1}^{m} \mathcal{A}_s(t) \vec{R}_s(\vec{k}) e^{\iota(\vec{k}\cdot\vec{x} - \omega_s(\vec{k})t)}.$$

Inserting this into (5.77), the eigenproblem (5.71) drops out, leaving the equation

$$\sum_{s=1}^{m} \frac{d\mathcal{A}_s}{dt} \vec{R}_s(\vec{k}) e^{\iota(\vec{k}\cdot\vec{x} - \omega_s(\vec{k})t)} = \vec{F} e^{\iota(\vec{k}\cdot\vec{x} - ct)}.$$

Taking the inner product with the left eigenvector and canceling the common phase terms yields simple scalar equations for the components $\mathcal{A}_s$ for $s = 1, 2, \ldots, m$,

$$(5.78) \qquad \frac{d\mathcal{A}_s}{dt} = F_s e^{\iota(\omega_s(\vec{k}) - c)t}, \quad F_s = \vec{L}_s \cdot \vec{F}.$$

The solution can be written immediately:

Case 1: $\omega_s(\vec{k}) - c \ne 0$, $\mathcal{A}_s = F_s \frac{e^{\iota(\omega_s(\vec{k}) - c)t}}{\omega_s(\vec{k}) - c}$, and

Case 2: $\omega_s(\vec{k}) - c = 0$, $\mathcal{A}_s = F_s t$.

For case 1, the solution is bounded in time, while for case 2, resonant forcing occurs and the solution grows linearly unless the coefficient is zero. This leads to the following:

PROPOSITION 5.6 *The PDE* (5.77) *has bounded behavior if and only if either*

$$c \ne \omega_s(\vec{k}) \text{ for all } 1 \le s \le m \quad \text{or} \quad c = \omega_{s'}(\vec{k}) \text{ and } \vec{L}_{s'} \cdot \vec{F} = 0.$$

Next, we utilize this proposition to study precisely when the solution to (5.76) has sublinear growth with the right-hand side determined by (5.74).

Resonant Interactions. Enforcing the conditions of the previous proposition as applied to our equation (5.76) will yield evolution equations for the wave amplitudes $\mathcal{A}_p$ on the slow scales $(\vec{X}, T)$. The nonlinear term in (5.76) looks like
(5.79)
$$
B\left(\vec{u}_1, \frac{\partial \vec{u}_1}{\partial x_j}\right) =
$$

$$
\sum_{p,q=1}^{\vec{r}} \mathcal{A}_p \mathcal{A}_q e^{\iota((\vec{k}_p + \vec{k}_q) \cdot \vec{x} - (\omega_{l_p}(\vec{k}_p) + \omega_{l_p}(\vec{k}_q))t)} \left(\iota k_q^j\right) B_j\left(\vec{R}_{l_q}(\vec{k}_p), \vec{R}_{l_q}(\vec{k}_q)\right)
$$

$$
+ \mathcal{A}_p \mathcal{A}_q^* e^{\iota((\vec{k}_p - \vec{k}_q) \cdot \vec{x} - (\omega_{l_q}(\vec{k}_p) - \omega_{l_q}(\vec{k}_q))t)} \left(-\iota k_q^j\right) B_j\left(\vec{R}_{l_q}(\vec{k}_p), \vec{R}_{l_q}^*(\vec{k}_q)\right) + \text{c.c.}
$$

Let
$$
\vec{k}_{p,q}^{\pm} = \vec{k}_p \pm \vec{k}_q, \quad \omega_{p,q}^{\pm} = \omega_{l_q}(\vec{k}_p) \pm \omega_{l_q}(\vec{k}_q).
$$

The nonlinear term creates forcing at the wavenumbers $\vec{k}_{p,q}^{\pm}$. These are the potential additional new waves ($\vec{r} > r$) that have to be included in the solution. From the auxiliary problem, we learned that resonance occurs if the forcing frequency is one of the eigenfrequencies of the system. This gives the *resonance condition*

$$
\omega_s(\vec{k}_{p,q}^{\pm}) = \omega_{p,q}^{\pm} \text{ for some } s, \quad \vec{k}_{p,q}^{\pm} = \vec{k}_p + \vec{k}_q.
$$

If this condition is not met, there is no resonance at this wavenumber. If resonance does occur, we need to insure the coefficient $\vec{F}_s$ in the auxiliary problem is zero, which leads to coupling of the wave amplitudes at different wavenumbers on the slow time and space scales and differential equations for the wave amplitudes, $\mathcal{A}_p(\vec{X}, T)$ from (5.74). Energy can be transferred between existing wavenumbers or new wavenumbers can be generated in this case. We already know from previous work in this chapter that the linear contribution in (5.76) from $\mathcal{A}_s \vec{R}_s$ to $\vec{L}_s \cdot \vec{F} = 0$ yields linear transport terms.

As a simple example, we ask if new phases can be created by a single wavenumber $\vec{k}_1$. The right-hand side forces at wavenumbers $\vec{k}_1 + \vec{k}_1 = 2\vec{k}_1$ and $\vec{k}_1 - \vec{k}_1 = 0$. The condition for resonance in the first case is $\omega_s(2\vec{k}_1) = 2\omega_s(\vec{k}_1)$. This is called *subharmonic instability* because a wave can generate another wave at half the wavelength. The other resonance condition is $\omega_s(0) = 0$ for some s and is called *wave–mean flow interaction*. If neither of these possibilities exist, the equations reduce to the linear WKB transport equations for each phase.

Wave–Mean Flow Equations. Suppose $\omega_0(0) = 0$ and that subharmonic instability does not occur for channel l. Consider a solution with only two phases, one at wavenumber $\vec{k}_1$ and a second at wavenumber 0, the mean flow, in other words,

$$
\vec{u}_1 = \bar{\mathcal{A}}(\vec{X}, T)\vec{R}_0 + \mathcal{A}_1(\vec{X}, T)e^{\iota(\vec{x} \cdot \vec{k}_1 - \omega_1(\vec{k}_1)t)}\vec{R}_1 + \mathcal{A}_1^*(\vec{X}, T)e^{-\iota(\vec{x} \cdot \vec{k}_1 - \omega_l(\vec{k}_1)t)}\vec{R}_1^*.
$$

Then we apply our proposition from the auxiliary problem at the wavenumbers $\vec{k} = 0$ and $\vec{k} = \vec{k}_1$ to obtain resonant interaction equations between a wave and the mean flow

(5.80)
$$\frac{\partial \bar{\mathcal{A}}}{\partial T} + \nabla_k \omega_0(0) \cdot \nabla_X \bar{\mathcal{A}} + \Gamma_0 |\mathcal{A}_1|^2 = 0 \,,$$
$$\frac{\partial \mathcal{A}_1}{\partial T} + \nabla_k \omega_l(\vec{k}_1) \cdot \nabla_X \mathcal{A}_1 + \Gamma_1 \mathcal{A}_1 \bar{\mathcal{A}} = 0 \,,$$

where

(5.81)
$$\Gamma_0 = \vec{L}_0(0) \cdot \left(\sum_{j=1}^{d} (-\iota k_1^j) B_j(\vec{R}_l, \vec{R}_l^*) + \text{c.c.} \right),$$
$$\Gamma_1 = \vec{L}_l(\vec{k}_1) \cdot \left(\sum_{j=1}^{d} (\iota k_1^j) B_j(\vec{R}_0, \vec{R}_l) \right).$$

From the equations, we see that if the mean flow is initially zero, it can nevertheless be generated by the interaction of a wave with itself in these equations.

Following the same lines, equations can be written for wave amplitudes at $\vec{k}_1$ and $2\vec{k}_1$ when subharmonic resonance occurs in the absence of wave-mean resonance. And in the case that both occur simultaneously, one finds equations for three coupled wave amplitudes.

EXERCISE 5.11. Does wave-mean flow interaction occurs for the midlatitude planetary equations. Compute the coefficients Γ_0 and Γ_1 in this case. Are they nonzero?

Three-Wave Interactions. Subharmonic resonance is something of a fluke; the generic case has two wavenumbers summing to make a third with the corresponding frequencies also summing. We can write this resonance condition as

(5.82) $$\vec{k}_1 + \vec{k}_2 + \vec{k}_3 = 0 \,, \quad \omega_{l_1}(\vec{k}_1) + \omega_{l_2}(\vec{k}_2) + \omega_{l_3}(\vec{k}_3) = 0 \,,$$

where $\omega_{l_j}(\vec{k}_j)$ are solutions to the eigenvalue problem (5.71). We assume that we have chosen the $\vec{k}_j$ to be nonzero and distinct. Also, we assume that there is no mean flow interaction and no subharmonic instability. These conditions generally keep the three-wave interaction pure, and we can form a three-wave solution

$$\vec{u}_1 = \sum_{p=1}^{3} \mathcal{A}_p \vec{R}_{l_p}(\vec{k}_p) e^{\iota(\vec{k}_p \cdot \vec{x} - \omega_{l_p}(\vec{k}_p)t)} + \text{c.c.}$$

Application of the auxiliary problem at each of the three wavenumbers yields the slow time evolution equations for the amplitudes.

Three-Wave Resonance Equations.

(5.83)
$$\frac{\partial \mathscr{A}_1}{\partial T} + \nabla_k \omega_{l_1}(\vec{k}_1) \cdot \nabla_X \mathscr{A}_1 + \iota \Gamma_1 \mathscr{A}_2^* \mathscr{A}_3^* = 0,$$
$$\frac{\partial \mathscr{A}_2}{\partial T} + \nabla_k \omega_{l_2}(\vec{k}_2) \cdot \nabla_X \mathscr{A}_2 + \iota \Gamma_2 \mathscr{A}_1^* \mathscr{A}_3^* = 0,$$
$$\frac{\partial \mathscr{A}_3}{\partial T} + \nabla_k \omega_{l_3}(\vec{k}_3) \cdot \nabla_X \mathscr{A}_3 + \iota \Gamma_3 \mathscr{A}_1^* \mathscr{A}_2^* = 0,$$

(5.84)
$$\Gamma_1 = -\vec{L}_{l_1}(\vec{k}_1) \cdot \left(k_3^j B_j(\vec{R}_{l_2}^*(\vec{k}_2), \vec{R}_{l_3}^*(\vec{k}_3)) + k_2^j B_j(\vec{R}_{l_3}^*(\vec{k}_3), \vec{R}_{l_2}^*(\vec{k}_2)) \right),$$
$$\Gamma_2 = -\vec{L}_{l_2}(\vec{k}_2) \cdot \left(k_3^j B_j(\vec{R}_{l_1}^*(\vec{k}_1), \vec{R}_{l_3}^*(\vec{k}_3)) + k_1^j B_j(\vec{R}_{l_3}^*(\vec{k}_3), \vec{R}_{l_1}^*(\vec{k}_1)) \right),$$
$$\Gamma_3 = -\vec{L}_{l_3}(\vec{k}_3) \cdot \left(k_2^j B_j(\vec{R}_{l_1}^*(\vec{k}_1), \vec{R}_{l_2}^*(\vec{k}_2)) + k_1^j B_j(\vec{R}_{l_2}^*(\vec{k}_2), \vec{R}_{l_1}^*(\vec{k}_1)) \right).$$

More discussion of asymptotic expansions of this type can be found in chapter 10 of the book by Anile et al. [1]. In the case of no space dependence, exact solutions can be written for the three wave ODEs; see chapter 5 of the book by Craik [3] and the references therein.

A physical example occurs in the rapidly rotating shallow water equations. Recall

$$\omega_\pm(\vec{k}) = \pm\sqrt{gH_0|\vec{k}|^2 + f^2} \quad \text{inertio-gravity waves,}$$
$$\omega_0 = 0 \qquad\qquad\qquad \text{quasi-geostrophic waves.}$$

Subharmonic instability does not occur, and the interaction coefficients for waves and the mean flow are zero, so the three wave equations do not become complicated by these other behaviors. Algebraically, it is impossible to obtain resonance among three inertio-gravity waves. A detailed discussion of this fact is postponed until Chapter 8. However, two oppositely directed gravity waves and one quasi-geostrophic (QG) wave can interact. Let $\vec{k}_3$ correspond to the QG mode:

$$\vec{k}_1 + \vec{k}_2 = \vec{k}_3, \quad \omega_+(\vec{k}_1) + \omega_-(\vec{k}_2) = 0, \quad (\Leftrightarrow |\vec{k}_1| = |\vec{k}_2|).$$

The three wave equations in this case are

(5.85)
$$\frac{\partial \mathscr{A}_1}{\partial T} + \nabla_k \omega_+(\vec{k}_1) \cdot \nabla_X \mathscr{A}_1 \cdot \iota \Gamma_1 \mathscr{A}_2^* \mathscr{A}_3^* = 0,$$
$$\frac{\partial \mathscr{A}_2}{\partial T} + \nabla_k \omega_-(\vec{k}_2) \cdot \nabla_X \mathscr{A}_2 + \iota \Gamma_2 \mathscr{A}_1^* \mathscr{A}_3^* = 0,$$
$$\frac{\partial \mathscr{A}_3}{\partial T} = 0.$$

The QG mode allows a passing gravity wave to scatter into an oppositely moving gravity wave, but the QG mode itself is not affected by the passing waves. This result will have important consequences when we consider the related issues of fast averaging of gravity waves later in Chapter 8.

The midlatitude planetary waves have many three-wave resonances; see section 3.26 of Pedlosky's book [29].

Let's summarize the nonlinear WKB analysis presented above for the constant-coefficient first-order system

(5.86)

$$\frac{\partial \vec{u}}{\partial t} + \sum_{j=1}^{d} A_j \frac{\partial \vec{u}}{\partial x_j} + B\vec{u} + \mathcal{B}(\vec{u}, \nabla \vec{u}) = 0,$$

$$\vec{u}(\vec{x}, t)\big|_{t=0} = \epsilon \sum_{p=1}^{r} \mathcal{A}_p^0(\epsilon \vec{x}) \vec{R}_{l_p}(\vec{k}_p) e^{\iota \vec{k}_p \cdot \vec{x}} + \text{c.c.}$$

The normal mode $\vec{u}(\vec{x}, t) = \exp(\iota(\vec{k} \cdot \vec{x} - \omega(\vec{k})t)\vec{R}(\vec{k})$ is a solution to the linear version of (5.86) provided

$$\left(-\iota \omega(\vec{k})I + \iota \sum_{j=1}^{d} A_j k_j + B \right) \vec{R}(\vec{k}) = 0$$

has eigenvalues $\omega_j(\vec{k})$ and eigenvectors $\vec{R}_j(\vec{k})$ for $j = 1, 2, \ldots, J$. The dual basis L_j is given by

(5.87)
$$\langle \vec{R}_j(\vec{k}), \vec{L}_{j'}(\vec{k}) \rangle = \delta_{jj'} = \begin{cases} 0, & j \neq j', \\ 1, & j = j', \end{cases}$$

where $\langle \vec{R}, \vec{L} \rangle = \vec{L}^* \cdot \vec{R}$. So, in fact, $\vec{L}^*$ is the left eigenvector of the system. With the long-time scales $\vec{X} = \epsilon \vec{x}$ and $T = \epsilon t$, the asymptotic solution of (5.86) is

(5.88)
$$\vec{u}^{\epsilon}(\vec{x}, t) = \epsilon \sum_{j=1}^{\tilde{r}} \mathcal{A}_p(\vec{X}, T) e^{\iota(\vec{k}_p \cdot \vec{x} - \omega_{l_p}(\vec{k}_p)t)} \vec{R}_{l_p}(\vec{k}_p) \bigg|_{\substack{\vec{X} = \epsilon \vec{x} \\ T = \epsilon t}} + \text{c.c.}$$

In general, $r < \tilde{r}$ as the nonlinearity causes energy exchange among waves and creation of waves not in the initial data; $\tilde{r}$ can be as large as the number of normal modes of the system, or just some subset. The condition we want is that the set $\{\vec{k}_p, \omega_{l_p}(\vec{k}_p) : p = 1, \ldots, \tilde{r}\}$ be complete, meaning that the sum of any two elements remains in the set. That is, if $\vec{k}_{p_1} + \vec{k}_{p_2} = \vec{k}$ and $\omega_{p_1}(\vec{k}_1) + \omega_{p_2}(\vec{k}_2) = \omega$, then there exists $1 \leq p_3 \leq \tilde{r}$ such that $\vec{k} = \vec{k}_{p_3}$ and $\omega = \omega_{l_{p_3}}(\vec{k}_{p_3})$.

Recall that to prevent linear growth in t of the second-order term, we enforce conditions on the slow evolution of the wave amplitude $\mathcal{A}_p$. This is just the Fredholm alternative

(5.89)
$$\frac{\partial \mathcal{A}_p}{\partial T} + \nabla_k \omega_{l_p}(\vec{k}_p) \cdot \nabla_X \mathcal{A}_p$$

$$+ \iota \sum_{\substack{\vec{k}_{p_1} + \vec{k}_{p_2} = \vec{k}_p \\ \omega_{l_{p_1}}(\vec{k}_{p_1}) + \omega_{l_{p_2}}(\vec{k}_{p_2}) = \omega_{l_p}(\vec{k}_p)}} \left\langle \mathcal{B}\left(\vec{R}_{l_{p_1}}(\vec{k}_{p_1}), \vec{k}_{p_2} \otimes \vec{R}_{l_{p_2}}(\vec{k}_{p_2}) \right), \vec{L}_{l_p}(\vec{k}_p) \right\rangle \mathcal{A}_{p_1} \mathcal{A}_{p_2} = 0.$$

A special case is the three-wave solution $\tilde{r} = 3$ of a resonant triad

(5.90)
$$\vec{k}_1 + \vec{k}_2 + \vec{k}_3 = 0, \qquad \omega_{l_1}(\vec{k}_1) + \omega_{l_2}(\vec{k}_2) + \omega_{l_3}(\vec{k}_3) = 0.$$

The three wave equations are given with $\vec{c}_{l_j} = \nabla_k \omega_{l_j}(\vec{k}_j)$ for $j = 1, 2, 3$,

$$\frac{\partial \mathcal{A}_1}{\partial T} + \vec{c}_{l_1}(\vec{k}_1) \cdot \nabla_X \mathcal{A}_1 + \iota \Gamma_1 \mathcal{A}_2^* \mathcal{A}_3^* = 0,$$

(5.91)
$$\frac{\partial \mathcal{A}_2}{\partial T} + \vec{c}_{l_2}(\vec{k}_2) \cdot \nabla_X \mathcal{A}_2 + \iota \Gamma_2 \mathcal{A}_3^* \mathcal{A}_1^* = 0,$$

$$\frac{\partial \mathcal{A}_3}{\partial T} + \vec{c}_{l_3}(\vec{k}_3) \cdot \nabla_X \mathcal{A}_3 + \iota \Gamma_3 \mathcal{A}_1^* \mathcal{A}_2^* = 0.$$

The interaction coefficients are

$$(5.92) \quad \Gamma_1 = -\langle \mathcal{B}(\vec{R}_{l_3}^*(\vec{k}_3), \vec{k}_2 \otimes \vec{R}_{l_2}^*(\vec{k}_2)) + \mathcal{B}(\vec{R}_{l_2}^*(\vec{k}_2), \vec{k}_3 \otimes \vec{R}_{l_3}^*(\vec{k}_3)), \vec{L}_{l_1}(\vec{k}_1) \rangle$$

with Γ_2 and Γ_3 given by a cyclic permutation of the indices.

5.8. Nonlinear WKB and the Boussinesq Equations

The Boussinesq equations discussed in earlier chapters do not have the straight-forward form of a first-order hyperbolic system. Here we show how to generalize the above procedure to these equations. Below, $\vec{v}_H = (v_1, v_2)$ is the horizontal velocity vector and w is the vertical (x_3) velocity, so the full velocity is $\vec{v} = (\vec{v}_H, w)$. ρ is the density perturbations around stably stratified equilibrium and ϕ is the pressure perturbations

$$\frac{\partial \vec{v}_H}{\partial t} + \nabla_H \phi + \vec{v} \cdot \nabla \vec{v}_H = 0, \qquad \frac{\partial w}{\partial t} + \rho + \frac{\partial \phi}{\partial x_3} + \vec{v} \cdot \nabla w = 0,$$

(5.93)
$$\frac{\partial \rho}{\partial t} - w + \vec{v} \cdot \nabla \rho = 0, \qquad \text{div } \vec{v} = 0.$$

For simplicity in exposition, the effects of rotation are ignored here. The nonlinear WKB theory that we have described cannot be directly applied to the Boussinesq equations because the divergence-free condition has no time dependence. The matrix multiplying the $\frac{\partial}{\partial t}$ term is singular. To proceed, we must either fix the theory or fix the equation, and we choose the latter. The pressure, which is the Lagrange multiplier of the divergence-free condition, can be eliminated by taking the divergence of the momentum equation

$$\Delta \phi = -\frac{\partial p}{\partial x_3} - \text{div}(\vec{v} \cdot \nabla \vec{v}).$$

This elliptic equation is solved by inverting the Laplacian. The details of this calculation are presented in Section 7.2. Think of doing this either in Fourier space or by using the Poisson kernel

$$\nabla \phi = \nabla \Delta^{-1} \left(-\frac{\partial \rho}{\partial x_3} - \text{div}(\vec{v} \cdot \nabla \vec{v}) \right).$$

This gives the nonlocal form of the Boussinesq equations

$$\frac{\partial \vec{v}_H}{\partial t} + \nabla_H \Delta^{-1}\left(-\frac{\partial \rho}{\partial x_3}\right) + \vec{v} \cdot \nabla \vec{v}_H - \nabla_H \Delta^{-1}(\mathrm{div}(\vec{v} \cdot \nabla \vec{v})) = 0\,,$$

$$(5.94)\quad \frac{\partial w}{\partial t} + \rho + \frac{\partial}{\partial x_3}\Delta^{-1}\left(-\frac{\partial \rho}{\partial x_3}\right) + \vec{v} \cdot \nabla w - \frac{\partial}{\partial x_3}\Delta^{-1}(\mathrm{div}(\vec{v} \cdot \nabla \vec{v})) = 0\,,$$

$$\frac{\partial \rho}{\partial t} - w + \vec{v} \cdot \nabla \rho = 0\,.$$

After eliminating the pressure, our equation is a nonlocal first-order system in the form (5.86) for the theory as given there. An eigenvalue problem will show us the structure of the linear system. The state vector

$$\vec{u} = \begin{pmatrix} \vec{v}_H \\ w \\ \rho \end{pmatrix}$$

solves (5.94), an equation of the form

$$(5.95)\qquad\qquad \vec{u}_t + \mathcal{L}\vec{u} + \mathcal{B}(\vec{u}, \vec{u}) = 0\,.$$

The linear term is

$$\mathcal{L}\vec{u} = \begin{pmatrix} -\nabla_H \Delta^{-1}\rho_{x_3} \\ \rho - \frac{\partial}{\partial x_3}\Delta^{-1}\rho_{x_3} \\ -w \end{pmatrix}\,,$$

and the bilinear term is

$$\mathcal{B}(\vec{u}, \vec{u}) = \begin{pmatrix} \vec{v} \cdot \nabla \vec{v}_H - \nabla_H \Delta^{-1}(\mathrm{div}(\vec{v} \cdot \nabla \vec{v})) \\ \vec{v} \cdot \nabla w - \frac{\partial}{\partial x_3}\Delta^{-1}(\mathrm{div}(\vec{v} \cdot \nabla \vec{v})) \\ \vec{v} \cdot \nabla \rho \end{pmatrix}\,.$$

REMARK. If $\mathrm{div}\,\vec{v}\big|_{t=0} = 0$, then $\mathrm{div}\,\vec{v} = 0$ for all times.

It is left to the student to prove this by taking the divergence of the momentum equation and finding that $(\mathrm{div}\,\vec{v})_t = 0$. Once again this procedure is discussed in more detail in Section 7.3 and Section 8.6.

We turn to the linear analysis for $\vec{u}_t + \mathcal{L}\vec{u} = 0$ and consider plane wave solutions $\vec{u}(\vec{x}, t) = \exp(\iota(\vec{k} \cdot \vec{x} - \omega(\vec{k})t))\vec{R}(\vec{k})$. This gives an eigenvalue problem

$$(5.96)\qquad\qquad (-\iota\omega I + \mathcal{L}(\vec{k}))\vec{R} = 0$$

where the symbol of the linear operator is

$$(5.97)\qquad\qquad \mathcal{L}(\iota\vec{k}) = \begin{pmatrix} 0 & 0 & 0 & -\frac{k_1 k_3}{|\vec{k}|^2} \\ 0 & 0 & 0 & -\frac{k_2 k_3}{|\vec{k}|^2} \\ 0 & 0 & 0 & \frac{|\vec{k}_H|^2}{|\vec{k}|^2} \\ 0 & 0 & -1 & 0 \end{pmatrix}\,.$$

The notation for the wavenumber vectors is $\vec{k}_H = (k_1, k_2)$ and $\vec{k} = (\vec{k}_H, k_3)$. We now solve the eigenvalue problem to find eigenvalues $\omega_j(\vec{k})$ for $j = 1, 2, 3, 4$ and the associated eigenvectors $\vec{R}_j(\vec{k})$. The former are the roots of the characteristic determinant

$$\det(-\iota\omega(\vec{k})I + \mathcal{L}(\iota\vec{k})) = \omega^2\left(\omega^2 - \frac{|\vec{k}_H|^2}{|\vec{k}|^2}\right) = 0\,.$$

There are two slow $\omega = 0$ modes and two fast $\omega = \pm|\vec{k}_H|/|\vec{k}|$ modes. For the slow modes, let's write the eigenvectors

$$\vec{R}^0(\vec{k}) = \begin{pmatrix} -k_2/|\vec{k}| \\ k_1/|\vec{k}| \\ 0 \\ 0 \end{pmatrix} \quad \text{shear mode,} \qquad \widetilde{\vec{R}}^0(\vec{k}) = \begin{pmatrix} k_1/|\vec{k}| \\ k_2/|\vec{k}| \\ 0 \\ 0 \end{pmatrix} \quad \text{not physical.}$$

To see that the last mode is not physical, recall that the divergence taken in Fourier space is the dot product with the wave vector

$$\widehat{\operatorname{div} \vec{v}} = \iota\vec{k} \cdot \hat{v}\,.$$

The velocity associated with $\vec{R}^0$ is divergence free, but that associated with $\widetilde{\vec{R}}^0$ is not. The mathematical framework we have chosen contains this unphysical field, but as we learned in an earlier remark, if we start with divergence-free fields, they will remain so. Thus, if we do not excite this artificial mode in the initial data, it will not grow. So we have only three real eigenmodes: the shear vortical mode $\omega^0 = 0$ and the two fast modes $\omega^\pm = \pm|\vec{k}_H|/|\vec{k}|$:

$$\vec{R}^\pm(\vec{k}) = \frac{1}{\sqrt{2}} \begin{pmatrix} \pm\frac{k_1 k_3}{|\vec{k}_H||\vec{k}|} \\ \pm\frac{k_2 k_3}{|\vec{k}_H||\vec{k}|} \\ \mp\frac{|\vec{k}_H|}{|\vec{k}|} \\ -\iota \end{pmatrix} \quad \text{gravity modes.}$$

Fact. $\{\vec{R}^0(\vec{k}), \vec{R}^+(\vec{k}), \vec{R}^-(\vec{k})\}$ form an orthonormal basis, so the dual basis is the same,

$$\vec{L}_j(\vec{k}) = \vec{R}_j(\vec{k}) \quad \text{for } j = 1, 2, 3.$$

When written in the basis where $\operatorname{div} \vec{v} = 0$, the matrix $\mathcal{L}(\iota\vec{k})$ appears skew-Hermitian, in agreement with the orthonormal eigenvectors.

Geometrically, the fast frequency is the cosine of the angle between the horizontal wave vector $\vec{k}_H$ and the total wave vector $\vec{k}$,

$$\omega^\pm = \pm\frac{|\vec{k}_H|}{|\vec{k}|} = \pm\cos\theta\,,$$

and we recognize the dispersion relation for internal gravity waves from Section 2.4. The horizontal wave vectors ($\theta = \frac{\pi}{2}$) move at the fastest frequency, the buoyancy frequency, which has been normalized to 1. The vertically oriented waves ($\theta = 0$) are steady. This case is degenerate, with only slow waves, and special formulas are needed for the eigenvectors (see [8]).

The fast frequency ω^+ is constant on the surface of a cone at an angle θ to the horizontal. The group velocity $\vec{c}^+ = \nabla_k \omega^+$ is perpendicular to the surface of constant ω and hence orthogonal to $\vec{k}$.

Explicit computation of the group velocity

$$(5.98) \qquad \vec{c}^\pm = \nabla_k \omega^\pm = \pm \frac{k_3}{|\vec{k}|^2} \left(\frac{k_1 k_3}{|\vec{k}_H| \, |\vec{k}|}, \frac{k_2 k_3}{|\vec{k}_H| \, |\vec{k}|}, -\frac{|\vec{k}_H|}{|\vec{k}|} \right)$$

shows that it is in the same direction as the fluid velocity in the associated plane wave.

Everything is now set up to do WKB asymptotics, which go through just as they did earlier in this chapter. For instance, we might consider a small amplitude three-wave solution

$$(5.99) \qquad \vec{u} = \epsilon \sum_{p=1}^{3} \mathcal{A}_p(\vec{X}, T) e^{\iota (\vec{k}_p \cdot \vec{x} - \omega_p(\vec{k}_p) t)} \vec{R}_p(\vec{k}_p) + \text{c.c.} + O(\epsilon^2)$$

where the three-wave resonance condition is satisfied,

$$(5.100) \qquad \vec{k}_1 + \vec{k}_2 + \vec{k}_3 = 0, \quad \omega_1(\vec{k}_1) + \omega(\vec{k}_2) + \omega_3(\vec{k}_3) = 0.$$

Then the wave amplitudes evolve on the slow time and space scales according to the three wave equations

$$(5.101) \qquad \begin{aligned} \frac{\partial \mathcal{A}_1}{\partial t} + \vec{c}_1(\vec{k}_1) \cdot \nabla_X \mathcal{A}_1 + \iota \Gamma_1 \mathcal{A}_2^* \mathcal{A}_3^* &= 0, \\ \frac{\partial \mathcal{A}_2}{\partial t} + \vec{c}_2(\vec{k}_2) \cdot \nabla_X \mathcal{A}_2 + \iota \Gamma_2 \mathcal{A}_3^* \mathcal{A}_1^* &= 0, \\ \frac{\partial \mathcal{A}_3}{\partial t} + \vec{c}_3(\vec{k}_3) \cdot \nabla_X \mathcal{A}_3 + \iota \Gamma_3 \mathcal{A}_1^* \mathcal{A}_2^* &= 0. \end{aligned}$$

The interaction coefficients are

$$(5.102) \qquad \begin{aligned} \Gamma_{\iota_1} = &-\left(\vec{v}_{i_3} \cdot \vec{k}_{i_2} \right) \langle \vec{R}_{i_2}(\vec{k}_{i_2}), \, B(\vec{k}_{i_1}) \vec{R}_{i_1}(\vec{k}_{i_1}) \rangle \\ &- \left(\vec{v}_{i_2} \cdot \vec{k}_{i_3} \right) \langle \vec{R}_{i_3}(\vec{k}_{i_3}), \, B(\vec{k}_{i_1}) \vec{R}_{i_1}(\vec{k}_{i_1}) \rangle \end{aligned}$$

where (i_1, i_2, i_3) is a cyclic permutation of the indices $(1,2,3)$ and $B(\vec{k})$ is the 4×4 matrix, written here in block form

$$(5.103) \qquad B(\vec{k}) + \begin{pmatrix} I - \frac{kk^\mathsf{T}}{|\vec{k}|^2} & 0 \\ 0 & 1 \end{pmatrix}.$$

$B(\vec{k})$ acts as a projection matrix onto the space of divergence-free fields, but since our eigenvectors are already divergence free, we have $B(\vec{k}) \vec{R}_i(\vec{k}) = \vec{R}_i(\vec{k})$ for $i = 1, 2, 3$. This simplifies the interaction coefficients, which are nevertheless quite complicated.

EXAMPLE 5.4 (Two Gravity Waves and One Vortical Mode). Let $\omega_3 = 0$ be the vortical mode with wavenumber $\vec{k}_3$. Then the frequency matching condition becomes $\omega_2 = -\omega_1$. This says that waves 1 and 2 are in opposite families and the resonance condition (let $\vec{k}_j = (\vec{k}_{jH}, k_j^3)$) is

$$(5.104) \qquad \vec{k}_1 + \vec{k}_2 + \vec{k}_3 = 0, \qquad \frac{|\vec{k}_{1H}|}{|\vec{k}_1|} = -\frac{|\vec{k}_{2H}|}{|\vec{k}_2|}.$$

The three wave equations are (5.101) except that $c_3 = 0$, and the third interaction coefficient is

$$(5.105) \qquad \Gamma_3 = \frac{1}{4} \frac{\vec{k}_{1H}^{\perp} \cdot \vec{k}_{2H}}{|\vec{k}_{1H}| \, |\vec{k}_{2H}| \, |\vec{k}_1| \, |\vec{k}_2| \, |\vec{k}_3|} \left(|\vec{k}_{1H}|^2 (k_2^3)^2 - |\vec{k}_{2H}|^2 (k_1^3)^2 \right) = 0$$

by the resonance condition (5.104). Analogous to the interaction of inertio-gravity waves with the QG mode in the rotating shallow water equations, the vortical mode is not influenced by the fast waves.

EXAMPLE 5.5 (Three Gravity Waves in a Vertical Plane). The general algebraic problem of identifying resonant triads is rather difficult. The problem is simplified when the three waves lie in the same vertical plane.

Instead of using the normal polar coordinates, we will write the wavenumber vectors $\vec{k}_j = k_j(\cos \theta_j, \sin \theta_j)$, restrict the angle θ_j to run from $-\frac{\pi}{2}$ to $\frac{\pi}{2}$, and allow the wave magnitude k_j to be either positive or negative. This set of wave vectors will be associated with the $\omega > 0$ fast waves. For the second set of waves we will let θ range from $\frac{\pi}{2}$ to $\frac{3\pi}{2}$. These will have $\omega < 0$. Note that the wave vectors (k, θ) and $(-k, \theta + \pi)$ are the same. With these definitions, we have the simple formula $\omega(\vec{k}_j) = \cos \theta_j$. The required construction is to find three angles satisfying the resonance condition

$$(5.106) \qquad \cos \theta_1 + \cos \theta_2 + \cos \theta_3 = 0.$$

Then by the law of sines

$$\frac{k_1}{\sin(\theta_3 - \theta_2)} = \frac{k_2}{\sin(\theta_1 - \theta_3)} = \frac{k_3}{\sin(\theta_2 - \theta_1)}.$$

There remains a free parameter to determine the overall magnitude of the wave vectors

$$(5.107) \qquad \begin{aligned} \vec{k}_1 &= k \sin(\theta_3 - \theta_2)(\cos \theta_1, \sin \theta_1), \\ \vec{k}_2 &= k \sin(\theta_1 - \theta_3)(\cos \theta_2, \sin \theta_2), \\ \vec{k}_3 &= k \sin(\theta_2 - \theta_1)(\cos \theta_3, \sin \theta_3). \end{aligned}$$

The reader may compute that $\vec{k}_1 + \vec{k}_2 + \vec{k}_3 = 0$. The interaction coefficients are given by

$$(5.108) \qquad \Gamma_1 = (k_2 - k_3)\Gamma, \qquad \Gamma_2 = (k_3 - k_1)\Gamma, \qquad \Gamma_3 = (k_1 - k_2)\Gamma,$$

where

$$(5.109) \qquad \Gamma = 2 \sin\left(\frac{\theta_3 - \theta_2}{2}\right) \sin\left(\frac{\theta_1 - \theta_3}{2}\right) \sin\left(\frac{\theta_2 - \theta_1}{2}\right).$$

The form of the coefficients (5.108) could be predicted by the following conserved quantities. The energy of the system is

$$E = \int (\rho^2 + |\vec{v}|^2) d\vec{x} = \epsilon^2 \sum_{j=1}^{3} |\mathcal{A}_j|^2 + O(\epsilon^3),$$

and there is a second conserved quantity for the plane case. Let ξ be the scalar vorticity. In the (x_1, x_3)-plane,

$$\xi = \frac{\partial v_1}{\partial x_3} - \frac{\partial w}{\partial x_1}, \qquad P = \int \rho \xi = \epsilon^2 \sum_{j=1}^{3} k_j |\mathcal{A}_j|^2 + O(\epsilon^3).$$

The conservation of these quantities to leading order gives two constraints on the interaction coefficients

$$\Gamma_1 + \Gamma_2 + \Gamma_3 = 0, \qquad k_1 \Gamma_1 + k_2 \Gamma_2 + k_3 \Gamma_3 = 0.$$

The vector $(\Gamma_1, \Gamma_2, \Gamma_3)$ is perpendicular to the vectors $(1,1,1)$ and $\vec{k}$. This implies

$$(\Gamma_1, \Gamma_2, \Gamma_3) = \Gamma \vec{k} \times (1, 1, 1).$$

In the deep ocean, governed by the stably stratified equations, there is a wave field that has certain universal features on length scales between 1 meter and 1 kilometer and time scales on the order of hours. This so-called *Garrett-Munk spectrum* has the following features: It is symmetric in vertical wavenumber and horizontally isotropic. In the time domain, it has an ω^{-2} power law with a strong peak near the inertial (rotation) frequency. And in the physical domain, there are certain critical wavenumber magnitudes that can be understood in terms of the linear modes of the system. The common belief is that this spectrum is maintained by the resonant interactions among gravity waves. The resonant set for the Boussinesq contains many triads involving waves on a variety of scales. McComas and Bretherton [27] identified certain classes of three-wave planar resonances that have some scale separation and that may be especially important in the energy exchange in Garrett-Munk. They came to this conclusion by using a statistical closure for the Boussinesq equations that allows you to compute energy fluxes within the spectrum for various triads. Those interested might start with one of the review papers on the subject, such as Muller et al. [28].

In *parametric instability* two wave vectors are nearly equal and opposite, with nearly equal frequency. They sum to make a very small magnitude wave vector ($\vec{k}_3$ of size $\delta \ll 1$) with twice the frequency. Omitting spatial variation, the three wave resonance equations are

(5.110) $\quad \dfrac{\partial \mathcal{A}_1}{\partial T} + \iota \Gamma \mathcal{A}_2^* \mathcal{A}_3^* = 0, \qquad \dfrac{\partial \mathcal{A}_2}{\partial T} + \iota \Gamma \mathcal{A}_3^* \mathcal{A}_1^* = 0, \qquad \dfrac{\partial \mathcal{A}_3}{\partial T} - 2\iota \Gamma \mathcal{A}_1^* \mathcal{A}_2^* = 0,$

where $\Gamma = O(\delta)$. If these equations are linearized about the state $\mathcal{A}_1 = \mathcal{A}_2 = 0$ and $\mathcal{A}_3 = \mathcal{A}_0$, then

$$\frac{\partial^2 \mathcal{A}_1}{\partial T^2} - \Gamma^2 |\mathcal{A}_0|^2 \mathcal{A}_1 = 0.$$

So the large-scale wave $\vec{k}_3$ is unstable to modulation at small scales with half the frequency. This is thought to be a mechanism for feeding energy into the near-inertial waves with small vertical scale, but is less important in transferring energy than the following two triads.

In *induced diffusion* two waves of nearly opposite frequency and wavenumber resonate with a large-wavelength, nearly vertical wave with low frequency $(\vec{k}_3, \omega_3 = O(\delta))$. The interaction coefficient Γ_3 is $O(\delta)$ while the others are size 1, so while the presence of the large wave allows the interaction to take place, it will change slowly while the others exchange energy. The name of this triad comes from the fact that if you examine wave action (rather than energy), you find that the resonance of nearby wavenumbers acts as a diffusion in wavenumber space of wave action.

The triad of *elastic scattering* is

$$\vec{k}_1 = \vec{k}_H - m\vec{e}_3, \qquad \omega_1 = \omega,$$
$$\vec{k}_2 = -\vec{k}_H - m\vec{e}_3, \qquad \omega_2 = -\omega,$$
$$\vec{k}_3 = 2m\vec{e}_3, \qquad \omega_3 = O(\delta).$$

This time the nearly vertical third wavenumber is not small magnitude, but again the interaction coefficient $\Gamma_3 = O(\delta)$ is small, and that wave evolves more slowly than the others. Thus the word *elastic* describes the way the fast waves might reflect off the vertical wave. This mechanism is said to symmetrize the wavenumber spectrum.

Additional references for this material are chapter 10 of the book by Anile et al. [1] and the paper by P. Ripa [32]. We have hinted here about the fact that resonant triads might be very important for dispersive wave turbulence in geophysical flows. A brief introduction to this topic, which contains additional references, can be found in the article by A. Majda [18].

CHAPTER 6

Simplified Equations for the Dynamics
of Strongly Stratified Flow

Several mesoscale regimes in the atmosphere with spatial scales on the order of 10 km to 500 km and in the regions of strong stable stratification in the upper troposphere or lower stratosphere involve fluid flow with strong stratification but weak rotation. There is an outstanding unsolved problem in atmospheric science to explain the $K^{-5/3}$ energy spectrum that occurs universally on these scales; the exponent is the same as Kolmogoroff's famous one but the physical phenomena are very different. An entire issue of *Theoretical and Computational Fluid Dynamics* [37] is devoted to the observations, numerical experiments, and theory attempting to explain these phenomena. There are also a large number of laboratory experiments for fluid flow with strong stable stratification that exhibit remarkable phenomena and also serve as idealized examples for behavior in the atmosphere or ocean at mesoscales. The experimental and theoretical efforts in understanding these issues have been surveyed recently by Riley and Lelong [31], where numerous references are given.

Motivated by the above issues, it is natural to ask whether there are simplified dynamic equations that describe fluid motion with strong stratification when rotational effects are weak or entirely absent. In the beginning of this chapter, we show how to derive the appropriate limiting dynamics formally from the rotating Boussinesq equations. While we will not carry out the details here, a rigorous mathematical justification of this limit is possible by utilizing the ideas in Chapter 4 for the shallow water equations or the ideas to be presented in Chapter 7 for the quasi-geostrophic limit of the Boussinesq equations. In fact, the equations formally derived below are the rigorous singular limit dynamics of strong stratification with general unbalanced initial data (see Embid and Majda [8] and related discussion in Chapter 8).

Instead, in this chapter, we emphasize how exact solutions and mathematical properties of the limiting dynamics can be utilized to give insight and a laminar analytic model for remarkable recent experimental observations in decaying strongly stratified turbulence. The experimental observations are reported by Fincham, Maxworthy, and Spedding [9]. The use of strongly stratified limiting dynamics as an analytical laminar model for these turbulence experiments can be found in the paper by A. Majda and M. Grote [20]; the presentation at the end of this chapter closely follows this work. More discussion and references can be found in the previously mentioned article by Riley and Lelong.

6.1. Nondimensionalization of the Boussinesq Equations for Stably Stratified Flow

Recall from Chapters 1 and 2 the Boussinesq equations for stably stratified flow.

Boussinesq Equations.

HORIZONTAL CONSERVATION OF MOMENTUM:

(6.1)
$$\frac{D^H \vec{v}_H}{Dt} + w \frac{\partial \vec{v}_H}{\partial x_3} = -\frac{1}{\rho_b} \nabla_H p .$$

VERTICAL CONSERVATION OF MOMENTUM:

(6.2)
$$\frac{D^H w}{Dt} + w \frac{\partial w}{\partial x_3} = -\frac{1}{\rho_b} \frac{\partial p}{\partial x_3} - \frac{g}{\rho_b} \rho .$$

CONSERVATION OF MASS:

(6.3)
$$\frac{D\rho}{Dt} = -w \frac{\partial \bar{\rho}}{\partial x_3} .$$

INCOMPRESSIBILITY:

(6.4)
$$\mathrm{div}_H \, \vec{v}_H + \frac{\partial w}{\partial x_3} = 0 ,$$

where the velocity field $\vec{v}$ is written in terms of its horizontal component $\vec{v}_H$ and its vertical component w, $\vec{v} = (\vec{v}_H, w)$, and the density $\tilde{\rho}$ is given by a mean $\bar{\rho}$ and the deviation ρ,

(6.5)
$$\tilde{\rho} = \bar{\rho} + \rho , \quad \bar{\rho} = \rho_b - bx_3 ,$$

where the stratified mean density $\bar{\rho}$ and $b > 0$ for stable stratification. The superscript "H" in equations (6.1)–(6.4) means that the operation is performed on the horizontal components only, so that the material derivative $\frac{D^H}{Dt}$, the horizontal gradient ∇_H, and the horizontal divergence div_H are given by

(6.6) $\dfrac{D^H}{Dt} = \dfrac{\partial}{\partial t} + \vec{v}_H \cdot \nabla_H , \quad \nabla_H = \left(\dfrac{\partial}{\partial x_1}, \dfrac{\partial}{\partial x_2} \right) , \quad \mathrm{div}_H \, \vec{v}_H = \dfrac{\partial v_1}{\partial x_1} + \dfrac{\partial v_2}{\partial x_2} .$

Next we nondimensionalize the Boussinesq equations (6.1)–(6.4) utilizing the following scales for length, time, velocity, density, and pressure:

(6.7)

$$
\begin{aligned}
&L \quad \text{length scale} \\
&U \quad \text{velocity scale} \\
&T_e = \tfrac{L}{U} \quad \text{eddy turnover time} \\
&T_N = N^{-1} \quad \text{buoyancy time} \\
&\rho_b \quad \text{mean density} \\
&\bar{p} \quad \text{mean pressure,}
\end{aligned}
$$

where N is the buoyancy (Brunt-Väisälä) frequency,

$$N = -\left(\frac{g}{\rho_b} \frac{\partial \bar{\rho}}{\partial x_3} \right)^{1/2} ,$$

and the ρ_b is the mean density in equation (6.5). With this choice of scales we introduce the following nondimensional variables:

Nondimensionalization.

(6.8) $\qquad \vec{x}' = \dfrac{\vec{x}}{L}, \quad t' = \dfrac{t}{T_e}, \quad \vec{v}' = \dfrac{\vec{v}}{U}, \quad \rho' = \dfrac{\rho}{\rho_b B}, \quad p' = \dfrac{p}{\bar{p}},$

where $B > 0$ is a numerical factor to be fixed when we discuss the distinguished asymptotic limit of the Boussinesq equations. In terms of the nondimensional variables, the Boussinesq equations (6.1)–(6.4) become

$$\frac{D^H \vec{v}'_H}{Dt'} + w' \frac{\partial \vec{v}'_H}{\partial x'_3} = -\left(\frac{\bar{p}}{\rho_b U^2}\right) \nabla_H p',$$

(6.9) $\qquad \dfrac{D^H w'}{Dt'} + w' \dfrac{\partial w'}{\partial x'_3} = -\left(\dfrac{\bar{p}}{\rho_b U^2}\right) \dfrac{\partial p'}{\partial x'_3} - \left(\dfrac{BgL}{U^2}\right) \rho',$

$$\frac{D\rho'}{Dt'} = \left(-\frac{\partial \bar{\rho}}{\partial x_3} \frac{L}{B\rho_b}\right) w', \quad \text{div}_H \vec{v}'_H + \frac{\partial w'}{\partial x'_3} = 0,$$

where the operators div_H and ∇_H now act on the primed variables.

Next we introduce the nondimensional numbers

(6.10) $\qquad \text{Fr} = \dfrac{T_N}{T_e} = \dfrac{U}{LN}, \qquad \overline{P} = \dfrac{\bar{p}}{\rho_b U^2}, \qquad \Gamma = \dfrac{NU}{gB}.$

As already discussed in Chapter 3, the Froude number Fr is the ratio of the buoyancy time to the advective time scale of the motion of eddies. If Fr $\ll 1$ we expect the flow to be highly stratified and the fluid motions to be essentially two-dimensional in horizontal planes. For example, for mesoscale atmospheric motions in the lower stratosphere or upper troposphere the buoyancy time is of the order of minutes, so $T_N = 10^2$ s, whereas the eddy turnover time is of the order of hours, $T_e = 10^4$ s. In this case the Froude number is Fr $= 0.01 \ll 1$, and the flow is highly stratified. The Euler number $\overline{P}$ is the ratio of the pressure forces to inertial forces. The remaining nondimensional number Γ in equation (6.10) is the ratio of the buoyancy time to the "gravity" time obtained from the fluid velocity U and gravity g, $U/(gB)$. Utilizing these nondimensional numbers in the equations in (6.9) and dropping the primes, we finally get the following:

Nondimensional Boussinesq Equations.

(6.11) $\qquad \dfrac{D^H \vec{v}_H}{Dt} + w \dfrac{\partial \vec{v}_H}{\partial x_3} = -\overline{P} \nabla_H p,$

(6.12) $\qquad \dfrac{D^H w}{Dt} + w \dfrac{\partial w}{\partial x_3} = -\overline{P} \dfrac{\partial p}{\partial x_3} - (\Gamma)^{-1} (\text{Fr})^{-1} \rho,$

(6.13) $\qquad \dfrac{D\rho}{Dt} = \Gamma (\text{Fr})^{-1} w,$

(6.14) $\qquad \text{div}_H \vec{v}_H + \dfrac{\partial w}{\partial x_3} = 0.$

6.1.1. Formal Asymptotic Derivation of the Layer Two-Dimensional Equations from the Boussinesq Equations in the Small Froude Number Limit.

As we mentioned earlier, for smaller values of the Froude number Fr we expect a higher degree of stratification in the fluid, with the flow increasingly constrained to two-dimensional horizontal motions. In the limit of small Froude number, $Fr \ll 1$, we expect the three-dimensional Boussinesq equations to collapse into a continuum "stack of pancakes" of two-dimensional incompressible flows, layered in the vertical direction. Here we will show that this is indeed the case via the formal asymptotic expansion of the Boussinesq equations in the small Froude number limit. For this derivation we need to make some assumptions about the nondimensional numbers Fr, $\overline{P}$, and Γ in the nondimensional Boussinesq equations in (6.11)–(6.14),

Assumption 1: The Froude number Fr is small,

$$(6.15) \qquad Fr = \epsilon \quad \text{where } \epsilon \ll 1 \,.$$

Assumption 2: The Euler number $\overline{P}$ is equal to 1,

$$(6.16) \qquad \overline{P} = 1 \,,$$

so that the pressure is in balance with the convective momentum transfer.

Assumption 3: The nondimensional number Γ is equal to 1,

$$(6.17) \qquad \Gamma = 1, \text{ or equivalently, } B = \frac{NU}{g} \,.$$

The condition in equation (6.17) imposes a balance in the interaction effects between the density ρ and the vertical velocity w in the Boussinesq equations in (6.12) and (6.13). Namely, the forcing effect that the vertical velocity w has on the density ρ in the density equation (6.13) is the same as the forcing effect that the density has on the vertical velocity in the vertical momentum equation (6.12). The mathematical reason for the choice of the nondimensional factor B in equation (6.17) is that Γ appears in the two singular terms involving Fr^{-1} in the Boussinesq equations (6.11)–(6.14). Setting the value of Γ equal to 1 guarantees that both singular terms in (6.12) and (6.13) have the same coefficient Fr^{-1}, and the combined contribution of these singular terms in the energy identity will cancel each other.

Summarizing, we have the following:

DISTINGUISHED ASYMPTOTIC LIMIT ASSUMPTION FOR THE BOUSSINESQ EQUATIONS:

$$(6.18) \qquad Fr = \epsilon \ll 1, \quad \overline{P} = 1, \quad \Gamma = 1, \text{ that is, } B = NU/g \,.$$

Finally, introducing the assumptions in equation (6.18) into the nondimensional Boussinesq equations in (6.11)–(6.14) we have the following:

DISTINGUISHED SCALING OF BOUSSINESQ EQUATIONS:

$$(6.19) \qquad \frac{D^H \vec{v}_H^{\,\epsilon}}{Dt} + w^\epsilon \frac{\partial \vec{v}_H^{\,\epsilon}}{\partial x_3} = -\nabla_H p^\epsilon \,,$$

$$(6.20) \qquad \frac{D^H w^\epsilon}{Dt} + w^\epsilon \frac{\partial w^\epsilon}{\partial x_3} = -\frac{\partial p^\epsilon}{\partial x_3} - \epsilon^{-1} \rho^\epsilon \,,$$

$$(6.21) \qquad \frac{D\rho^\epsilon}{Dt} = \epsilon^{-1} w^\epsilon ,$$

$$(6.22) \qquad \mathrm{div}_H \, \vec{v}_H^\epsilon + \frac{\partial w^\epsilon}{\partial x_3} = 0 .$$

For the derivation of the incompressible two-dimensional layered equations from the Boussinesq equations we assume that the velocity $\vec{v}^\epsilon$, the density ρ^ϵ and the pressure p^ϵ have asymptotic expansions of the form

$$(6.23) \qquad \begin{aligned} \vec{v}_H^\epsilon(\vec{x}, t) &= \vec{v}_H^{(0)}(\vec{x}, t) + \epsilon \vec{v}_H^{(1)}(\vec{x}, t) + O(\epsilon^2) , \\ w^\epsilon(\vec{x}, t) &= w^{(0)}(\vec{x}, t) + \epsilon w^{(1)}(\vec{x}, t) + O(\epsilon^2) , \\ \rho^\epsilon(\vec{x}, t) &= \rho^{(0)}(\vec{x}, t) + \epsilon \rho^{(1)}(\vec{x}, t) + O(\epsilon^2) , \\ p^\epsilon(\vec{x}, t) &= \epsilon^{-1} p^{(-1)}(\vec{x}, t) + \epsilon^0 p^{(0)}(\vec{x}, t) + O(\epsilon) . \end{aligned}$$

Introduce these asymptotic expansions into the Boussinesq equations (6.19)–(6.22), and solve for the powers of ϵ. To the leading-order $O(\epsilon^{-1})$ in the horizontal momentum equation (6.19), we get

$$(6.24) \qquad \nabla_H p^{(-1)} = 0 ,$$

so that the leading-order term $p^{(-1)}$ of the pressure is independent of the horizontal variables and only depends on the vertical variable

$$(6.25) \qquad p^{(-1)} = p^{(-1)}(x_3, t) .$$

On the other hand, the leading-order $O(\epsilon^{-1})$ in the vertical momentum equation (6.20) yields

$$(6.26) \qquad \rho^{(0)}(x_3, t) = \frac{\partial p^{(-1)}}{\partial x_3}(x_3, t) ,$$

so that to leading order the density and the pressure are in hydrostatic balance. This is a trivial equation. The remaining $O(\epsilon^{-1})$ contribution comes from the density equation (6.21), and it implies that to the leading-order $O(\epsilon^0)$, the vertical component of the velocity is identically zero,

$$(6.27) \qquad w^{(0)}(\vec{x}, t) = 0 .$$

Next we derive the equations satisfied by the leading term $\vec{v}_H^{(0)}$ of the horizontal velocity. The leading-order contribution of order $O(\epsilon^0)$ in the incompressibility condition of the horizontal momentum equation (6.19) yields

$$(6.28) \qquad 0 = \mathrm{div}_H \, \vec{v}_H^{(0)} + \frac{\partial w^{(0)}}{\partial x_3} = \mathrm{div}_H \, \vec{v}_H^{(0)} ,$$

where we have utilized equation (6.27) for $w^{(0)}$. This shows that the horizontal velocity field $\vec{v}_H^{(0)}$ is incompressible to leading order in ϵ. On the other hand, the leading contribution of order $O(\epsilon^0)$ in the horizontal momentum equation (6.19) yields

$$(6.29) \qquad \frac{D^H \vec{v}_H^{(0)}}{Dt} = -\nabla_H p^{(0)} ,$$

where we have again utilized equation (6.27) to eliminate the leading-order contribution of the vertical velocity. Collecting the equations in (6.25)–(6.29), we conclude that the singular low Froude limit of the Boussinesq equations is given by

Layered Two-Dimensional Incompressible Flow Equations.

HYDROSTATIC BALANCE:

$$(6.30) \qquad \rho^{(0)}(x_3, t) = \frac{\partial p^{(-1)}}{\partial x_3}(x_3, t) .$$

ZERO VERTICAL COMPONENT OF THE VELOCITY:

$$(6.31) \qquad w^{(0)}(\vec{x}, t) = 0 .$$

TWO-DIMENSIONAL INCOMPRESSIBLE EULER EQUATIONS FOR THE HORIZONTAL COMPONENT OF THE VELOCITY WITH VERTICAL VARIATION:

$$(6.32) \qquad \mathrm{div}_H \, \vec{v}_H^{(0)} = 0 ,$$

$$(6.33) \qquad \frac{D^H \vec{v}_H^{(0)}}{Dt} = -\nabla_H p^{(0)} .$$

This derivation shows formally that in the limit of low Froude number the motions of the stably stratified flow are constrained severely in the vertical direction; effectively there is no vertical motion but only the hydrostatic balance of forces between the density and the pressure. The flow is thus constrained to two-dimensional horizontal motions where in each plane the fluid flow is governed by the incompressible two-dimensional Euler equations. In this extreme regime of stratification given by the limit of low Froude number, there is decoupling in the motions of the fluid masses at the different heights, creating the "stack of pancakes" of two-dimensional horizontal flows mentioned earlier, where the stacking in the vertical direction is dictated solely by the hydrostatic balance.

6.1.2. The Small Froude Number Asymptotic Limit for a Viscous Boussinesq Equation with Rotation.
In the previous section we saw that in the limit of small Froude number, the Boussinesq equations for three-dimensional stably stratified flow become highly stratified and collapse into a family of two-dimensional incompressible Euler flows, with no interaction between the fluid in different layers, and where the stratification in the vertical direction is dictated by hydrostatic balance. In that discussion we did not consider the effect of viscosity in the Boussinesq equations. If we take into account the viscosity of the fluid, then we no longer expect that the motion of the fluid in the horizontal layers will take place independently of what is happening in neighboring layers. On the contrary, in the presence of viscosity, we know that there will be transfer of momentum between neighboring layers of fluid moving at different velocities as a result of friction between the layers. Consequently, there must be coupling terms for the neighboring layers in the equations for momentum transfer. In laboratory experiments such viscosity is not negligible, and often for the atmosphere or ocean the term *viscosity* is used to represent "turbulent diffusivity" in models for larger scales without complete justification. In this section we want to consider the asymptotic limit of small Froude

number for the viscous Boussinesq equations and to show how viscosity introduces coupling between the different two-dimensional layered incompressible flow solutions. We will see that the coupling takes the form of a forcing term depending on the viscosity and the changes of the horizontal velocity as one moves from layer to layer in the vertical direction. The asymptotic derivation of the limiting equations is similar to the one just given for the nonviscous case in the previous section. First we will present the distinguished scaling for the viscous Boussinesq equations, and then proceed with the derivation of the corresponding limit equations, while concentrating on the new effects due to the introduction of viscosity into the Boussinesq equations.

We start with the Boussinesq equations in (6.1)–(6.4) with the addition of viscosity in the momentum equations

VISCOUS BOUSSINESQ EQUATIONS:

(6.34)
$$\frac{D^H \vec{v}_H}{Dt} + w\frac{\partial \vec{v}_H}{\partial x_3} + f v_H^\perp = -\nabla_H p + \mu \Delta \vec{v}_H ,$$

$$\frac{D^H w}{Dt} + w\frac{\partial w}{\partial x_3} = -\frac{\partial p}{\partial x_3} - \frac{g}{\rho_b}\rho + \mu\Delta w ,$$

$$\frac{D\rho}{Dt} = -w\frac{\partial \bar{\rho}}{\partial x_3} , \quad \text{div}_H \vec{v}_H + \frac{\partial w}{\partial x_3} = 0 ,$$

where μ is the viscosity of the fluid and f is the constant rotation frequency. We use the same nondimensionalization given in (6.8), and introduce additional nondimensional parameters, the Reynolds number Re, defined by

(6.35)
$$\text{Re} = \frac{T_\mu}{T_e} = \frac{\rho_b L^2 \mu^{-1}}{LU^{-1}} = \frac{\rho_b U L}{\mu} ,$$

so that the Reynolds number Re is interpreted as the ratio of the viscous dissipation time to the eddy turnover time. The nondimensional number $\text{Ro} = \frac{v}{Lf}$ is the Rossby number representing the ratio of the rotation time T_f to the eddy turnover time T_e. The resulting nondimensional form of the equations in (6.34) is

NONDIMENSIONAL VISCOUS ROTATING BOUSSINESQ EQUATIONS:

(6.36)
$$\frac{D^H \vec{v}_H}{Dt} + w\frac{\partial \vec{v}_H}{\partial x_3} + (\text{Ro})^{-1}\vec{v}_H^\perp = -(\overline{P})\nabla_H p + (\text{Re})^{-1}\Delta \vec{v}_H ,$$

$$\frac{D^H w}{Dt} + w\frac{\partial w}{\partial x_3} = -(\overline{P})\frac{\partial p}{\partial x_3} - (\Gamma)^{-1}(\text{Fr})^{-1}\rho + (\text{Re})^{-1}\Delta w ,$$

$$\frac{D\rho}{Dt} = \Gamma(\text{Fr})^{-1}w , \quad \text{div}_H \vec{v}_H + \frac{\partial w}{\partial x_3} = 0 .$$

In addition to the three assumptions concerning the nondimensional numbers Fr, $\overline{P}$, and Γ in equations (6.15)–(6.17), we add one more assumption regarding the Reynolds number Re.

Assumption 4:

(6.37)   The Reynolds number Re is independentof the Froude number $\text{Fr} = \epsilon$.

On mesoscales in the atmosphere, the Rossby number is $O(1)$; i.e., the effects of rotation are weak. Thus we make the following assumption here:

Assumption 5: The Rossby number Ro $= O(1)$ with Fr $= \epsilon$.

With the assumptions in equation (6.18), together with assumptions 4 and 5 in equation (6.37), the equations in (6.34)–(6.36) become:

DISTINGUISHED SCALING OF THE VISCOUS ROTATING BOUSSINESQ EQUATIONS:

(6.38)
$$\frac{D^H \vec{v}_H^\epsilon}{Dt} + w^\epsilon \frac{\partial \vec{v}_H^\epsilon}{\partial x_3} + (\text{Ro})^{-1} \vec{v}_H^{\perp \epsilon} = -\nabla_H p^\epsilon + (\text{Re})^{-1} \Delta_H \vec{v}_H^\epsilon + \frac{\partial^2 \vec{v}_H^\epsilon}{\partial x_3^2},$$

(6.39)
$$\frac{D^H w^\epsilon}{Dt} + w^\epsilon \frac{\partial w^\epsilon}{\partial x_3} = -\frac{\partial p^\epsilon}{\partial x_3} - \epsilon^{-1} \rho^\epsilon + (\text{Re})^{-1} \Delta_H w^\epsilon + \frac{\partial^2 w^\epsilon}{\partial x_3^2},$$

(6.40)
$$\frac{D\rho^\epsilon}{Dt} = \epsilon^{-1} w^\epsilon,$$

(6.41)
$$\text{div}_H \vec{v}_H^\epsilon + \frac{\partial w^\epsilon}{\partial x_3} = 0,$$

where we split the Laplacian operator appearing in the momentum equations (6.38) and (6.39) into their horizontal and vertical components

(6.42)
$$\Delta = \Delta_H + \frac{\partial^2}{\partial x_3^2}, \quad \Delta_H = \frac{\partial^2}{\partial x_1^2} + \frac{\partial^2}{\partial x_2^2}.$$

The derivation of the asymptotic limit of the equations in (6.38)–(6.39) in the limit of small Froude number follows very closely the derivation done previously for the nonviscous Boussinesq equations in (6.19)–(6.22). We assume an asymptotic expansion as in equation (6.23) for the horizontal and vertical velocity components $\vec{v}_H^\epsilon$ and w^ϵ, the density ρ^ϵ, and the pressure p^ϵ, plug into the viscous Boussinesq equations (6.38)–(6.41), and solve for the powers of ϵ.

We observe that the same derivation leading to equations (6.24)–(6.28) in the nonviscous Boussinesq equations applies verbatim to the viscous case. The only equation that is changed by the introduction of viscosity is the horizontal momentum equation for the leading-order term $\vec{v}_H^{(0)}$ of the horizontal velocity. In fact, the leading-order contribution of order $O(\epsilon^0)$ in the horizontal momentum equation (6.38) yields

(6.43)
$$\frac{D^H \vec{v}_H^{(0)}}{Dt} + (\text{Ro})^{-1} \vec{v}_H^{\perp (0)} = -\nabla_H p^{(0)} + (\text{Re})^{-1} \Delta_H \vec{v}_H^{(0)} + (\text{Re})^{-1} \frac{\partial^2 \vec{v}_H^{(0)}}{\partial x_3^2},$$

where we utilized equation (6.27) to eliminate the leading-order contribution of the vertical velocity $w^{(0)}$. Therefore the small Froude asymptotic limit of the viscous Boussinesq equations in (6.38)–(6.41) is given by equations (6.24)–(6.28) and (6.43):

Layered Two-Dimensional Incompressible Flow Equations in the Limit of Strong Stratification.

HYDROSTATIC BALANCE:

$$(6.44) \qquad \rho^{(0)}(x_3, t) = \frac{\partial p^{(-1)}}{\partial x_3}(x_3, t) \,.$$

ZERO VERTICAL COMPONENT OF THE VELOCITY:

$$(6.45) \qquad w^{(0)}(\vec{x}, t) = 0 \,.$$

TWO-DIMENSIONAL NAVIER-STOKES EQUATIONS WITH VISCOUS FORCING OF THE HORIZONTAL COMPONENT OF THE VELOCITY:

$$(6.46) \qquad \text{div}_H\, \vec{v}_H^{(0)} = 0 \,,$$

$$(6.47) \qquad \frac{D^H \vec{v}_H^{(0)}}{Dt} + (\text{Ro})^{-1} \vec{v}_H^{\perp} = -\nabla_H p^{(0)} + (\text{Re})^{-1} \Delta_H \vec{v}_H^{(0)} + (\text{Re})^{-1} \frac{\partial^2 \vec{v}_H^{(0)}}{\partial x_3^2} \,.$$

We recall that in the two-dimensional layered incompressible nonviscous equations in (6.30)–(6.33), the fluid motion in each horizontal layer is independent of the fluid motion in the other layers, where the motion in each layer is governed by the two-dimensional incompressible Euler equations in (6.32) and (6.33). In contrast, introducing the viscosity into the Boussinesq equations induces the interactive transfer of momentum between different horizontal layers in the fluid. This transfer of momentum in the Navier-Stokes equations in (6.47) is the result of the weak coupling between the layers produced by the forcing term

$$(\text{Re})^{-1} \frac{\partial^2 \vec{v}_H^{(0)}}{\partial x_3^2} \,.$$

In conclusion, the layered two-dimensional viscous incompressible flow equations in (6.43)–(6.47) show that in the asymptotic limit of small Froude number, the flow is layered horizontally so that the vertical velocity is zero, the vertical stacking of the layers is governed by hydrostatic balance, and the viscosity induces transfer of horizontal momentum between the layers through vertical variations of the horizontal velocity.

6.2. The Vorticity Stream Formulation and Elementary Properties of the Limit Equations for Strongly Stratified Flow

We begin with the dissipation of kinetic energy, $E(t) = \frac{1}{2}\|\vec{v}_H\|_2^2$, where for any function f, $\|f\|_2^2 = \int f^2\, dx\, dy\, dz$. By multiplying equation (6.47) by $\vec{v}_H$ and integrating by parts, it is straightforward to derive the identity,

$$(6.48) \qquad \frac{d}{dt} E = -(\text{Re})^{-1} \left\| \frac{\partial \vec{v}_H}{\partial z} \right\|_2^2 - (\text{Re})^{-1} \|\nabla_H \vec{v}_H\|_2^2 \equiv \varepsilon_z + \varepsilon_{2D} \,.$$

In the above formula, we have split the dissipation of kinetic energy into vertical and horizontal pieces.

We consider solutions of (6.47) that are periodic in both the vertical and horizontal. Under these circumstances the horizontal velocity field $\vec{v}_H$ satisfying the equations in (6.47) has the unique decomposition

$$(6.49) \qquad \vec{v}_H = \vec{V}_H + \nabla_H^\perp \psi ,$$

where $\vec{V}_H(z, t) = \int \vec{v}_H(x, y, z, t) \, dx \, dy$ is a vertically sheared periodic flow and ψ is the periodic horizontal stream function with $\nabla_H^\perp \psi = (-\partial_y \psi, \partial_x \psi, 0)$. Although the velocity field is purely horizontal in the limiting dynamics, the total vorticity $\vec{\omega}$ is three-dimensional and

$$(6.50) \qquad \vec{\omega} = \text{curl} \begin{pmatrix} \vec{V}_H + \nabla_H^\perp \psi \\ 0 \end{pmatrix} = \begin{pmatrix} \partial_z \vec{v}_H^\perp \\ 0 \end{pmatrix} + \begin{pmatrix} \vec{\omega}_H' \\ \omega \end{pmatrix} ,$$

where ω and $\vec{\omega}_H'$ are given by

$$(6.51) \qquad \omega = \Delta_H \psi , \qquad \vec{\omega}_H' = \frac{\partial}{\partial z} \nabla_H \psi .$$

The horizontal components of vorticity and the vertical component of energy dissipation are linked in the limiting dynamics. Indeed, since

$$(6.52) \qquad \frac{\partial \vec{v}_H}{\partial z} = \frac{\partial \vec{v}_H}{\partial z} + \frac{\partial}{\partial z} \nabla_H^\perp \psi ,$$

we infer using (6.50) that

$$(6.53) \qquad \left| \frac{\partial \vec{v}_H}{\partial z} \right| = \left| \vec{\omega}_H' + \frac{\partial \vec{V}_H^\perp}{\partial z} \right| \quad \text{pointwise.}$$

Next, we derive the vorticity-stream form of the equations for the limiting slow dynamics in (6.47). We recall the decomposition $\vec{v}_H = \vec{V}_H(z, t) + \nabla_H^\perp \psi$ from (6.7) and also from (6.51) that the vertical component of the vorticity is related to the stream function ψ by $\omega = \Delta_H \psi$. Therefore, computing the horizontal average (6.47) and the vertical component of the vorticity from the curl of the horizontal momentum equations in the limiting dynamics in (6.47) yields:

Vorticity-Stream Form of the Limiting Dynamics for Low Froude and Finite Rossby Numbers. The horizontal velocity is $\vec{v}_H = \vec{V}_H + \nabla_H^\perp \psi$, where the vertical shear $\vec{V}_H$ satisfies

$$(6.54) \qquad \frac{\partial \vec{V}_H}{\partial t} + (\text{Ro})^{-1} \vec{V}_H^\perp = (\text{Re})^{-1} \frac{\partial^2}{\partial z^2} \vec{V}_H .$$

The vertical vorticity ω and the stream function ψ satisfy the vorticity-stream equations

$$(6.55) \quad \frac{\partial \omega}{\partial t} + \vec{V}_H \cdot \nabla_H \omega + J_H(\psi, \omega) = (\text{Re})^{-1} \Delta_H \omega + (\text{Re})^{-1} \frac{\partial^2 \omega}{\partial z^2} , \qquad \Delta_H \psi = \omega ,$$

where $J_H(\psi, \omega) = \nabla_H^\perp \psi \cdot \nabla_H \omega$ is the Jacobian of ψ and ω in the horizontal variables.

We leave the detailed derivation of (6.54) and (6.55) from (6.46) and (6.47) as an interesting, straightforward exercise for the reader.

An Exact Solution Procedure for the Limit Dynamics. We build exact solutions of the limiting dynamics in (6.54) and (6.55), which we utilize below to give qualitative insight into recent remarkable experiments on turbulence in strongly stratified flows.

From equation (6.54) it is clear that the vertical shear component $\vec{V}_H$ evolves independently of the horizontal velocity $\nabla_H^\perp \psi$. In fact, the dynamics of $\vec{V}_H$ are described by a linear equation with constant coefficients, which we can solve immediately via Fourier series. Let $R(t)$ be the rotation matrix

$$(6.56) \qquad R(t) = \begin{pmatrix} \cos(\mathrm{Ro})^{-1}t & \sin(\mathrm{Ro})^{-1}t \\ -\sin(\mathrm{Ro})^{-1}t & \cos(\mathrm{Ro})^{-1}t \end{pmatrix}.$$

Then the solution of (6.54) is

$$(6.57) \qquad \vec{V}_H(z,t) = \sum_{k=-\infty}^{\infty} e^{ikz} e^{-(\mathrm{Re})^{-1}k^2 t} R(t) \widehat{\vec{V}_H}(k).$$

From (6.57) we see that $\vec{V}_H(z,t)$ represents a time-decaying vertical shear flow, which rotates with uniform angular velocity $(\mathrm{Ro})^{-1}$. We note that if there is no rotation and $(\mathrm{Ro})^{-1} = 0$, then $R(t) = I$.

Next, we construct *special exact solutions* of (6.55). We select a *single fixed energy shell* $\Lambda \geq 1$ and consider a superposition of *horizontal* Fourier modes $\vec{k}_H = (k_1, k_2)$, k_1 and k_2 integers, within the shell,

$$(6.58) \qquad \psi(x,y,z,t) = \sum_{|\vec{k}_H^j|^2 = \Lambda} A_j(z,t) e^{i\vec{k}_H^j \cdot \vec{x}_H} e^{-(\mathrm{Re})^{-1}\Lambda t}.$$

To ensure that ψ is real, we must include for each Fourier component A_j its complex conjugate $A_i = A_j^*$, with $\vec{k}_H^i = -\vec{k}_H^j$. With this particular choice of ψ, we immediately derive the two formulas

$$(6.59) \qquad \omega = \Delta_H \psi = -\Lambda\psi, \qquad \vec{\omega}_H' = \nabla_H \frac{\partial}{\partial z}(-\Lambda^{-1}\omega) = -\Lambda^{-1}\left(\nabla_H \frac{\partial \omega}{\partial z}\right).$$

The first property in (6.59) causes the nonlinear term in (6.55), $J_H(\psi, \omega)$, to vanish so that (6.55) becomes a *linear* advection-diffusion equation; moreover, this condition is self-consistently satisfied by the solution of the advection-diffusion equation as time evolves independently of z and t because we restrict all horizontal wavenumbers $\vec{k}_H$ to satisfy $|\vec{k}_H|^2 = \Lambda$. By introducing (6.58) into (6.55), we see that the complex amplitudes $A_j(z,t)$ solve the linear equation

$$(6.60) \quad \frac{\partial}{\partial t}A_j(z,t) + i\vec{k}_H^j \cdot \vec{V}_H(z,t)A_j(z,t) = (\mathrm{Re})^{-1}\frac{\partial^2}{\partial z^2}A_j(z,t), \quad |\vec{k}_H^j|^2 = \Lambda,$$

for some given initial conditions $A_j(z,t) = A_j^0(z)$ at $t = 0$.

In other words, if we define $\mathcal{A}$ by

$$\mathcal{A}(\vec{x}_H, z, t) = \sum_{|\vec{k}_H^j|^2 = \Lambda} A_j(z,t) e^{i\vec{k}_H^j \cdot \vec{x}_H},$$

then $\mathcal{A}$ acts as a passive scalar and is the solution of the advection-diffusion equation

$$\frac{\partial \mathcal{A}}{\partial t} + \vec{V}_H(z, t) \cdot \nabla \mathcal{A} = (\mathrm{Re})^{-1} \frac{\partial^2 \mathcal{A}}{\partial z^2}.$$

Note that while the equation for $\vec{V}_H(z, t)$ is linear and the equation for $\mathcal{A}$ is also linear, the coefficients for the equation for $\mathcal{A}$ depend on the solution $\vec{V}_H$. Thus, the exact solution procedure described above actually depends on nonlinear interaction and is called a *quasi-linear exact solution*.

6.3. Solutions of the Limit Dynamics with Strong Stratification as Models for Laboratory Experiments

Recent laboratory observations in the decaying wake region for stratified flows at long time display a remarkable transition from columnar vortex structures with dominant vertical vorticity to layered "pancake" vortex sheets with dominant horizontal vorticity. The dynamic process, which achieves this radical reorganization of the vorticity field, exhibits a concurrent dominance of vertical dissipation of kinetic energy as compared with horizontal dissipation. Recent numerical simulations in idealized periodic geometry with strongly stratified flows also exhibit a dominance of layered "pancake" vortex sheets with dominant horizontal vorticity.

The detailed experimental observations of Fincham et al. [9] mentioned in the introduction are extremely revealing regarding the details of the collapse process mentioned above. These authors utilize high resolution Digital Particle Image Velocimetry (DPIV) in decaying stratified grid turbulence at large times (many buoyancy times) with Reynolds numbers varying from Re $= 760$ to Re $= 12000$. The salient features observed in these experiments, especially at the lower Reynolds numbers, are the following:

Experimental Observations.

    A. The flow remains predominantly horizontal during the transition process with Froude number Fr $\ll 1$ and Richardson number Ri $\gg 1$. This condition for approximately horizontal flow means that $\vec{v} = \vec{v}_v + \vec{v}_H$ with $|\vec{v}_v| \ll |\vec{v}_H|$.

    B. At the initial stage, the flow looks like a "complex sea" of columnar dipole vortices in a weakly vertically sheared background flow. Thus, if $\vec{\omega} = (\vec{\omega}_H, \omega)$ is the three-dimensional vorticity, at the initial stage there is much larger vertical vorticity

(6.61)
$$|\omega| \gg |\vec{\omega}_H|.$$

    C. At the later stages of the decay process, the flow field remains horizontal while the vorticity is mostly horizontal and arranged in vertically layered pancakes of horizontal vorticity with

(6.62)
$$|\vec{\omega}_H| \gg |\omega|.$$

D. Consider the (horizontal) velocity field $\vec{v}_H(x, y, z, t)$ from A. Fincham et al. split the contributions to the dissipation of kinetic energy

$$\frac{d}{dt} \frac{1}{2} \int |\vec{v}_H|_2^2 = \varepsilon$$

into horizontal and vertical dissipation components $\varepsilon = \varepsilon_{2D} + \varepsilon_z$ with

$$(6.63) \qquad \varepsilon_{2D} = -v \int |\nabla_H \vec{v}|_2^2, \quad \varepsilon_z = -v \int \left| \frac{\partial \vec{v}}{\partial z} \right|_2^2.$$

Then in the transition process from B to C, the vertical dissipation of energy dominates over horizontal dissipation of kinetic energy, i.e.,

$$(6.64) \qquad |\varepsilon_z| \gg |\varepsilon_{2D}|.$$

The interested reader can consult figures 3, 4, and 7 in Fincham et al. for quantitative confirmation of the features of the experimental observations, which we have outlined briefly in A through D above; in particular, typically 90% or more of the contribution to the dissipation arises from $|\varepsilon_z|$ as compared with $|\varepsilon_{2D}|$.

The development of simplified models that simultaneously capture the features in A through D in a qualitative fashion is an important theoretical issue. From A, these flows are strongly stratified with Fr $\ll$ 1; thus, the starting point for the theoretical developments presented here are the simplified equations for low Froude number limiting dynamics, developed earlier in this chapter. Here we use the exact solutions of the dynamics in the limit of strong stratification developed in the last section. We demonstrate below that they simultaneously have all of the requirements in A, B, C, and D. The solutions built here are laminar models, of course, without explicit turbulence as in the actual experiments.

Exact Solutions Exhibiting Vertical Collapse. In all of the solutions presented in this section utilizing the exact solution procedure in (6.57)–(6.60), for initial data we use the periodic dipole flow configuration with stream function

$$(6.65) \qquad \psi(x, y, z, 0) = \sin(x) + \sin(y).$$

The initial condition for the background vertically sheared motion, $\vec{V}_H(z, t)$, is given by

$$(6.66) \qquad \vec{V}_H(z, 0) = \begin{pmatrix} V_{\max} \sin(z) \\ 0 \end{pmatrix}.$$

The constant $V_{\max}$ controls the strength of the shear; typically, we set $V_{\max} = 0.1$ so that the magnitude of the vertical shear is 5% of the size of the vorticity magnitude in the dipole. Clearly, these initial data mimic the experimentally observed initial stage described in B, involving a sea of dipole vortices in a weakly vertically sheared background flow. Below, we also vary the strength of the vertical shear through the moderate amplitudes for $V_{\max}$ between 0.05 and 0.5.

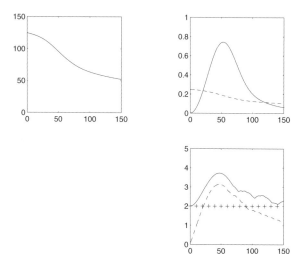

FIGURE 6.1. The exact solutions described in Section 6.3 without rotation $(\text{Ro})^{-1} = 0$, with Reynolds number $\text{Re} = 1000$ and with maximal vertical shear strength $V_{\max} = 0.1$, are monitored versus time: the total energy [top left]; the horizontal and vertical energy dissipation rates, ε_{2D} (dashed) and ε_z (plain) [top right]; the maximal horizontal vorticity $\vec{\omega}_H$ (dashed), the maximal vertical vorticity $\vec{\omega}_z$ (plus), and the maximal total vorticity $\vec{\omega}$ (plain) [bottom right]. From *Physics of Fluids*, vol. 9 (1997), no. 10, p. 2936, fig. 1. Copyright © American Institute of Physics.

Vertical Collapse Dynamics at Low Froude Numbers. Here we show that the dynamical evolution of the exact solutions from (6.57)–(6.60) with the initial data from (6.65) qualitatively captures all of the features from the experiments outlined in C and D from Section 6.3.

At Reynolds number $\text{Re} = 1000$ we set $V_{\max} = 0.1$ in (6.66) for the initial data and suppress rotation with $(\text{Ro})^{-1} = 0$. In Figure 6.1 we graph the energy decay as a function of time in the first frame, while the second frame follows the horizontal and vertical dissipation rate defined earlier in (6.48). The third frame in Figure 6.1 follows the vertical, horizontal, and total vorticity amplitudes as a function of time. Clearly by time $t = 50$, the vertical dissipation rate swamps the horizontal dissipation rate and contributes nearly 80% of the dissipation. Also, simultaneously, the horizontal vorticity maximum dominates the vertical vorticity maximum. These two graphs give strong evidence that both effects from C and D in the introduction are occurring simultaneously in the exact solutions from (6.57)–(6.60) with the above initial data.

To further corroborate the behavior of these exact solutions as a qualitative model for the process in C and D, we present snapshots of the corresponding vorticity field at times $t = 0, 10, 20, 30, 40, 50$ in Figure 6.2. Here we plot the set of points where the vorticity is within 90% of its maximum. Clearly at time $t = 0$, we see essentially the columnar vortex dipole pair since the vertical shear is very

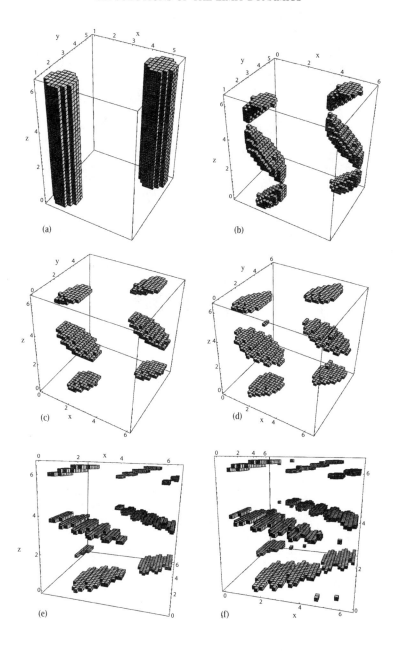

FIGURE 6.2. The exact solutions described in Section 6.3 without rotation $(Ro)^{-1} = 0$, with Reynolds number $Re = 1000$, and with maximal vertical shear strength $V_{max} = 0.1$. The locations where the total vorticity $|\vec{\omega}|$ is within 90% of its maximum are shown at various times: (a) $t = 0$, (b) $t = 10$, (c) $t = 20$, (d) $t = 30$, (e) $t = 40$, (f) $t = 50$. From *Physics of Fluids*, vol. 9 (1997), no. 10, pp. 2937–38, fig. 2(a). Copyright © American Institute of Physics.

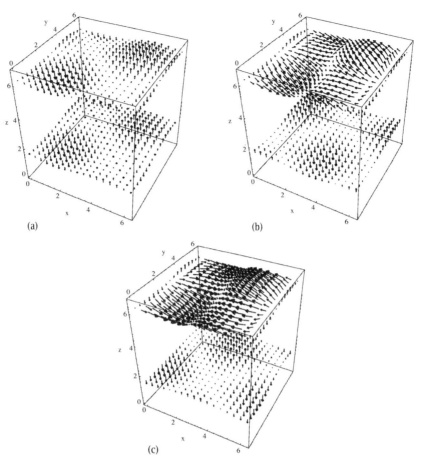

FIGURE 6.3. Snapshots of the vorticity field at two different heights, $z = \frac{\pi}{2}$ and $z = 2\pi$, are shown in various times: (a) $t = 0$, (b) $t = 20$, (c) $t = 50$. The amplitude of the vorticity is scaled to 25% of its actual strength. From *Physics of Fluids*, vol. 9 (1997), no. 10, pp. 2938–39, fig. 2(b). Copyright © American Institute of Physics.

weak. The dynamic formation of the layered pancake sheets with large horizontal vorticity is already evident at times $t = 20, 30$ and continues in time until $t = 50$. We also note that the initial breakup of the vertical vortex tubes at $t = 10$ occurs, as one would expect, in the vicinity of the locations at $z = 0, \pi, 2\pi$, where the vertical shear has its maximum gradient so that the viscous dissipation is more effective.

In Figure 6.3 we display the vorticity field at two different heights, $z = \frac{\pi}{2}$ and $z = 2\pi$, at times $t = 0, 20, 50$. These snapshots of the vorticity clearly demonstrate the strong increase with time in the horizontal component of the vorticity at locations of maximal gradient in the vertical shear.

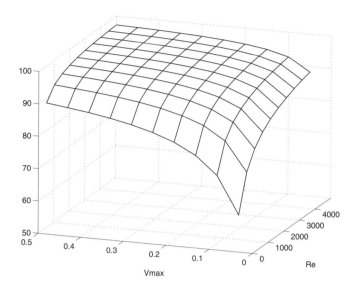

FIGURE 6.4. For the exact solutions described in Section 6.3 without rotation $(Ro)^{-1} = 0$, the vertical energy dissipation ε_z as a percentage of total dissipation ε is shown at the temporal maximum of vertical dissipation as the Reynolds number Re and the vertical shear strength V_{max} vary. From *Physics of Fluids*, vol. 9 (1997), no. 10, p. 2939, fig. 3. Copyright © American Institute of Physics.

The graphical data presented in Figures 6.1, 6.2, and 6.3 demonstrate that the exact solutions of the low Froude number limiting dynamics described in Section 6.2 are a qualitative mode for all of the features observed experimentally and summarized in B, C, and D above. We emphasize that no vertical overturning was utilized in building these solutions.

To see the effect of Reynolds number and the maximum of the vertical shear on the vertical dissipation, we varied these parameters systematically for the exact solutions in (6.57)–(6.60), with the initial data in (6.65) and (6.66) for Reynolds numbers $500 \leq Re \leq 5000$ and V_{max} with $0.05 \leq V_{max} \leq 0.5$. In Figure 6.4 we plot the vertical dissipation as a percentage of total dissipation at the temporal maximum of vertical dissipation as the parameters Re and V_{max} vary. Note that even with $V_{max} = 0.05$, 85% of the dissipation rate is vertical at Re = 5000. Not surprisingly, the larger values of V_{max} have even more efficient vertical dissipation.

The Effect of a Finite Rossby Number on the Model Solution for Vertical Collapse. Here we present the exact solutions with the initial data in (6.54)–(6.55) for the equations in (6.51)–(6.52) that apply in the low Froude number limit at finite Rossby numbers. We will see below that fairly weak rotation can prevent the vertical collapse process in the laminar solutions even at moderate Rossby numbers, such as Ro = 5, 1. We utilize the fixed Reynolds number Re = 1000 and vary the Rossby number through the four moderately large values Ro = 50, 10, 5, 1. In

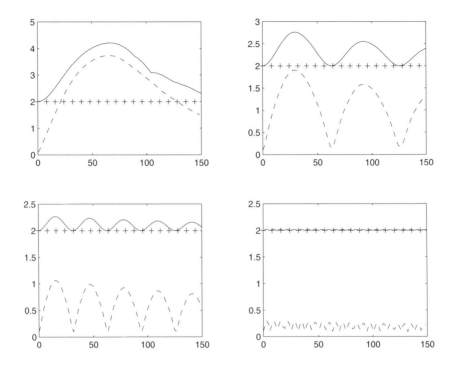

FIGURE 6.5. The exact solutions with rotation, with Reynolds number Re $= 1000$ and with maximal vertical shear strength $V_{\max} = 0.1$, are monitored versus time. The maximal horizontal vorticity $\vec{\omega}_H$ (dashed), the maximal vertical vorticity $\vec{\omega}_z$ (plus), and the maximal total vorticity $\vec{\omega}$ (plain), are shown for varying Rossby numbers (upper left: Ro $= 50$; upper right: Ro $= 10$; lower left: Ro $= 5$; lower right: Ro $= 1$).

Figure 6.5, we graph the time history of the horizontal and vertical vorticity maxima for Ro $= 50, 10, 5, 1$, respectively, while in Figures 6.6, 6.7, and 6.8 we graph the exact solution at times $t = 10, 20, 30$ for the four Rossby numbers Ro $= 50$, $10, 5, 1$. In Figures 6.5, 6.6, 6.7, and 6.8, Ro $= 50$ is the upper left panel, Ro $= 10$ is the upper right, Ro $= 5$ is the lower left, and Ro $= 1$ is the lower right. Figure 6.5 clearly demonstrates the decreasing production of horizontal vorticity $|\vec{\omega}_H|$ for decreasing Rossby numbers with negligible horizontal vorticity at Ro $= 1$. It is evident from Figures 6.6 through 6.8 that even a mild tendency toward vertical collapse is prevented by rotation at Ro $= 1$ while the rotation is strong enough, even at Ro $= 5$, to completely prevent vertical collapse. We note, however, that vertical collapse with horizontal vortex sheets occurs for Ro $= 10$. Thus, even moderate Rossby numbers can prevent vertical collapse in laminar exact solutions in the low Froude number limit. Of course, these laminar vortex columns can be unstable to other perturbations in the initial data, and this is an interesting issue not addressed here.

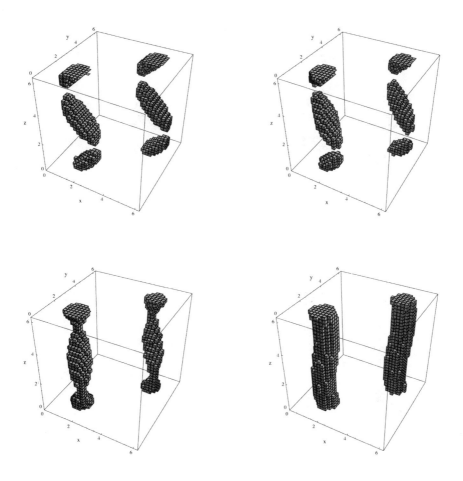

FIGURE 6.6. The exact solutions at time $t = 10$ with Reynolds number Re $= 1000$ and with maximal vertical shear strength $V_{max} = 0.1$. The locations where the total vorticity $|\vec{\omega}|$ is within 90% of its maximum are shown for varying Rossby numbers (upper left: Ro $= 50$; upper right: Ro $= 10$; lower left: Ro $= 5$; lower right: Ro $= 1$).

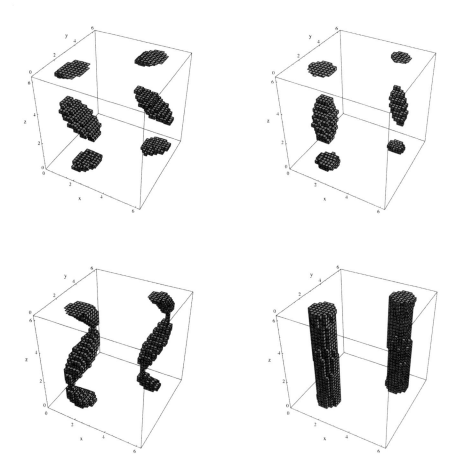

FIGURE 6.7. The exact solutions at time $t = 20$ with Reynolds number Re $= 1000$ and with maximal vertical shear strength $V_{max} = 0.1$. The locations where the total vorticity $|\vec{\omega}|$ is within 90% of its maximum are shown for varying Rossby numbers (upper left: Ro $= 50$; upper right: Ro $= 10$; lower left: Ro $= 5$; lower right: Ro $= 1$).

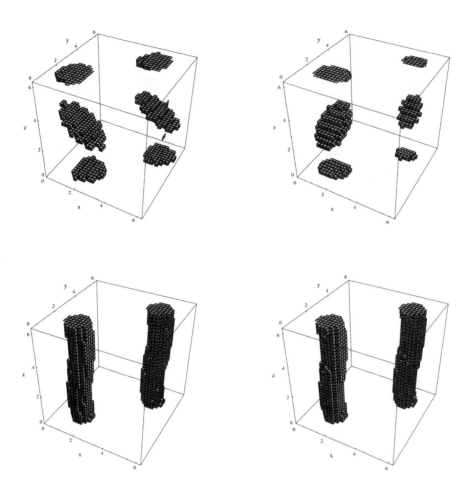

FIGURE 6.8. The exact solutions at time $t = 30$ with Reynolds number Re $= 1000$ and with maximal vertical shear strength $V_{max} = 0.1$. The locations where the total vorticity $|\vec{\omega}|$ is within 90% of its maximum are shown for varying Rossby numbers (upper left: Ro $= 50$; upper right: Ro $= 10$; lower left: Ro $= 5$; lower right: Ro $= 1$).

CHAPTER 7

The Stratified Quasi-Geostrophic Equations as a Singular Limit of the Rotating Boussinesq Equations

7.1. Introduction

In Chapters 4 and 6 we have investigated separately the effects that rapid rotation or strong stratification have on the fluid flow. In Chapter 4 we studied the rotating shallow water equations with no density stratification and showed that for small Rossby number (rapid rotation) the rotating shallow water equations converge to the quasi-geostrophic equations. On the other hand, in Chapter 6 we studied the Boussinesq equations for stably stratified flow and showed that when the stratification is strong (small Froude number) then the Boussinesq equations converge to the simplified two-dimensional layered incompressible Euler equations.

However, formally many realistic geophysical flows are subject to the combined effects of rotation and density stratification. In this chapter we want to study flows that are influenced by both rotation and stratification. The governing equations are the rotating Boussinesq equations introduced in Chapter 1. Of special interest is the situation when the flow is under the action of rapid rotation (small Rossby number) and strong stratification (small Froude number). We will show that when the Rossby and Froude numbers are both small and comparable in magnitude, then the rapidly rotating Boussinesq equations for stably stratified flow converge to the celebrated stratified quasi-geostrophic equations. In Section 7.2 we re-introduce the rotating Boussinesq equations and extend Ertel's vorticity from Chapter 2 to include the effect of the rotating reference frame. In Section 7.3 we nondimensionalize the rotating Boussinesq equations and introduce the relevant nondimensional numbers. This analysis is literally the sum of the previous ones done for the rotating shallow water equations in Chapter 4 and the Boussinesq equations without rotation in Chapter 6. In Section 7.4 we consider the distinguished asymptotic limit when both the Rossby and Froude numbers are small and of the same order of magnitude, and then we give the formal asymptotic derivation of the quasi-geostrophic equations from the rotating Boussinesq equations in the limit of small Rossby and Froude numbers. The remaining sections, Sections 7.5 through 7.7, are devoted to the rigorous proof of convergence of the rotating Boussinesq equations to the quasi-geostrophic equations in the limit of small Rossby and Froude numbers. The proof of the theorem follows the same basic strategy already utilized in Chapter 4. First we derive energy estimates for the solution and the time derivative of the solution of the rotating Boussinesq equations, uniformly in the small expansion parameter that represents the Rossby and

Froude numbers. The main new twist here is the need to derive estimates for the pressure. To accomplish this we show that the pressure is given in terms of the other variables by a Poisson equation, and then we use elliptic regularity estimates for the pressure. Then we combine these uniform estimates with a compactness argument to establish convergence of the solutions as the Rossby and Froude numbers tend to zero. Finally, we identify the limit equations as the quasi-geostrophic equations.

7.2. The Rotating Boussinesq Equations

Recall from Chapter 1 the rotating Boussinesq equations for stably stratified flow (we ignore the effects of viscous and heat dissipation):

HORIZONTAL CONSERVATION OF MOMENTUM:

$$(7.1) \qquad \frac{D^H \vec{v}_H}{Dt} + w \frac{\partial \vec{v}_H}{\partial x_3} + f \vec{v}_H^{\perp} = -\rho_b^{-1} \nabla_H \phi \,.$$

VERTICAL CONSERVATION OF MOMENTUM:

$$(7.2) \qquad \frac{D^H w}{Dt} + w \frac{\partial w}{\partial x_3} = -\rho_b^{-1} \frac{\partial \phi}{\partial x_3} - \frac{g}{\rho_b} \rho \,.$$

CONSERVATION OF MASS:

$$(7.3) \qquad \frac{D\rho}{Dt} = -w \frac{\partial \bar{\rho}}{\partial x_3} \,.$$

INCOMPRESSIBILITY:

$$(7.4) \qquad \mathrm{div}_H \, \vec{v}_H + \frac{\partial w}{\partial x_3} = 0 \,,$$

where the velocity field $\vec{v}$ is written in terms of its horizontal component $\vec{v}_H$ and its vertical component w, $\vec{v} = (\vec{v}_H, w)$, and the density $\tilde{\rho}$ is given by a mean $\bar{\rho}$ and the deviation ρ

$$(7.5) \qquad \tilde{\rho} = \bar{\rho} + \rho \,, \quad \bar{\rho} = \rho_b - bx_3 \,,$$

where the stratified mean density $\bar{\rho}$ has $b > 0$ for stable stratification. The pressure is denoted by ϕ. Equations (7.1)–(7.4) are written in a horizontally rotating reference frame with rotation frequency f. While the planetary β-plane effects allowing for variation of this rotation frequency with latitude are important at large scale, for simplicity in exposition we assume that f is constant here. Such effects have been discussed briefly in Section 1.2 and Chapter 5 for simplified approximations; see [6, 26, 29] for more discussion. The superscript "H" in equations (7.1)–(7.4) means that the operation is performed on the horizontal components only, so that the material derivative $\frac{D^H}{Dt}$, the horizontal gradient ∇_H, and the horizontal divergence div_H are given by

$$(7.6) \quad \frac{D^H}{Dt} = \frac{\partial}{\partial t} + \vec{v}_H \cdot \nabla_H \,, \quad \nabla_H = \left(\frac{\partial}{\partial x_1}, \frac{\partial}{\partial x_2} \right), \quad \mathrm{div}_H \, \vec{v}_H = \frac{\partial v_1}{\partial x_1} + \frac{\partial v_2}{\partial x_2} \,.$$

In this rotating frame of reference we define the absolute vorticity $\vec{\omega}_a$ as the sum of the fluid dynamic vorticity $\vec{\omega}$ and the vorticity $\vec{\Omega}$ induced by the reference frame

$$(7.7) \qquad \vec{\omega}_a = \vec{\omega} + \vec{\Omega}, \quad \vec{\omega} = \operatorname{curl} \vec{v}, \quad \vec{\Omega} = f\vec{e}_3.$$

In terms of the absolute vorticity $\vec{\omega}_a$ we have the following extension of Ertel's theorem discussed in Chapter 2 for the nonrotating Boussinesq equations to the case of rotating reference frames:

THEOREM 7.1 (Ertel's Theorem) *Let the absolute vorticity $\vec{\omega}_a$ be defined by equation (7.7), and the density $\tilde{\rho}$ by equation (7.5). Then for any solution of the rotating Boussinesq equations (7.1)–(7.4) we have*

$$(7.8) \qquad \frac{D}{Dt}(\vec{\omega}_a \cdot \nabla \tilde{\rho}) = 0.$$

PROOF: Evaluation of the left side of equation (7.8) yields

$$(7.9) \qquad \frac{D}{Dt}(\vec{\omega}_a \cdot \nabla \tilde{\rho}) = \vec{\omega}_a \cdot \frac{D}{Dt}(\nabla \tilde{\rho}) + \frac{D\vec{\omega}_a}{Dt} \cdot \nabla \tilde{\rho}.$$

To evaluate the terms in the right side of equation (7.9), we utilize the rotating Boussinesq equations in (7.1)–(7.4). First we rewrite the density equation (7.3) as

$$(7.10) \qquad \frac{D\tilde{\rho}}{Dt} = 0.$$

Taking the gradient of equation (7.10) yields

$$(7.11) \qquad \frac{D}{Dt}(\nabla \tilde{\rho}) + ({}^{\mathsf{T}}\nabla \vec{v})\nabla \tilde{\rho} = 0,$$

where the gradient velocity matrix is given by $(\nabla \vec{v})_{ij} = \partial v_i/\partial x_j$. To evaluate the advective derivative of the total vorticity $\vec{\omega}_a$ in equation (7.9), we start with the momentum equations in (7.1) and (7.2),

$$(7.12) \qquad \frac{D\vec{v}}{Dt} + \vec{\Omega} \times \vec{v} = \rho_b^{-1}\nabla \phi - \frac{g\rho}{\rho_b}\vec{e}_3.$$

Next we take the curl of equation (7.12). In Proposition 2.1, we actually showed that

$$(7.13) \qquad \operatorname{curl}\left(\frac{D\vec{v}}{Dt}\right) = \frac{D\vec{\omega}}{Dt} - (\nabla \vec{v})\vec{\omega}.$$

On the other hand, it is straightforward to check that for $\vec{\Omega} = f\vec{e}_3$,

$$(7.14) \qquad \operatorname{curl}(\vec{\Omega} \times \vec{v}) = (\vec{v} \cdot \nabla)\vec{\Omega} - (\nabla \vec{v})\vec{\Omega}.$$

Therefore, if we take the curl of the momentum equations in (7.12), it follows that

$$(7.15) \qquad \frac{D\vec{\omega}}{Dt} - (\nabla \vec{v})\vec{\omega} + (\vec{v} \cdot \nabla)\vec{\Omega} - (\nabla \vec{v})\vec{\Omega} = \frac{g}{\rho_b}\begin{pmatrix} \nabla_H^\perp \rho \\ 0 \end{pmatrix}$$

and $\nabla_H^\perp = (-\frac{\partial}{\partial x_2}, \frac{\partial}{\partial x_1})$ is the horizontal perpendicular gradient. Recalling the definition of $\vec{\omega}_a$ in equation (7.7), we conclude that

$$(7.16) \qquad \frac{D\vec{\omega}_a}{Dt} + (\nabla \vec{v})\vec{\omega}_a + \frac{g}{\rho_b}\begin{pmatrix} \nabla_H^\perp \rho \\ 0 \end{pmatrix}.$$

Plugging equations (7.11) and (7.16) back into equation (7.9) yields

$$(7.17) \qquad \frac{D}{Dt}(\vec{\omega}_a \cdot \nabla \tilde{\rho}) = -\vec{\omega}_a \cdot (({}^{\mathsf{T}}\nabla \vec{v})\nabla \tilde{\rho}) + \nabla_H \tilde{\rho} \cdot \nabla_H^{\perp} \tilde{\rho} + \nabla \tilde{\rho} \cdot ((\nabla \vec{v})\vec{\omega}_a) \,.$$

Clearly the dot product of ∇_H and $\nabla_H^{\perp}$ is zero. It is also clear that the two quadratic contributions from $\nabla \vec{v}$ and its transpose are the same, so they cancel each other in equation (7.17). Therefore the right side of equation (7.17) is zero and we conclude that

$$(7.18) \qquad \frac{D}{Dt}(\vec{\omega}_a \cdot \nabla \tilde{\rho}) = 0 \,.$$

This proves Ertel's theorem. $\qquad\qquad\qquad\qquad\qquad\qquad\qquad\qquad\qquad\qquad\qquad\qquad\square$

We conclude this section by showing how the pressure ϕ in the Boussinesq equations can be recovered from the other state variables by solving a Poisson equation. The fact that the pressure ϕ can be obtained as the solution of an elliptic equation will be a key new ingredient needed for the rigorous proof of convergence of the rotating Boussinesq equations to the quasi-geostrophic equations in Section 7.7.

PROPOSITION 7.2 (Elliptic Equation for the Pressure) *The pressure ϕ in the rotating Boussinesq equations (7.1)–(7.4) satisfies the Poisson equation*

$$(7.19) \qquad -\rho_b^{-1}\Delta\phi = (\nabla \vec{v}) : ({}^{\mathsf{T}}\nabla \vec{v}) + \frac{g}{\rho_b}\frac{\partial\rho}{\partial x_3} - f\omega_3 \,,$$

where

$$(7.20) \qquad (\nabla \vec{v}) : ({}^{\mathsf{T}}\nabla \vec{v}) = \sum_{i,j=1}^{3} \frac{\partial v_i}{\partial x_j}\frac{\partial v_j}{\partial x_i} \,.$$

Note that $A : {}^{\mathsf{T}}A$ is short-hand notation for the trace of the matrix $A^{\mathsf{T}}A$.

PROOF: We start with the momentum equations in (7.1) and (7.2),

$$(7.21) \qquad \frac{D\vec{v}}{Dt} + f\vec{e}_3 \times \vec{v} = -\rho_b^{-1}\nabla\phi - \frac{g\rho}{\rho_b}\vec{e}_3 \,.$$

Applying the divergence to equation (7.21) yields

$$(7.22) \quad \operatorname{div}\vec{v}_t + \operatorname{div}(\vec{v} \cdot \nabla\vec{v}) + \operatorname{div}(f\vec{e}_3 \times \vec{v}) = -\rho_b^{-1}\operatorname{div}\nabla\phi - \operatorname{div}\left(\frac{g\rho}{\rho_b}\vec{e}_3\right).$$

Let us evaluate the different terms in equation (7.22). Since the velocity field $\vec{v}$ is divergence free, then $\operatorname{div}\vec{v}_t = 0$, and the divergence of the advective derivative reduces to

$$\operatorname{div}(\vec{v} \cdot \nabla\vec{v}) = \sum_{i=1}^{3}\frac{\partial}{\partial x_i}\left(\sum_{j=1}^{3} v_j \frac{\partial v_i}{\partial x_j}\right) = \vec{v} \cdot \nabla(\operatorname{div}\vec{v}) + \sum_{i,j=1}^{3}\frac{\partial v_j}{\partial x_i}\frac{\partial v_i}{\partial x_j}$$

$$(7.23)$$

$$= \sum_{i,j=1}^{3}\frac{\partial v_j}{\partial x_i}\frac{\partial v_i}{\partial x_j} = (\nabla\vec{v}) : ({}^{\mathsf{T}}\nabla\vec{v}) \,,$$

where $(\nabla \vec{v}) : ({}^{\mathsf{T}}\nabla \vec{v})$ is defined by equation (7.20) as the trace of the matrix $\nabla \vec{v}^{\mathsf{T}} \nabla \vec{v}$. The remaining terms in equation (7.22) are easily evaluated:

(7.24)
$$\operatorname{div}(f \vec{e}_3 \times \vec{v}) = -f \left(\frac{\partial v_2}{\partial x_1} - \frac{\partial v_1}{\partial x_2} \right) = -f \omega_3 \,,$$

$$\operatorname{div}\left(\frac{g\rho}{\rho_b} \vec{e}_3 \right) = \frac{g}{\rho_b} \frac{\partial \rho}{\partial x_3} \,, \quad \operatorname{div} \nabla \phi = \Delta \phi \,.$$

Plugging the results from equations (7.23) and (7.24) back into equation (7.22) yields the Poisson equation for ϕ,

(7.25)
$$-\rho_b^{-1} \Delta \phi = (\nabla \vec{v}) : ({}^{\mathsf{T}}\nabla \vec{v}) + \frac{g}{\rho_b} \frac{\partial \rho}{\partial x_3} - f \omega_3 \,,$$

which is the same equation as in (7.19). $\square$

Note that the formula for the pressure in Proposition 7.2 in terms of the velocity and density gradients can be inverted, differentiated, and substituted back into equations (7.1), (7.2), and (7.3) for the velocity $(\vec{v}_H, w)$ and the density perturbation ρ. In this fashion the Boussinesq equations become four nonlinear, nonlocal equations with time derivatives for four unknowns with the pressure eliminated and the incompressibility constraint satisfied automatically at later times when this is true for the initial data. This is the analogue of Leray's projection for incompressible flow; see chapter 1 of the book by Majda and Bertozzi [19].

7.3. The Nondimensional Rotating Boussinesq Equations

Next we nondimensionalize the Boussinesq equations (7.1)–(7.4) utilizing the following scales for length, time, velocity, density, and pressure:

(7.26)

L	horizontal length scale
U	mean horizontal advective velocity
$T_e = \frac{L}{U}$	eddy turnover time
$T_R = f^{-1}$	rotation time
$T_N = N^{-1}$	buoyancy time
ρ_b	mean density
$\bar{p}$	mean pressure

where N is the buoyancy (Brunt-Väisälä) frequency

$$N = \left(-\frac{g}{\rho_b} \frac{\partial \bar{p}}{\partial x_3} \right)^{1/2} \,,$$

and the ρ_b is the mean density in equation (7.5). With this choice of scales we introduce the nondimensional variables that follow.

Nondimensionalization.

(7.27)
$$\vec{x}'_H = \frac{\vec{x}_H}{L} \,, \quad x'_3 = \frac{x_3}{L} \,, \quad t' = \frac{t}{T_e} \,, \quad \vec{v}'_H = \frac{\vec{v}_H}{U} \,,$$

$$w' = \frac{w}{U} \,, \quad \rho' = \frac{\rho}{\rho_b B} \,, \quad \phi' = \frac{\phi}{\bar{p}} \,,$$

where $\vec{x}_H = (x_1, x_2)$ denotes the horizontal space variables. The quantity $B > 0$ in the density equation in (7.27) is a numerical factor to be fixed later when we discuss the distinguished asymptotic limit for the rotating Boussinesq equations. In terms of the nondimensional variables, the rotating Boussinesq equations (7.1)–(7.4) become

(7.28)
$$\frac{D^H \vec{v}'_H}{Dt'} + w' \frac{\partial v'_H}{\partial x'_3} + \left(\frac{fL}{U}\right) \vec{v}'^{\perp}_H = -\left(\frac{\bar{p}}{\rho_b U^2}\right) \nabla_H \phi',$$

$$\frac{D^H w'}{Dt'} + w' \frac{\partial w'}{\partial x'_3} = -\left(\frac{\bar{p}}{\rho_b U^2}\right) \frac{\partial \phi'}{\partial x'_3} - \left(\frac{BgL}{U^2}\right) \rho',$$

$$\frac{D\rho'}{Dt'} = \left(\frac{L}{gB}\right) \left(-\frac{g}{\rho_b} \frac{\partial \bar{\rho}}{\partial x_3}\right) w', \quad \text{div}_H \vec{v}'_H + \frac{\partial w'}{\partial x'_3} = 0,$$

where the operators div_H and ∇_H now act on the primed variables.

Next we introduce the nondimensional numbers

(7.29) $$\text{Ro} = \frac{T_R}{T_e} = \frac{U}{Lf}, \quad \text{Fr} = \frac{T_N}{T_e} = \frac{U}{LN}, \quad \overline{P} = \frac{\bar{p}}{\rho_b U^2}, \quad \Gamma = \frac{BgL}{U^2}.$$

The Rossby number Ro is the ratio of the rotation time to the time scale of horizontal motion of the fluid. The Froude number Fr is the ratio of the buoyancy time to the time scale of vertical motion of the fluid. We are interested in situations where both the Rossby and Froude numbers are small, so that the fluid will be highly stratified and confined to essentially horizontal motion, and the rotation is fast enough to influence the dynamics of the fluid motion. For example, for large-scale motions of atmospheric weather patterns, the time scale of motion is of several days, whereas the rotation time is one day and the buoyancy time is of the order of only a few minutes. In this case it is clear that the Rossby and Froude numbers are small. The Euler number $\overline{P}$ is the ratio of the pressure forces to inertial forces. The nondimensional number Γ can be interpreted as the ratio of the mean potential energy to the mean vertical kinetic energy in the fluid. The reader can easily observe that the nondimensional numbers appearing in equation (7.29) include the nondimensional numbers introduced earlier in Chapters 4 and 6, where we studied the simpler problems of the rotating shallow water equations with no stratification and the Boussinesq equations with no rotation. Utilizing these nondimensional numbers in the equations in (7.28), and dropping the primes, we finally get the

Nondimensional Rotating Boussinesq Equations.

(7.30) $$\frac{D^H \vec{v}_H}{Dt} + w \frac{\partial \vec{v}_H}{\partial x_3} + (\text{Ro})^{-1} \vec{v}^{\perp}_H = -\overline{P} \nabla_H \phi,$$

(7.31) $$\frac{D^H w}{Dt} + w \frac{\partial w}{\partial x_3} = -\overline{P} \frac{\partial \phi}{\partial x_3} - \Gamma \rho,$$

(7.32) $$\frac{D\rho}{Dt} = (\Gamma)^{-1} (\text{Fr})^{-2} w,$$

$$(7.33) \qquad \text{div}_H \, \vec{v}_H + \frac{\partial w}{\partial x_3} = 0.$$

Next we derive the nondimensional version of Ertel's theorem. If in addition to the nondimensionalized variables in (7.27) we define the nondimensional vorticity $\vec{\omega}'$ in terms of the eddy rotation frequency T_e^{-1}, $\vec{\omega}' = T_e \vec{\omega}$, then Ertel's theorem in (7.8) is written in the nondimensional variables as

$$(7.34) \qquad \begin{aligned} &\frac{D}{Dt'}\left\{ \left(\vec{\omega}' + \left(\frac{fL}{U} \right) \vec{e}_3 \right) \right. \\ &\left. \cdot \left[\left(\frac{gB}{L} \right) \left(-\frac{g}{\rho_b} \frac{\partial \bar{\rho}}{\partial x_3} \right)^{-1} \left(-\nabla_H \rho' - \frac{\partial \rho'}{\partial x_3'} \vec{e}_3 + \vec{e}_3 \right) \right] \right\} = 0. \end{aligned}$$

After introducing the nondimensional numbers in (7.29) and dropping the primes, equation (7.34) yields

NONDIMENSIONAL ERTEL'S THEOREM:

$$(7.35) \qquad \frac{D}{Dt}\left\{ \omega_3 - (\text{Ro})^{-1} \Gamma (\text{Fr})^2 \frac{\partial \rho}{\partial x_3} - \Gamma (\text{Fr})^2 \left(\vec{\omega}_H \cdot \nabla_H \rho + \omega_3 \frac{\partial \rho}{\partial x_3} \right) \right\} = 0,$$

where $\vec{\omega}_H$ denotes the horizontal component of the vorticity, $\vec{\omega}_H = (\omega_1, \omega_2)$. Finally, we write the nondimensional version of the Poisson equation (7.19) for the pressure. In terms of the nondimensional variables given in (7.27), equation (7.19) becomes

$$(7.36) \qquad -\left(\frac{\bar{p}}{\rho_b U^2} \right) \left(\Delta_H \phi' + \frac{\partial^2 \phi'}{\partial x_3'^2} \right) =$$

$$(\nabla \vec{v}') : ({}^T\!\nabla \vec{v}') + \left(\frac{gBL}{U^2} \right) \frac{\partial \rho'}{\partial x_3'} - \left(\frac{fL}{U} \right) \omega_3',$$

and after the introduction of the nondimensional numbers in (7.29) and dropping the primes, equation (7.36) reduces to

NONDIMENSIONAL POISSON EQUATION FOR THE PRESSURE:

$$(7.37) \qquad -\overline{P}\left(\Delta_H \phi + \frac{\partial^2 \phi}{\partial x_3^2} \right) = (\nabla \vec{v}) : ({}^T\!\nabla \vec{v}) + \Gamma \frac{\partial \rho}{\partial x_3} - (\text{Ro})^{-1} \omega_3.$$

7.4. Formal Asymptotic Derivation of the Quasi-Geostrophic Equations as a Distinguished Asymptotic Limit of Small Rossby and Froude Numbers

In Chapter 4 we showed that the equations for constant density rotating shallow water converged to the quasi-geostrophic equations in the limit of rapid rotation (Ro $\ll$ 1), and that one of the key features of the resulting quasi-geostrophic equations was the balance between rotation and pressure forces. On the other hand, in Chapter 6 we studied the Boussinesq equations of stably stratified fluid and showed formally that these equations converged to two-dimensional layered equations in the limit of high stratification (Fr $\ll$ 1); these layered equations involved zero vertical velocity and essentially no fluid dynamical interaction among the layers (in the inviscid case). Here we want to combine the effects of both high rotation and high stratification for the rotating Boussinesq equations. We will show that the rotating Boussinesq equations converge to the quasi-geostrophic equations with stratification for a distinguished asymptotic limit of small Rossby and Froude numbers.

We start with the discussion of the distinguished asymptotic limit conditions necessary for the derivation of the quasi-geostrophic equations. We make the following assumptions on the nondimensional numbers Ro, Fr, Γ, and $\overline{P}$ in (7.29):

> **Assumption 1:** The Rossby number Ro is small,

$$(7.38) \qquad \text{Ro} = \epsilon \quad \text{where } \epsilon \ll 1 \,,$$

> so that the time scale of motion of the fluid is large compared with the period of rotation of the reference frame, and therefore the rotation affects the motion of the flow.
> **Assumption 2:** Geostrophic Balance. The rotation and the pressure forces are in balance,

$$(7.39) \qquad \overline{P} = (\text{Ro})^{-1} \,,$$

> and this implies that in the horizontal momentum equation (7.30), the Coriolis force and the horizontal gradient pressure force are in balance.
> **Assumption 3:** The Froude number Fr is small and proportional to the Rossby number Ro,

$$(7.40) \qquad \text{Fr} = F\text{Ro} \,,$$

> where F is a constant (although it can be small). This assumption implies that the Froude number is small; therefore the buoyancy time is small compared with the eddy turnover time, and the fluid is highly stratified.
> **Assumption 4:** The nondimensional number Γ is in balance with the inverse of the Froude number,

$$(7.41) \qquad \Gamma = (\text{Fr})^{-1}, \quad \text{or equivalently,} \quad B = \frac{NU}{g} \,.$$

> The condition in equation (7.41) enforces the balance of the buoyancy force in the vertical momentum equation (7.31) and the density changes from buoyant convection in equation (7.32). We also remark that a mathematical motivation for the balance assumption 4 is that it will permit us to obtain stability estimates for the energy of the system, *uniformly* in ϵ, in spite of the presence of singular terms of order ϵ^{-1} in the differential equations (7.30)–(7.33).

Summarizing, we have the following:

QUASI-GEOSTROPHIC (QG) SCALING ASSUMPTIONS:

$$(7.42) \qquad \text{Ro} = \epsilon \ll 1, \quad \overline{P} = \epsilon^{-1}, \quad \text{Fr} = \epsilon F, \quad \Gamma = \epsilon^{-1}F^{-1} \,.$$

Finally, introducing the scaling assumptions from (7.42) back into the nondimensional rotating Boussinesq equations in (7.30)–(7.33) yields

QG DISTINGUISHED LIMIT OF THE ROTATING BOUSSINESQ EQUATIONS:

$$(7.43) \qquad \frac{D^H \vec{v}_H^\epsilon}{Dt} + w^\epsilon \frac{\partial \vec{v}_H^\epsilon}{\partial x_3} = -\epsilon^{-1}(\vec{v}_H^{\epsilon \perp} + \nabla_H \phi^\epsilon),$$

$$(7.44) \qquad \frac{D^H w^\epsilon}{Dt} + w^\epsilon \frac{\partial w^\epsilon}{\partial x_3} = -\epsilon^{-1} \frac{\partial \phi^\epsilon}{\partial x_3} - \epsilon^{-1} F^{-1} \rho^\epsilon,$$

$$(7.45) \qquad \frac{D\rho^\epsilon}{Dt} = \epsilon^{-1} F^{-1} w^\epsilon,$$

$$(7.46) \qquad \mathrm{div}_H \vec{v}_H^\epsilon + \frac{\partial w^\epsilon}{\partial x_3} = 0.$$

Next we write the formulas for Ertel's theorem and the Poisson equation for the pressure in terms of the quasi-geostrophic scaling in (7.42). In terms of the scaling in (7.42) Ertel's theorem yields

QG SCALING OF ERTEL'S THEOREM:

$$(7.47) \qquad \frac{D}{Dt} \left\{ \omega_3 - F \frac{\partial \rho}{\partial x_3} - \epsilon F \left(\vec{\omega}_H \cdot \nabla_H \rho + \omega_3 \frac{\partial \rho}{\partial x_3} \right) \right\} = 0.$$

Similarly, introducing the scaling from equation (7.42) into the Poisson equation for the pressure in (7.37) yields

QG SCALING OF THE POISSON EQUATION FOR THE PRESSURE:

$$(7.48) \qquad -\left(\Delta_H \phi + \frac{\partial^2 \phi}{\partial x_3^2} \right) = \epsilon (\nabla \vec{v}) : ({}^T \nabla \vec{v}) + F^{-1} \frac{\partial \rho}{\partial x_3} - \omega_3.$$

Derivation of the QG equations. Next we give the formal derivation of the quasi-geostrophic equations for stratified flow from the rotating Boussinesq equations in the asymptotic limit of small Rossby and Froude numbers summarized in (7.42). We assume that the velocity $\vec{v}^\epsilon$, the density ρ^ϵ, and the pressure ϕ^ϵ have asymptotic expansions of the form

$$(7.49) \qquad \begin{aligned} \vec{v}_H^\epsilon(\vec{x}, t) &= \vec{v}_H^{(0)}(\vec{x}, t) + \epsilon \vec{v}_H^{(1)}(\vec{x}, t) + O(\epsilon^2), \\ w^\epsilon(\vec{x}, t) &= w^{(0)}(\vec{x}, t) + \epsilon w^{(1)}(\vec{x}, t) + O(\epsilon^2), \\ \rho^\epsilon(\vec{x}, t) &= \rho^{(1)}(\vec{x}, t) + \epsilon \rho^{(1)}(\vec{x}, t) + O(\epsilon^2), \\ \phi^\epsilon(\vec{x}, t) &= \phi^{(0)}(\vec{x}, t) + \epsilon \phi^{(1)}(\vec{x}, t) + O(\epsilon^2). \end{aligned}$$

We introduce these asymptotic expansions into the rotating Boussinesq equations (7.43)–(7.46) and solve for the powers of ϵ. Let us start with the terms associated with the leading power ϵ^{-1}. The leading contribution of order $O(\epsilon^{-1})$ in the horizontal momentum equation (7.43) yields the geostrophic balance equation for the horizontal component $\vec{v}_H^{(0)}$ of the velocity field and the horizontal gradient of the pressure

$$(7.50) \qquad \vec{v}_H^{(0)} = \nabla_H^\perp \phi^{(0)},$$

and taking the horizontal curl of (7.50) we obtain

$$(7.51) \qquad \Delta_H \phi^{(0)} = \omega_3^{(0)}.$$

On the other hand, the leading-order $O(\epsilon^{-1})$ in the vertical momentum equation (7.44) yields the hydrostatic balance of the buoyancy forces from the density and the vertical gradient of the pressure

$$(7.52) \qquad \frac{\partial \phi^{(0)}}{\partial x_3} = -F^{-1} \rho^{(0)} \, .$$

Finally, the leading-order $O(\epsilon^{-1})$ in the density equation yields the strong stratification restriction of zero vertical component $w^{(0)}$ of the velocity field

$$(7.53) \qquad w^{(0)}(\vec{x}, t) = 0 \, .$$

Notice that the equations (7.50) and (7.52) for the horizontal and vertical components of the velocity field $\vec{v}^{(0)}$ automatically satisfy the incompressibility restriction in (7.46),

$$(7.54) \qquad \mathrm{div}_H \, \vec{v}_H^{(0)} + \frac{\partial w^{(0)}}{\partial x_3} = \mathrm{div}_H \, \nabla_H^\perp \phi^{(0)} = 0 \, .$$

To close the system we need to derive an equation for the vertical component of the vorticity $\omega_3^{(0)}$. This is done with the help of Ertel's theorem. If we introduce the asymptotic expansions in (7.49) into Ertel's equation in (7.47) and collect the leading contribution of order $O(\epsilon^0)$, we obtain

$$(7.55) \qquad \frac{D^H}{Dt} \left(\omega_3^{(0)} - F \frac{\partial \rho^{(0)}}{\partial x_3} \right) = 0 \, ,$$

where the material derivative is evaluated at the velocity $\vec{v}_H^{(0)}$,

$$(7.56) \qquad \frac{D^H}{Dt} = \frac{\partial}{\partial t} + \vec{v}_H^{(0)} \cdot \nabla_H \, .$$

From equation (7.51) we know that the pressure and the vertical component of the vorticity are related by a Poisson equation. On the other hand, if we differentiate the hydrostatic balance equation (7.52) in the vertical variable, we obtain

$$(7.57) \qquad \frac{\partial \rho^{(0)}}{\partial x_3} = -F \frac{\partial^2 \phi^{(0)}}{\partial x_3^2} \, .$$

Finally, introducing equations (7.51) and (7.57) into equation (7.55) for the vertical component of the vorticity yields

$$(7.58) \qquad \frac{D^H}{Dt} \left(\Delta_H \phi^{(0)} + F^2 \frac{\partial^2 \phi^{(0)}}{\partial x_3^2} \right) = 0 \, .$$

In conclusion, collecting equations (7.50), (7.52), (7.53), and (7.58), we have the celebrated equations for quasi-geostrophic flow.

QG equations for stably stratified flow.

GEOSTROPHIC BALANCE IN THE HORIZONTAL DIRECTION:

(7.59)
$$\vec{v}_H^{(0)} = \nabla_H^\perp \phi^{(0)} \, .$$

HYDROSTATIC BALANCE IN THE VERTICAL DIRECTION:

(7.60)
$$\frac{\partial \phi^{(0)}}{\partial x_3} = -F^{-1} \rho^{(0)} \, .$$

ZERO VERTICAL COMPONENT OF THE VELOCITY:

(7.61)
$$w^{(0)}(\vec{x}, t) = 0 \, .$$

CONSERVATION OF POTENTIAL VORTICITY:

(7.62)
$$\frac{D^H}{Dt}\left(\Delta_H \phi^{(0)} + F^2 \frac{\partial^2 \phi^{(0)}}{\partial x_3{}^2} \right) = 0 \, .$$

The quasi-geostrophic equations were originally derived by Charney in the late 1940s. Historically they represented the first realistic equations capable of describing some of the main features of large-scale atmospheric motion. The reader can notice many similarities between the quasi-geostrophic for rapidly rotating and highly stratified flow just derived and both the quasi-geostrophic equations in Chapter 4, and the two-dimensional layered equations in Chapter 6. However, we remark that whereas in the two-dimensional layered equations in Chapter 6 there was no exchange mechanism among the layers, the equations for the potential vorticity (7.50)–(7.51) and (7.57)–(7.58) show that geostrophic balance allows for communication between layers through pressure variations in the vertical direction. In the remaining part of this chapter we will prove rigorously the convergence of the rotating Boussinesq equations to the quasi-geostrophic equations in the asymptotic limit of small Rossby and Froude numbers. The interested reader can consult the books of Gill and Pedlosky for interesting detailed applications of these equations to geophysical flows.

7.5. Rigorous Convergence of the Rotating Boussinesq Equations to the Quasi-Geostrophic Equations

Next we study rigorously the convergence of the rotating Boussinesq equations (7.43)–(7.46) to the quasi-geostrophic equations (7.59)–(7.62) in the asymptotic limit of small Rossby and Froude numbers as given by the assumptions in (7.42). For simplicity in the exposition we work with periodic boundary conditions, that is, we assume that the spatial domain is the three-dimensional torus $\mathbb{T}^3$.

THEOREM 7.3 *Consider the rotating Boussinesq equations for stably stratified flow for* $\vec{u}^\epsilon = (\vec{v}^\epsilon, \rho^\epsilon) = (\vec{v}_H^\epsilon, w^\epsilon, \rho^\epsilon)$ *and small Rossby and Froude numbers* $\mathrm{Fr} = F\mathrm{Ro}$, $\mathrm{Ro} = \epsilon \ll 1$,

(7.63)
$$\frac{D^H \vec{v}_H^\epsilon}{Dt} + w^\epsilon \frac{\partial \vec{v}_H^\epsilon}{\partial x_3} = -\epsilon^{-1}(\vec{v}_H^{\epsilon\perp} + \nabla_H \phi^\epsilon) \, ,$$

(7.64)
$$\frac{D^H w^\epsilon}{Dt} + w^\epsilon \frac{\partial w^\epsilon}{\partial x_3} = -\epsilon^{-1}\left(\frac{\partial \phi^\epsilon}{\partial x_3} + F^{-1} \rho^\epsilon \right) \, ,$$

$$(7.65) \qquad \frac{D\rho^\epsilon}{Dt} = \epsilon^{-1} F^{-1} w^\epsilon \,,$$

$$(7.66) \qquad \mathrm{div}_H \, \vec{v}_H^\epsilon + \frac{\partial w^\epsilon}{\partial x_3} = 0 \,,$$

where the initial data $\vec{u}_0 = (\vec{v}_{H0}, w_0, \rho_0)$ is given by

$$(7.67) \qquad \begin{aligned} \vec{v}_H^\epsilon(\vec{x}, t)\big|_{t=0} &= \vec{v}_{H0}(\vec{x}) + \epsilon \vec{v}_{H1}(\vec{x}) \,, \\ w^\epsilon(\vec{x}, t)\big|_{t=0} &= w_0(\vec{x}) + \epsilon w_1(\vec{x}) \,, \\ \rho^\epsilon(\vec{x}, t)\big|_{t=0} &= \rho_0(\vec{x}) + \epsilon \rho_1(\vec{x}) \,, \end{aligned}$$

with $\vec{v}_{H0}, \vec{v}_{H1}, w_0, w_1, \rho_0,$ and ρ_1 in $H^s(\mathbb{T}^3)$, $s \geq 4$ (so that $s > \frac{N}{2} + 2$), and where the velocity field satisfies initially the incompressibility constraint in (7.66)

$$(7.68) \qquad \mathrm{div}_H \, \vec{v}_{Hj} + \frac{\partial w_j}{\partial x_3} = 0 \quad \text{for } j = 0, 1 \,.$$

Then we have

(i) *If $\vec{u}_0$ is in $H^s(\mathbb{T}^3)$, then the classical solution $\vec{u}^\epsilon$ of equations (7.63)–(7.66) exists on a common time interval $[0, T]$ independent of ϵ and satisfies the uniform H^s estimate*

$$(7.69) \qquad \max_{0 \leq t \leq T} \left\{ \|\vec{u}^\epsilon(t)\|_s + \|\phi^\epsilon(t)\|_{s+1} \right\} \leq C \,,$$

where C is independent of ϵ.

(ii) *If the initial data $\vec{u}_0$ satisfies to order $O(\epsilon)$ the additional restrictions of geostrophic balance, zero vertical velocity, and zero density deviation from the hydrostatic balance state*

$$(7.70) \qquad \vec{v}_{H0} = \nabla_H^\perp \phi_0 \,, \qquad \frac{\partial \phi_0}{\partial x_3} = -F^{-1} \rho_0 \,, \qquad w_0(\vec{x}) \equiv 0 \,,$$

then the time derivative $\vec{u}_t^\epsilon$ of the solution $\vec{u}^\epsilon$ satisfies the uniform estimate

$$(7.71) \qquad \max_{0 \leq t \leq T} \left\{ \|\vec{u}_t^\epsilon(t)\|_1 + \|\phi_t^\epsilon\|_2 \right\} \leq C \,,$$

where C is independent of ϵ. Moreover, when the initial data satisfies the additional restrictions in (7.70), then there is a function $\vec{u}^{(0)} = (\vec{v}_H^{(0)}, w^{(0)}, \rho^{(0)})$ in $C([0, T], H^s(\mathbb{T}^3))$, and $\phi^{(0)}$ in $C([0, T], H^{s+1}(\mathbb{T}^3))$ so that

$$(7.72) \qquad \max_{0 \leq t \leq T} \left\{ \|\vec{u}^\epsilon(t) - \vec{u}^{(0)}(t)\|_{C^1} + \|\phi^\epsilon(t) - \phi^{(0)}(t)\|_{C^2} \right\} \to 0$$

as $\epsilon \to 0$, and $\vec{u}^{(0)}$ and $\phi^{(0)}$ solve the quasi-geostrophic equations

$$(7.73) \qquad \vec{v}_H^{(0)} = \nabla_H^\perp \phi^{(0)} \,,$$

$$(7.74) \qquad \frac{\partial \phi^{(0)}}{\partial x_3} = -F^{-1} \rho^{(0)} \,,$$

$$(7.75) \qquad w^{(0)}(\vec{x}, t) = 0 \,,$$

$$(7.76) \qquad \frac{D^H}{Dt} \left(\Delta_H \phi^{(0)} + F^2 \frac{\partial^2 \phi^{(0)}}{\partial x_3^2} \right) = 0 \,.$$

In particular, this theorem gives as a corollary a constructive proof of the existence of solutions for the quasi-geostrophic equations in (7.73)–(7.76).

COROLLARY 7.4 *If the initial data $\vec{u}_0$ is in $H^s(\mathbb{T}^3)$ with $s \geq 4$, and $\vec{u}_0$ satisfies the balance conditions in (7.70), then there is a classical solution of the quasi-geostrophic equations (7.73)–(7.76) for a finite time interval $[0, T]$.*

It is worthwhile to remark that global existence for all times of solutions of the stratified quasi-geostrophic equations with the same smooth initial data follows directly as in chapter 3 of [**19**] for two-dimensional fluid flow.

PROOF OF THEOREM 7.3: The proof of the theorem follows the same strategy presented earlier in the context of the rotating shallow water equations in Chapter 4. This strategy consists in the derivation of energy estimates for the solution $\vec{u}^\epsilon$, the pressure ϕ^ϵ, and their time derivatives $\vec{u}_t^\epsilon$, ϕ_t^ϵ, uniformly in ϵ, followed by a compactness argument to prove the convergence of the sequences $\{\vec{u}^\epsilon(t)\}$ and $\{\phi^\epsilon(t)\}$ in the limit of $\epsilon \to 0$. We refer the reader back to the discussion in Chapter 4 for some of the tools that we will, once again, use here such as Sobolev's lemma, Lions-Aubin compactness, etc. The new twist in this proof is the derivation of elliptic regularity estimates for the pressure that can be derived from the Poisson equation in (7.48). The basic regularity estimate for periodic solutions of the Poisson equation is given in Proposition 7.7 in the next section. The assumption that $s \geq \frac{N}{2} + 2$ in the theorem is done for convenience to avoid the use of Sobolev spaces with negative exponent in the derivation of the regularity estimates for the pressure. The theorem is still true with the more natural conditions $s > \frac{N}{2} + 1$, but the proof then requires the use of negative Sobolev spaces like $H^{-1}(\mathbb{T}^3)$.

Next we summarize the main steps involved in the proof of the theorem:

Step 1: Derivation of nonlinear estimates for the high-order spatial derivatives of the solution $\vec{u}^\epsilon$ and the pressure ϕ^ϵ; these estimates must be valid for a fixed time interval $[0, T]$ and must be uniform in ϵ. These high norm estimates, with $s \geq 3$, combine energy estimates for the solution $\vec{u}^\epsilon$ and elliptic regularity estimates for the pressure ϕ^ϵ.

Step 2: Derivation of nonlinear energy estimates for the time derivative of the solution $\vec{u}^\epsilon$ and the pressure ϕ^ϵ in a low-order Sobolev norm. In order to guarantee uniform estimates in ϵ over the time interval $[0, T]$, the initial data must satisfy the additional balance constraints in (7.70). The additional L^2 derivative required by the condition $s \geq 4$ is needed to avoid the use of Sobolev spaces with negative exponents in the derivation of the elliptic regularity estimates leading to (7.71).

Step 3: Passage to the limit by a compactness argument. Once we have derived uniform estimates for the solution $\vec{u}^\epsilon$ in H^s and ϕ^ϵ in H^{s+1}, and for their time derivatives $\vec{u}_t^\epsilon$ in H^1 and ϕ_t^ϵ in H^2, we utilize the Lions-Aubin compactness lemma to guarantee the convergence of the sequence $\{\vec{u}^\epsilon(t)\}$ and $\{\phi^\epsilon(t)\}$ as $\epsilon \to 0$.

Step 4: Identification of the limit problem. In this final step we show that the limit $\vec{u}^{(0)}$ satisfies the three-dimensional quasi-geostrophic equations.

The proof exploits the uniform estimates satisfied by the converging sequence with passage to the limit in the rotating Boussinesq equations (7.63)–(7.66) and in Ertel's theorem (7.47) to derive the transport equation for the potential vorticity.

$\square$

7.6. Preliminary Mathematical Considerations

Let us start with the derivation of the basic energy identity and energy estimate needed for the proof of the theorem. For that purpose we consider the *rotating Boussinesq equations with forcing*:

$$(7.77) \qquad \frac{D\vec{v}}{Dt} = -\epsilon^{-1}\nabla\phi - \epsilon^{-1}F^{-1}\rho\vec{e}_3 - \epsilon^{-1}\vec{v}_H^{\perp} + \vec{\mathcal{F}}_v \,,$$

$$(7.78) \qquad \frac{D\rho}{Dt} = \epsilon^{-1}F^{-1}w + \mathcal{F}_\rho \,,$$

$$(7.79) \qquad \operatorname{div} \vec{v} = 0 \,.$$

Recall the conservation of energy principle for the stratified Boussinesq equations from Chapter 2. Here we use a precise form of such principles to derive estimates in the limit as $\varepsilon \to 0$. We define the energy density $\mathcal{E}$ by

$$(7.80) \qquad \mathcal{E} = \frac{1}{2}\vec{v}\cdot\vec{v} + \frac{1}{2}\rho^2 \,;$$

then the conservation of energy for the rotating Boussinesq equations with forcing is given in differential form by the following:

LEMMA 7.5 (Differential Form of the Conservation of Energy) *If $\vec{u} = (\vec{v}, \rho)$ satisfies equations (7.77)–(7.79), then the energy density $\mathcal{E}$ satisfies the differential equation*

$$(7.81) \qquad \frac{\partial\mathcal{E}}{\partial t} = \operatorname{div}\left((\mathcal{E} + \epsilon^{-1}\phi)\vec{v}\right) + \vec{u}\cdot\vec{\mathcal{F}} \,,$$

where $\vec{\mathcal{F}} = (\vec{\mathcal{F}}_v, \mathcal{F}_\rho)$.

PROOF: Differentiate the energy density $\mathcal{E}$ and utilize the rotating Boussinesq equations with forcing (7.77)–(7.78) to get

$$\begin{aligned}
\frac{\partial \mathcal{E}}{\partial t} &= \vec{v} \cdot \frac{\partial \vec{v}}{\partial t} + \rho \frac{\partial \rho}{\partial t} \\
&= \vec{v} \cdot \left(-\vec{v} \cdot \nabla \vec{v} - \epsilon^{-1} \nabla \phi - \epsilon^{-1} F^{-1} \rho \vec{e}_3 - \epsilon^{-1} \vec{v}_H^{\perp} + \vec{\mathcal{F}}_v \right) \\
&\quad + \rho \left(-\vec{v} \cdot \nabla \rho + \epsilon^{-1} F^{-1} w + \mathcal{F}_\rho \right) \\
&= -\vec{v} \cdot \nabla \left(\frac{1}{2} |\vec{v}|^2 + \frac{1}{2} \rho^2 + \epsilon^{-1} \phi \right) + \epsilon^{-1} (F^{-1} w \rho - F^{-1} w \rho) \\
&\quad - \epsilon^{-1} \vec{v} \cdot \vec{v}_H^{\perp} + \vec{v} \cdot \vec{\mathcal{F}}_v + \rho \mathcal{F}_\rho \\
&= -\vec{v} \cdot \nabla (\mathcal{E} + \epsilon^{-1} \phi) + \vec{u} \cdot \vec{\mathcal{F}} .
\end{aligned} \tag{7.82}$$

In addition, the incompressibility condition for the velocity field $\vec{v}$ in (7.79) implies that for any function f we have $\operatorname{div}(f\vec{v}) = \vec{v} \cdot \nabla f$. In particular, (7.82) reduces to

$$\frac{\partial \mathcal{E}}{\partial t} = -\operatorname{div}\left((\mathcal{E} + \epsilon^{-1} \phi) \vec{v} \right) + \vec{u} \cdot \vec{\mathcal{F}}, \tag{7.83}$$

which is the differential form of the conservation of energy stated in (7.81). $\qquad\square$

The energy identity given in (7.81) is essentially the same energy identity derived in Chapter 2 for the Boussinesq equations without rotation (except for the factor ϵ^{-1} multiplying the pressure term). Note from (7.81) that the dangerous singular terms of order $O(\varepsilon^{-1})$ are a perfect divergence as regards the energy density and thus will not make a singular contribution to the change of total energy in time. Therefore the derivations of the integral form of the conservation of energy and the basic L^2 energy estimate mimic line by line the discussion given there. If we define the total energy $E(t)$ as the integral of the energy density $\mathcal{E}$,

$$E(t) = \int_{\mathbb{T}^3} \left(\frac{1}{2} \vec{v} \cdot \vec{v} + \frac{1}{2} \rho^2 \right) dx , \tag{7.84}$$

then the integral form of the conservation of energy for the rotating Boussinesq equations with forcing is as follows:

LEMMA 7.6 (Integral Form of the Conservation of Energy) *If $\vec{u} = (\vec{v}, \rho)$ satisfy equations (7.77)–(7.79), then the total energy $E(t)$ satisfies the differential equation*

$$\frac{dE}{dt} = \int_{\mathbb{T}^3} \vec{u} \cdot \vec{\mathcal{F}} \, dx , \quad \text{where } \vec{\mathcal{F}} = (\vec{\mathcal{F}}_v, \mathcal{F}_\rho) . \tag{7.85}$$

The basic energy estimate for the rotating Boussinesq equations with forcing is

PROPOSITION 7.7 (Energy Estimate) *If $\vec{u} = (\vec{v}, \rho)$ satisfy equations (7.77)–(7.79), then the total energy $E(t)$ satisfies the differential inequality*

$$(7.86) \qquad \frac{dE}{dt} \leq \|\vec{\mathcal{F}}\|_0 (E(t))^{1/2},$$

where $\| \cdot \|_0$ is the L^2 norm, $| \cdot |_{L^\infty}$ is the L^∞ norm, and $\vec{\mathcal{F}}$ is the vector $\vec{\mathcal{F}} = (\vec{\mathcal{F}}_v, \mathcal{F}_\rho)$.

Next we consider the derivation of elliptic regularity estimates for the pressure ϕ, where ϕ is given by a Poisson equation as in (7.48). Before going into this discussion, it is convenient to remind the reader about an alternative characterization of a function u in the Sobolev space $H^s(\mathbb{T}^N)$ in terms of its Fourier coefficients $\hat{u}_{\vec{k}}$. The next lemma quantifies in a precise way the well-known fact that the smoother a periodic function is, the faster its Fourier coefficients decay.

LEMMA 7.8 (Equivalent Norm for $H^s(\mathbb{T}^N)$) *A function $u(\vec{x})$ is in $H^s(\mathbb{T}^N)$ if and only if*

$$(7.87) \qquad |u|_s^2 = \sum_{\vec{k} \in \mathbb{Z}^N} \left(1 + |2\pi\vec{k}|^2\right)^s |\hat{u}_{\vec{k}}|^2 < \infty,$$

where $\{\hat{u}_{\vec{k}}\}$ are the Fourier coefficients of u. Moreover, the norms $| \cdot |_s$ and $\| \cdot \|_s$ are equivalent,

$$(7.88) \qquad C^{-1}|u|_s \leq \|u\|_s \leq C|u|_s, \quad \text{where } C > 1 \text{ is a constant.}$$

PROOF: It is enough to prove the equivalence of the two norms in (7.88). Applying Plancherel's theorem to the definition of the Sobolev norm $\|u\|_s$ yields

$$(7.89) \qquad \|u\|_s^2 = \sum_{|\alpha| \leq s} \int_{\mathbb{T}^3} |D^\alpha u|^2 \, dx = \sum_{\vec{k} \in \mathbb{Z}^N} \left(\sum_{|\alpha| \leq s} |(2\pi\vec{k})^\alpha|^2 \right) |\hat{u}_{\vec{k}}|^2,$$

where $\vec{k}^\alpha = k_1^{\alpha_1} k_2^{\alpha_2} \cdots k_N^{\alpha_N}$. Let $P_s(\vec{k})$ be the polynomial

$$(7.90) \qquad P_s(\vec{k}) = \sum_{|\alpha| \leq s} \left|(2\pi\vec{k})^\alpha\right|^2.$$

Since the polynomial P_s is of degree $2s$, it is clear that it can be bounded above and below by the polynomial $(1 + |2\pi\vec{k}|^2)^s$,

$$(7.91) \quad C^{-1}(1 + |2\pi\vec{k}|^2)^s \leq P_s(\vec{k}) \leq C(1 + |2\pi\vec{k}|^2)^s \quad \text{with } C > 1 \text{ a constant.}$$

Introducing the inequalities in (7.91) back into equation (7.89) yields

$$(7.92) \qquad \begin{aligned} C^{-1} \sum_{\vec{k} \in \mathbb{Z}^N} \left(1 + |2\pi\vec{k}|^2\right)^s |\hat{u}_{\vec{k}}|^2 &\leq \sum_{\vec{k} \in \mathbb{Z}^N} P_s(\vec{k}) |\hat{u}_{\vec{k}}|^2 \\ &\leq C \sum_{\vec{k} \in \mathbb{Z}^N} \left(1 + |2\pi\vec{k}|^2\right)^s |\hat{u}_{\vec{k}}|^2, \end{aligned}$$

and this proves the equivalence of the norms $| \cdot |_s$ and $\| \cdot \|_s$ in (7.78). $\qquad \square$

LEMMA 7.9 (Elliptic Regularity Estimate for Poisson's Equation) *Let F be a periodic function with zero average, $\langle F \rangle = 0$. Then the unique solution ϕ of the Poisson equation*

$$(7.93) \qquad -\Delta\phi = F, \quad \langle \phi \rangle = 0,$$

satisfies the elliptic regularity estimate

$$(7.94) \qquad \|\phi\|_{s+1} \le C\|F\|_{s-1}, \quad \text{where } C \text{ is a constant.}$$

REMARK. The restriction $\langle F \rangle = 0$ for the inhomogeneous term F is the necessary solvability condition for the Poisson equation in (7.93). It turns out that the solution of the Poisson equation in (7.93) with periodic boundary conditions is only unique up to an additive constant. In order to single out a solution we add the zero-average restriction in (7.93). The restriction of having the average of a function u equal to zero is equivalent to the requirement that the Fourier coefficient $\hat{u}_0 = 0$.

PROOF: The proof is done by direct computation of the solution ϕ by expansion in Fourier series, followed by the estimate of $\|\phi\|_{s+1}$ in Fourier space with the help of the equivalent norm $|\cdot|_{s+1}$ in Lemma 7.8. If we expand both ϕ and F in Fourier series, substitute the resulting expansions into the Poisson equation in (7.93), and solve for the Fourier coefficients $\hat{\phi}_{\vec{k}}$ of ϕ, we obtain

$$(7.95) \qquad |2\pi\vec{k}|^2\hat{\phi}_{\vec{k}} = \widehat{F}_{\vec{k}}$$

for every Fourier mode. Equation (7.95) is readily solved for $\hat{\phi}_{\vec{k}}$ when $\vec{k} \ne 0$. When $\vec{k} = 0$, equation (7.95) cannot be solved unless F satisfies the solvability condition $\widehat{F}_0 = \langle F \rangle = 0$. In this case any value of $\hat{\phi}_0$ is a solution of (7.95) and we enforce uniqueness by requiring that $\hat{\phi}_0 = \langle \phi \rangle = 0$. Then the solution of the Poisson equation with periodic boundary conditions and zero average in (7.95) is given by

$$(7.96) \qquad \phi(\vec{x}) = \sum_{\vec{k}\ne 0} |2\pi\vec{k}|^{-2}\widehat{F}_{\vec{k}}e^{2\pi i\vec{k}\cdot\vec{x}},$$

and we can estimate the norm $\|\phi\|_{s+1}$ with the help of Lemma 7.8

$$(7.97) \qquad \|\phi\|_{s+1} \le C\sum_{\vec{k}\ne 0} \left(1 + |2\pi\vec{k}|^2\right)^{s+1}|2\pi\vec{k}|^{-4}|\widehat{F}_{\vec{k}}|^2.$$

Next, the bound $(1 + |2\pi\vec{k}|^2)^2 \le C|2\pi\vec{k}|^4$ is valid for $\vec{k} \in \mathbb{Z}^N$, and $\vec{k} \ne 0$. Introducing this bound back into the inequality in (7.97) and utilizing again Lemma 7.8, we obtain

$$(7.98) \qquad \|\phi\|_{s+1} \le C\sum_{\vec{k}\ne 0} \left(1 + |2\pi\vec{k}|^2\right)^{s-1}|\widehat{F}_{\vec{k}}|^2 \le \|F\|_{s-1}$$

and this concludes the proof of the lemma. $\qquad \qquad \square$

7.7. Proof of the Convergence Theorem

Next we provide the rigorous proof of the convergence of the solution of the rotating Boussinesq equations to the solution of the quasi-geostrophic equations in the limit of small Rossby and Froude numbers. The proof follows the strategy outlined in Section 7.5. We will be somewhat terse in those parts of the proof that are similar to the previous proof of convergence discussed in Chapter 4 for rotating shallow water. Instead we will concentrate greater attention on the new issues in the proof, for example, the derivation of the elliptic estimates for the pressure.

Step 1. *Derivation of energy estimates for the higher-order spatial derivatives of the solution $\vec{u}^\epsilon$ and the pressures ϕ^ϵ.*

To derive energy estimates for the derivative $D^\alpha \vec{u}^\epsilon$ for $|\alpha| \leq s$, we take the D^α space derivative of the rotating Boussinesq equations (7.63)–(7.66) and rearrange terms to obtain

$$(7.99) \quad \frac{D}{Dt}(D^\alpha \vec{u}^\epsilon) = -\epsilon^{-1}\nabla(D^\alpha \phi^\epsilon) - \epsilon^{-1}F^{-1}(D^\alpha \rho^\epsilon)\vec{e}_3 - \epsilon^{-1}(D^\alpha \vec{v})^\perp_H + \vec{\mathcal{F}}^\alpha_v,$$

$$(7.100) \quad \frac{D}{Dt}(D^\alpha \rho^\epsilon) = \epsilon^{-1}F^{-1}(D^\alpha w^\epsilon) + \mathcal{F}^\alpha_\rho,$$

$$(7.101) \quad \text{div}(D^\alpha \vec{v}^\epsilon) = 0,$$

where the forcing function $\vec{\mathcal{F}}^\alpha = (\vec{\mathcal{F}}^\alpha_v, \mathcal{F}^\alpha_\rho)$ consists of commutator terms involving the advection operator $\vec{v} \cdot \nabla$ and D^α,

$$(7.102) \quad \begin{aligned} \vec{\mathcal{F}}^\alpha_v &= \vec{v}^\epsilon \cdot \nabla(D^\alpha \vec{v}^\epsilon) - D^\alpha(\vec{v}^\epsilon \cdot \nabla \vec{v}^\epsilon), \\ \vec{\mathcal{F}}^\alpha_v &= \vec{v}^\epsilon \cdot \nabla(D^\alpha \rho^\epsilon) - D^\alpha(\vec{v}^\epsilon \cdot \nabla \rho^\epsilon). \end{aligned}$$

Applying the energy estimate in Proposition 7.7 to the function $D^\alpha \vec{u}^\epsilon$, followed by the estimate of the forcing term $\vec{\mathcal{F}}^\alpha$ with the sharp calculus inequality, we obtain, in the same fashion as done previously in Chapter 4, the energy estimate

$$(7.103) \quad \frac{d}{dt}\|\vec{u}^\epsilon\|_s \leq C_s |\nabla \vec{u}^\epsilon|_{L^\infty}\|\vec{u}^\epsilon\|_s,$$

where C_s is a constant independent of ϵ. Also proceeding in the same fashion as we did in Chapter 4, we can utilize the a priori estimate in (7.103) to extend the solution $\vec{u}^\epsilon$ to a fixed time interval $[0, T]$, uniformly in ϵ, and we can also derive for that time interval the a priori estimate

$$(7.104) \quad \max_{0 \leq t \leq T}\|\vec{u}^\epsilon(t)\|_s \leq C,$$

where C is independent of ϵ. Next we concentrate on the derivation of the elliptic regularity estimate for the pressure ϕ^ϵ. We know that the pressure ϕ^ϵ can be obtained as the solution of the Poisson equation in (7.48),

$$(7.105) \quad -\Delta\phi^\epsilon = \epsilon(\nabla \vec{v}^\epsilon) : (^T\nabla \vec{v}^\epsilon) + F^{-1}\frac{\partial \rho^\epsilon}{\partial x_3} - \omega^\epsilon_3, \quad \langle \phi^\epsilon \rangle = 0.$$

To be able to apply the elliptic estimate in (7.94), we must verify that the right side of (7.105) has zero average. The second and third terms in the right of (7.105) are perfect derivatives, so their average over the period cell is zero. The first term can be rewritten as $(\nabla \vec{v}^\epsilon) : ({}^{\mathrm{T}}\nabla \vec{v}^\epsilon) = \mathrm{div}(\vec{v}^\epsilon \cdot \nabla \vec{v}^\epsilon)$, so that the first term is also a perfect derivative and has zero average. Therefore the elliptic regularity estimate (7.94) in Lemma 7.9 yields

$$
(7.106) \quad \|\phi^\epsilon\|_{s+1} \leq C \left\| \epsilon(\nabla \vec{v}^\epsilon) : ({}^{\mathrm{T}}\nabla \vec{v}^\epsilon) + F^{-1}\frac{\partial \rho^\epsilon}{\partial x_3} - \omega_3^\epsilon \right\|_{s-1}
$$
$$
\leq C(1 + \epsilon \|\vec{u}^\epsilon\|_s)\|\vec{u}^\epsilon\|_s \,,
$$

where we estimated the right side of (7.106) with the help of the calculus inequalities for functions in Sobolev spaces $H^s(\mathbb{T}^N)$ with $s > \frac{N}{2} + 1$. Combining the estimate in (7.106) for ϕ^ϵ with the estimate in (7.104) for the solution $\vec{u}^\epsilon$ over the time interval $[0, T]$ yields

$$
(7.107) \quad \max_{0 \leq t \leq T} \|\phi^\epsilon(t)\|_{s+1} \leq C \,,
$$

and collecting the estimates for $\vec{u}^\epsilon$ and ϕ^ϵ in (7.104) and (7.107) finally yields the estimate in (7.69),

$$
(7.108) \quad \max_{0 \leq t \leq T} \left\{ \|\vec{u}^\epsilon(t)\|_s + \|\phi^\epsilon(t)\|_{s+1} \right\} \leq C \,,
$$

where C is a constant independent of ϵ.

Step 2. *Derivation of energy estimates for a low-order Sobolev norm of the time derivative of the solution $\vec{u}_t^\epsilon$ and the time derivative of the pressure ϕ_t^ϵ.*

First we show that if the initial data $\vec{u}^\epsilon(0)$ and the initial pressure $\phi^\epsilon(0)$ are in balance as required by the equations in (7.70), then the time derivative $\vec{u}_t^\epsilon$ is bounded initially, uniformly in ϵ. In fact, the rotating Boussinesq equations (7.63)–(7.65) imply that initially the time derivative $\vec{u}_t^\epsilon$ satisfies

$$
(7.109) \quad \vec{v}_t^\epsilon(0) = -\vec{v}_0 \cdot \nabla \vec{u}_0 - \epsilon^{-1}\nabla \phi_0 - \epsilon^{-1}F^{-1}\rho_0 \vec{e}_3 - \epsilon^{-1}\vec{v}_{H0}^\perp \,,
$$
$$
(7.110) \quad \rho_t^\epsilon(0) = -\vec{v}_0 \cdot \nabla \rho_0 + \epsilon^{-1}F^{-1}w_0 \,,
$$

and from the balance conditions for the initial data in (7.70) it follows that all the contributions of order ϵ^{-1} in equations (7.109) and (7.110) cancel. Therefore equations (7.109) and (7.110) reduce to

$$
(7.111) \quad \vec{u}_t^\epsilon(0) = -\vec{u}_0 \cdot \nabla \vec{u}_0 \,,
$$

and therefore the time derivative $\vec{u}_t^\epsilon$ of the solution is initially bounded in $H^1(\mathbb{T}^3)$

$$
(7.112) \quad \|\vec{u}_t^\epsilon(0)\|_1 = \|\vec{u}_0 \cdot \nabla \vec{u}_0\|_1 \leq C\|\vec{u}_0\|_s^2 \,.
$$

Next we derive the energy estimate for the time derivative of the solution $\vec{u}_t^\epsilon$ in the Sobolev space $H^1(\mathbb{T}^3)$. First we differentiate the rotating Boussinesq equations

(7.63)–(7.66) with respect to time and obtain

$$
(7.113) \qquad \frac{D\vec{v}_t^\epsilon}{Dt} = -\epsilon^{-1}\nabla\phi_t^\epsilon - \epsilon^{-1}F^{-1}\rho_t^\epsilon \vec{e}_3 - \epsilon^{-1}\vec{v}_{tH}^{\epsilon\perp} + \vec{\mathcal{F}}_v \,,
$$

$$
(7.114) \qquad \frac{D\rho_t^\epsilon}{Dt} = \epsilon^{-1}F^{-1}w_t^\epsilon + \mathcal{F}_\rho \,,
$$

$$
(7.115) \qquad \operatorname{div} \vec{v}_t^\epsilon = 0 \,,
$$

where the forcing term $\vec{\mathcal{F}} = (\vec{\mathcal{F}}_v, \mathcal{F}_\rho)$ is given by

$$
(7.116) \qquad \vec{\mathcal{F}} = \vec{v}^\epsilon \cdot \nabla \vec{u}_t^\epsilon - (\vec{v}^\epsilon \cdot \nabla \vec{u}^\epsilon)_t = -\vec{v}_t^\epsilon \cdot \nabla \vec{u}^\epsilon \,.
$$

Then the energy estimate in Proposition 7.7 (7.86) applied to $\vec{u}_t^\epsilon$ yields

$$
(7.117) \qquad \frac{d}{dt}\|\vec{u}_t^\epsilon\|_0 \le C|\nabla\vec{u}^\epsilon|_{L^\infty}\|\vec{u}_t^\epsilon\|_0 \,.
$$

Next we derive energy estimates for the mixed space-time derivative $\vec{u}_{tx_i}^\epsilon$ of the solution $\vec{u}^\epsilon$. Differentiating the rotating Boussinesq equations (7.63)–(7.66) with respect to t and x_i yields

$$
(7.118) \qquad \frac{D\vec{u}_{tx_i}^\epsilon}{Dt} = -\epsilon^{-1}\nabla\phi_{tx_i}^\epsilon - \epsilon^{-1}F^{-1}\rho_{tx_i}^\epsilon \vec{e}_3 - \epsilon^{-1}(\vec{v}_{tx_i}^\epsilon)_H^\perp + \vec{\mathcal{F}}_v \,,
$$

$$
(7.119) \qquad \frac{D\rho_{tx_i}^\epsilon}{Dt} = e^{-1}F^{-1}w_{tx_i}^\epsilon + \mathcal{F}_\rho \,,
$$

$$
(7.120) \qquad \operatorname{div} \vec{v}_{tx_i}^\epsilon = 0 \,,
$$

where the forcing term $\vec{\mathcal{F}} = (\vec{\mathcal{F}}_v, \mathcal{F}_\rho)$ is given by

$$
(7.121) \quad \vec{\mathcal{F}} = \vec{v}^\epsilon \cdot \nabla \vec{u}_{tx_i}^\epsilon - (\vec{v}^\epsilon \cdot \nabla \vec{u}^\epsilon)_{tx_i} = -\vec{v}_{tx_i}^\epsilon \cdot \nabla \vec{u}^\epsilon - \vec{v}_t^\epsilon \cdot \nabla \vec{u}_{x_i}^\epsilon - \vec{v}_{x_i}^\epsilon \cdot \nabla \vec{u}_t^\epsilon \,.
$$

Applying the energy estimate in (7.86) to $\vec{u}_{tx_i}^\epsilon$ in equations (7.118)–(7.120), followed with the calculus inequalities estimates of the forcing term $\vec{\mathcal{F}}$ in (7.121) yields

$$
(7.122) \qquad \frac{d}{dt}\|\vec{u}_{tx_i}^\epsilon\|_0 \le C\|\nabla\vec{u}^\epsilon\|_{C^1}\|\vec{u}_t^\epsilon\|_1 \le C\|\vec{u}^\epsilon\|_s\|\vec{u}_t^\epsilon\|_1 \,,
$$

where the last inequality follows from Sobolev's lemma with $s > \frac{N}{2}+2$. Collecting the energy estimates in (7.117) and (7.122) we obtain

$$
(7.123) \qquad \frac{d}{dt}\|\vec{u}_t^\epsilon\|_1 \le C\|\vec{u}^\epsilon\|_s\|\vec{u}_t^\epsilon\|_1 \,.
$$

Integrating the differential inequality in (7.123) and utilizing the uniform bound for $\vec{u}^\epsilon$ already derived in (7.108), we conclude that

$$
(7.124) \qquad \max_{0 \le t \le T} \|\vec{u}_t^\epsilon(t)\|_1 \le C \quad \text{uniformly in } \epsilon.
$$

Next we derive the bound for the time derivative of the pressure ϕ_t^ϵ. To derive the Poisson equation satisfied by the time derivative of the pressure ϕ_t^ϵ, we first apply the divergence operator to equation (7.113) for the time derivative of the velocity $\vec{v}_t^\epsilon$, with $\vec{\mathcal{F}}$ given by (7.116), and obtain

$$(7.125) \quad \text{div } \vec{v}_t^\epsilon + \text{div}(\vec{v}^\epsilon \cdot \nabla \vec{v}_t^\epsilon) =$$
$$- \epsilon^{-1} \Delta \phi_t^\epsilon - \epsilon^{-1} F^{-1} \frac{\partial \rho_t^\epsilon}{\partial x_3} - \epsilon^{-1} \omega_{3t}^\epsilon - \text{div}(\vec{v}_t^\epsilon \cdot \nabla \vec{v}^\epsilon),$$

and since both $\vec{v}^\epsilon$ and $\vec{v}_t^\epsilon$ are divergence free, then equation (7.125) reduces to

$$(7.126) \quad -\Delta \phi_t^\epsilon = \epsilon(\nabla \vec{v}^\epsilon) : ({}^\mathsf{T}\nabla \vec{v}_t^\epsilon) + \epsilon(\nabla \vec{v}_t^\epsilon) : ({}^\mathsf{T}\nabla \vec{v}^\epsilon) + F^{-1} \frac{\partial \rho_t^\epsilon}{\partial x_3} - \omega_{3t}^\epsilon.$$

Since the pressure ϕ^ϵ has zero average in (7.105), then we differentiate in time and conclude that the time derivative ϕ_t^ϵ also has zero average. In addition, the right side of (7.126) is a perfect divergence because it was obtained from (7.113) by applying the divergence operator. Therefore we can apply the elliptic regularity estimate in Lemma 7.9 (7.94) to the Poisson equation (7.126) for the time derivative of the pressure ϕ_t^ϵ and obtain

$$(7.127) \qquad \|\phi_t^\epsilon\|_2 \leq C(1 + \epsilon |\nabla \vec{v}^\epsilon|_{L^\infty}) \|\vec{u}_t^\epsilon\|_1.$$

The elliptic estimate for ϕ_t^ϵ in (7.127) and the energy estimate in (7.124) for $\vec{u}_t^\epsilon$ imply that

$$(7.128) \qquad \max_{0 \leq t \leq T} \|\phi_t^\epsilon(t)\|_2 \leq C,$$

and combining the estimates obtained in (7.124) and (7.128), we obtain the uniform estimate in (7.71) for $\vec{u}_t^\epsilon$ and ϕ_t^ϵ.

Step 3. *Passage to the limit by a compactness argument.*

In steps 1 and 2 we establish the validity of the estimates in equations (7.69) and (7.71) of the theorem. These estimates imply that $\{\vec{u}^\epsilon\}$ is a bounded sequence in $C([0, T], H^s(\mathbb{T}^3))$, $\{\vec{u}_t^\epsilon\}$ is a bounded sequence in $C([0, T], H^1(\mathbb{T}^3))$, $\{\phi^\epsilon\}$ is a bounded sequence in $C([0, T], H^{s+1}(\mathbb{T}^3))$, and $\{\phi_t^\epsilon\}$ is a bounded sequence in $C([0, T], H^2(\mathbb{T}^3))$. By the Lions-Aubin compactness lemma we know that there is a subsequence $\{\vec{u}^\epsilon\}$, $\{\phi^\epsilon\}$, so that $\vec{u}^\epsilon$ converges to $\vec{u}^{(0)}$ in $C([0, T], H^r(\mathbb{T}^3))$ for $0 \leq r < s$, and ϕ^ϵ converges to $\phi^{(0)}$ in $C([0, T], H^r(\mathbb{T}^3))$ for $0 \leq r < s + 1$. In particular, since $\frac{N}{2} + 2 < s$, we can pick r so that $\frac{N}{2} + 2 < r < s$ and utilize Sobolev's lemma to conclude that

$$(7.129) \quad \max_{0 \leq t \leq T} \left\{ \|\vec{u}^\epsilon(t) - \vec{u}^{(0)}(t)\|_{C^1} + \|\phi^\epsilon(t) - \phi^{(0)}(t)\|_{C^2} \right\} \leq$$
$$C \max_{0 \leq t \leq T} \left\{ \|\vec{u}^\epsilon(t) - \vec{u}^{(0)}(t)\|_r + \|\phi^\epsilon(t) - \phi^{(0)}(t)\|_r \right\} \to 0 \quad \text{as } \epsilon \to 0.$$

This proves the convergence statement in (7.72). Finally, we remark that although the Lions-Aubin lemma only gives a convergent subsequence, one can prove that the sequence itself is convergent because the solution of the quasi-geostrophic equations (7.73)–(7.76) is unique.

Step 4. *Identification of the limit problem.*

In this last step of the proof we want to show that the limits $\vec{u}^{(0)}$ and $\phi^{(0)}$ satisfy the quasi-geostrophic equations. First we establish the quasi-geostrophic balance condition in equation (7.73). From (7.63) it follows that

$$(7.130) \qquad \epsilon \left\| \frac{D\vec{v}_H^\epsilon}{Dt} \right\|_0 = \| \vec{v}_H^{\epsilon\perp} + \nabla_H \phi^\epsilon \|_0 .$$

From equations (7.69) and (7.71) it follows that the term $\| D\vec{v}_H^\epsilon / Dt \|_0$ in the left of (7.130) is bounded. Therefore, if we let $\epsilon \to 0$ in (7.130), and utilize the fact in (7.72) that $\vec{u}^\epsilon \to \vec{u}^{(0)}$ and $\phi^\epsilon \to \phi^{(0)}$ as $\epsilon \to 0$, it follows that

$$(7.131) \qquad 0 = \vec{v}_H^{(0)\perp} + \nabla_H \phi^{(0)} ,$$

which proves the quasi-geostrophic balance of the horizontal velocity and the horizontal gradient of the pressure in (7.73).

Next we establish the hydrostatic balance condition in (7.74). Equation (7.64) for the vertical component of momentum yields

$$(7.132) \qquad \epsilon \left\| \frac{Dw^\epsilon}{Dt} \right\|_0 = \left\| \frac{\partial \phi^\epsilon}{\partial x_3} + F^{-1} \rho^\epsilon \right\|_0 .$$

If we let $\epsilon \to 0$ as before, the left side again converges to zero whereas the right side is convergent, so that we obtain

$$(7.133) \qquad 0 = \frac{\partial \phi^{(0)}}{\partial x_3} + F^{-1} \rho^{(0)} .$$

This proves the hydrostatic balance condition between the density and the vertical gradient of the pressure in (7.73).

Next we prove that there is no vertical component in the velocity. From the density equation in (7.65) it follows that

$$(7.134) \qquad \epsilon \left\| \frac{D\rho^\epsilon}{Dt} \right\|_0 = \| F^{-1} w^\epsilon \|_0 .$$

Again, if we let $\epsilon \to 0$, the left side of equation (7.134) converges to zero, so that (7.134) yields

$$(7.135) \qquad w^{(0)} = 0 ,$$

and this proves the zero vertical velocity condition in (7.75).

The final step is to derive (7.76) for the potential vorticity. First we utilize the Poisson equation for the pressure to relate the vorticity and the pressure in the limit of $\epsilon \to 0$. The Poisson equation for the pressure in (7.48) is

$$(7.136) \qquad -\left(\Delta_H \phi^\epsilon + \frac{\partial^2 \phi^\epsilon}{\partial x_3{}^2} \right) = \epsilon (\nabla \vec{v}^\epsilon) : ({}^T \nabla \vec{v}^\epsilon) + F^{-1} \frac{\partial \rho^\epsilon}{\partial x_3} - \omega_3^\epsilon ,$$

and if we let $\epsilon \to 0$, then (7.136) reduces to

$$(7.137) \qquad -\left(\Delta_H \phi^{(0)} + \frac{\partial^2 \phi^{(0)}}{\partial x_3{}^2} \right) = F^{-1} \frac{\partial \rho^{(0)}}{\partial x_3} - \omega_3^{(0)} .$$

Differentiating the hydrostatic balance equation (7.133) in the vertical direction yields

$$\frac{\partial \rho^{(0)}}{\partial x_3} = -F \frac{\partial^2 \phi^{(0)}}{\partial x_3{}^2},$$
(7.138)

and if we introduce equation (7.138) into (7.137), then equation (7.137) reduces to

$$\omega_3^{(0)} = \Delta_H \phi^{(0)}.$$
(7.139)

The next step in the derivation of the equation for the potential vorticity is to let $\epsilon \to 0$ in the distribution formulation of Ertel's theorem. First we recall Ertel's theorem equation (7.47),

$$\frac{D}{Dt}\left\{ \omega_3^\epsilon - F \frac{\partial \rho^\epsilon}{\partial x_3} - \epsilon F \left(\vec{\omega}_H^\epsilon \cdot \nabla_H \rho^\epsilon + \omega_3^\epsilon \frac{\partial \rho^\epsilon}{\partial x_3} \right) \right\} = 0.$$
(7.140)

If $\psi(\vec{x}, t)$ is a test function in $C_0^\infty(\mathbb{T}^3 \times [0, T])$, then the distribution formulation of Ertel's theorem is

(7.141)
$$\int_0^\infty \int_{\mathbb{T}^3} (\psi_t + \vec{v}^\epsilon \cdot \nabla \psi) \left[\omega_3^\epsilon - F \frac{\partial \rho^\epsilon}{\partial x_3} \right.$$
$$\left. - \epsilon F \left(\vec{\omega}_H^\epsilon \cdot \nabla_H \rho^\epsilon + \omega_3^\epsilon \frac{\partial \rho^\epsilon}{\partial x_3} \right) \right] dx\, dt = 0.$$

By equation (7.72) we know that the integrand in (7.139) converges uniformly in the limit of $\epsilon \to 0$. Therefore we apply the dominated convergence theorem to conclude that in the limit of $\epsilon \to 0$, (7.141) reduces to

$$\int_0^\infty \int_{\mathbb{T}^3} (\psi_t + \vec{v}_H^{(0)} \cdot \nabla \psi) \left(\omega_3^{(0)} - F \frac{\partial \rho^{(0)}}{\partial x_3} \right) dx\, dt = 0,$$
(7.142)

and reversing the integrations by parts in (7.142), we deduce that the equation

$$\frac{D^H}{Dt} \left(\omega_3^{(0)} - F \frac{\partial \rho^{(0)}}{\partial x_3} \right) = 0$$
(7.143)

is satisfied in the distribution sense. Finally, introducing equations (7.138) and (7.139) into equation (7.143) yields

$$\frac{D^H}{Dt} \left(\Delta_H \phi^{(0)} + F^2 \frac{\partial^2 \phi^{(0)}}{\partial x_3{}^2} \right) = 0.$$
(7.144)

This establishes the equation for the transport of potential vorticity in (7.76) and concludes the proof of the theorem.

CHAPTER 8

Introduction to Averaging over Fast Waves
for Geophysical Flows

8.1. Introduction

Many problems in geophysical flows involve the interaction of different fluid dynamic mechanisms that evolve over two different disparate time scales. In this chapter I discuss these phenomena for the rapidly rotating shallow water equations (RSWE) and rapidly rotating Boussinesq equations described extensively in Chapters 4 and 7. First, I motivate these issues for RSWE; then I develop an abstract, formal two-time averaging principle motivated by this example. Next, this principle is applied to a general family of ODEs to illustrate the general structure of the limit. Then some amusing applications of this principle are presented for the ODEs generated by the exact local solution procedure for the Boussinesq equations developed in Chapter 2. In the next section, the systematic theory of fast-wave averaging is applied to the quasi-geostrophic limit of RSWE with the motivation of understanding nonlinear Rossby adjustment (see Gill [**11**, chap. 7]) in a fundamental mathematical fashion. In the final section of this Chapter, extensive applications of these ideas for unbalanced initial data are mentioned for both the low Froude number and low Froude, low Rossby number limits for the Boussinesq equations discussed earlier in Chapters 6 and 7 for unbalanced initial data. Although formal calculations are often emphasized here, many of the results presented below are mathematically rigorous. In particular Section 8.6 contains an amusing rigorous application of "weak convergence" ideas for the Boussinesq equations.

The interested reader can consult papers by Pedro Embid and Andrew Majda [**6, 7, 8**] for more mathematical details as well as extensive references to other theoretical and more applied work in these areas. The important mathematical paper by S. Schochet [**34**] is especially recommended to the reader interested in the rigorous mathematical details. Many more details and results on the interaction of fast and slow dynamics for the rotating stably stratified Boussinesq equations (RSSBE) can be found in [**8**], which also contains extensive references to more applied work for the atmosphere and ocean that motivates these developments.

8.2. Motivation for Fast-Wave Averaging

Low Froude Number, Low Rossby Number Limit for the Rotating Shallow Water Equations. We consider the RSWE with the familiar quasi-geostrophic scaling motivated in Chapter 4,

$$\text{(8.1)} \qquad\qquad \text{Ro} = \varepsilon, \quad \text{Fr} = F^{1/2}\varepsilon, \quad \theta = F\varepsilon,$$

with fixed $F > 0$ and $\varepsilon \ll 1$. As in Chapter 4, by utilizing (8.1) together with the rescaling, $h_{\text{old}} = F^{-1/2}h_{\text{new}}$, we obtain the rapidly rotating shallow water equations

$$\text{(8.2)} \qquad \begin{aligned} & \frac{\partial \vec{v}}{\partial t} + \varepsilon^{-1}\vec{v}^{\perp} + \varepsilon^{-1}F^{-1/2}\nabla h + \vec{v} \cdot \nabla \vec{v} = 0, \\ & \frac{\partial h}{\partial t} + \varepsilon^{-1}F^{-1/2}\operatorname{div}\vec{v} + \vec{v} \cdot \nabla h + h\operatorname{div}\vec{v} = 0. \end{aligned}$$

The equations in (8.2) have the general

ABSTRACT FORM OF RAPIDLY RSWE:

$$\text{(8.3)} \qquad\qquad \frac{\partial \vec{u}}{\partial t} + \varepsilon^{-1}\mathcal{L}(\vec{u}) + \mathcal{B}(\vec{u}, \vec{u}) = 0$$

where

$$\text{(8.4)} \quad \vec{u} = {}^{\mathsf{T}}(\vec{v}, h), \quad \mathcal{L}(\vec{u}) = \begin{pmatrix} \vec{v}^{\perp} + F^{-1/2}\nabla h \\ F^{-1/2}\operatorname{div}\vec{v} \end{pmatrix}, \quad \mathcal{B}(\vec{u}, \vec{u}) = \begin{pmatrix} \vec{v} \cdot \nabla \vec{v} \\ \operatorname{div}(h\vec{v}) \end{pmatrix}.$$

The linear operator $\mathcal{L}(\vec{u})$ in (8.3) is skew-symmetric and has three imaginary eigenvalues for any wavenumber $\vec{k} = (k_1, k_2)$ given by $i\omega(\vec{k})$ where

$$\text{(8.5)} \qquad \begin{aligned} & \omega^0(\vec{k}) = 0 && \text{potential vortical mode} \\ & && \text{with geostrophic balance} \\ & \omega^{\pm}(\vec{k}) = \pm\left(1 + F^{-1}|\vec{k}|^2\right)^{1/2} && \text{inertio-gravity modes.} \end{aligned}$$

Thus, looking at (8.3), we see that the rapidly RSWE have slow-mode propagation on advective time scales given by the potential vortical mode and fast modes of propagation given by the inertio-gravity modes.

As presented earlier in Chapter 4, a standard formal power series expansion of (8.3) and Ertel's potential vorticity theorem yields the familiar formal leading-order behavior given by the

QUASI-GEOSTROPHIC EQUATIONS:

$$\text{(8.6)} \qquad\qquad \vec{v} = F^{-1/2}\nabla^{\perp}h, \quad q = \Delta h - Fh, \quad \frac{Dq}{Dt} = 0.$$

A rigorous mathematical justification of this limiting procedure for initial data in geostrophic balance, i.e., $\vec{v}_0 = F^{-1/2}\nabla^{\perp}h_0$, has been presented in Chapter 4.

What happens to solutions of rapidly RSWE as $\varepsilon \to 0$ for general unbalanced initial data? When the initial data contains both geostrophically balanced initial data and also nontrivial gravity waves, i.e., *general unbalanced initial data, nonlinear Rossby adjustment* is said to occur for $\varepsilon \ll 1$ if the gravity waves average or radiate away and the local dynamics is still governed by the quasi-geostrophic equations (see Gill, chap. 7). Typically this problem in the geophysical community is studied in unbounded domains where the gravity waves can radiate to infinity despite the fact that the earth is a compact region. One question that motivates the work to be presented in this chapter is the following:

QUESTION. The geostrophically balanced portion of a flow is more easily measured than the random bath of fast gravity waves. Does nonlinear Rossby adjustment occur in a suitable sense even for general unbalanced initial data? In other words, even for unbalanced initial data, do the potential vortical modes move independently of the noisy sea of gravity modes and still continue to satisfy the quasi-geostrophic equations in some averaged sense?

Surprisingly, the answers to these questions are yes in the quasi-geostrophic limit in periodic geometry. I remark here that utilizing spatially periodic initial data is a severe test since the gravity waves cannot scatter and radiate away to infinity as in the classical Rossby adjustment (Gill, chap. 7). Next, I discuss general principles of fast-wave averaging for the equations in (8.3).

8.3. A General Framework for Averaging over Fast Waves

The example of the limiting behavior of RSWE motivates a study of the limiting behavior as $\varepsilon \to 0$ of the abstract problem

$$(8.7) \qquad \frac{\partial \vec{u}^\epsilon}{\partial t} + \varepsilon^{-1} \mathcal{L}(\vec{u}^\epsilon) + B(\vec{u}^\epsilon, \vec{u}^\epsilon) = 0, \quad \vec{u}^\epsilon\big|_{t=0} = \vec{u}_0(x).$$

I assume that (8.7) is defined on a Hilbert space X with inner product $(\vec{u}, \vec{w})_X$. In the elementary applications below, the Hilbert space X is either Euclidean space $\mathbb{R}^N$, or $\mathbb{C}^N$ with the standard Euclidean inner product. In the more sophisticated applications presented below for RSWE and the rotating Boussinesq equations in this chapter, the space X consists of the vector-valued square-integrable spatial periodic functions with the L^2 inner product or X is the Sobolev space H^s utilized extensively earlier in Chapters 4 and 7.

The main requirement on the linear operator $\mathcal{L}$ is that $\mathcal{L}$ is skew-symmetric on the Hilbert space X, i.e.,

$$(8.8) \qquad (\mathcal{L}\vec{u}, \vec{w})_X = -(\vec{u}, \mathcal{L}\vec{w})_X,$$

for all vectors $\vec{u}$ and $\vec{w}$ in a suitable domain of definition. It is a nice exercise for the reader to check that the operator for RSWE,

$$(8.9) \qquad \mathcal{L}(\vec{u}) = \begin{pmatrix} \vec{v}^\perp + F^{-1/2}\nabla h \\ F^{-1/2} \operatorname{div} \vec{v} \end{pmatrix}$$

with $\vec{u} = {}^\mathsf{T}(\vec{v}, h)$, is skew-symmetric in the L^2 inner product

$$(\vec{u}, \vec{w})_0 = \int_{\mathbb{T}^2} \vec{u} \cdot \vec{w} \, dx.$$

The goal here is to derive systematically a simplified averaged equation for the limiting dynamics of (8.7) as $\varepsilon \to 0$, valid on a time interval $0 < t < T$ with T fixed. I utilize the method of multiple scales. Recall that this method was also used in the last two sections of Chapter 5 in discussing weakly nonlinear dispersive waves. I assume that the solution $\vec{u}^\epsilon(t)$ for (8.7) depends on the fast scale, $\tau = \frac{t}{\epsilon}$, and on the slow scale t, i.e., $\vec{u}^\epsilon(t) = \vec{u}(t, \frac{t}{\epsilon})$. (Note that the $\vec{x}$-dependence in RSWE is actually absorbed in the operator theoretic form in (8.7), which should be regarded here as a finite- or infinite-dimensional system of ODEs.) Thus, for $\varepsilon \ll 1$ the solution $\vec{u}^\epsilon$ has the formal expansion

$$(8.10) \qquad \vec{u}^\epsilon(t) = \vec{u}^0(t, \tau)\big|_{\tau=\frac{t}{\epsilon}} + \epsilon u^1(t, \tau)\big|_{\tau=t/\epsilon}.$$

In order to have a formally valid asymptotic expansion, we need to guarantee formally that the second term on the right-hand side of (8.11) is always weaker than the leader term, $\vec{u}^0(t, \tau)$. Thus, we require the sublinear growth condition for the fast variable,

$$(8.11) \qquad \left|\vec{u}^1(t, \tau)\right| = o(\tau) \quad \text{uniformly for } 0 \leq \tau \leq \frac{T}{\epsilon}.$$

Substituting the ansatz from (8.10) collecting powers of ϵ, at the order $O(\epsilon^{-1})$ we have

$$(8.12) \qquad \frac{\partial \vec{u}^0}{\partial \tau} + \mathcal{L}(\vec{u}^0) = 0$$

with the solution

$$(8.13) \qquad \vec{u}^0(t, \tau) = e^{-\mathcal{L}\tau}\bar{u}(t).$$

Note that in the leading-order solution of the asymptotic equation in (8.13), the fast scales and the slow scales are factored in a special form that we exploit below.

With the ansatz from (8.10) substituted into (8.7), the terms of order $O(\epsilon^0)$ vanish provided that

$$(8.14) \qquad \frac{\partial \vec{u}^1}{\partial \tau} + \mathcal{L}(\vec{u}^1) = -\left(\frac{\partial \vec{u}^0}{\partial t} + \mathcal{B}(\vec{u}^0, \vec{u}^0)\right)$$

where $\vec{u}^0$ has the form in (8.13). The solution of (8.14) is readily given by the Duhamel formula

$$(8.15) \qquad e^{\tau\mathcal{L}}\vec{u}^1 = \vec{u}^1(t, \tau)\big|_{t=0} - \tau\frac{\partial \bar{u}}{\partial t}(t) - \int_0^\tau e^{\mathcal{L}s}\mathcal{B}(e^{-\mathcal{L}s}\bar{u}, e^{-\mathcal{L}s}\bar{u})ds.$$

Since $\mathcal{L}$ is a skew-symmetric operator on X, the operator $e^{\tau\mathcal{L}}$ preserves the norm in X, so $e^{\tau\mathcal{L}}\vec{u}^1$ satisfies the sublinear growth condition in (8.11) if and only if $\vec{u}^1$ does. First, without loss of generality, one can set the initial data $\vec{u}^1(t, \tau)|_{\tau=t/\epsilon} = 0$. From the explicit formula in (8.15), we observe that (8.11) is satisfied provided that $\bar{u}(\vec{x}, t)$ from (8.13) satisfies the following:

AVERAGED EQUATION OVER FAST WAVES:

$$(8.16) \quad \frac{\partial \bar{u}}{\partial t}(t) + \lim_{\tau \to \infty} \frac{1}{\tau}\int_0^\tau e^{s\mathcal{L}}\mathcal{B}(e^{-s\mathcal{L}}\bar{u}(t), e^{-s\mathcal{L}}\bar{u}(t))ds = 0, \quad \bar{u}(t)\big|_{t=0} = \vec{u}_0.$$

This completes our formal asymptotic derivation of the averaged equations for the reduced dynamics. Thus, the leading-order behavior is given by

$$u^\epsilon = e^{-\mathcal{L}(\frac{t}{\epsilon})}\bar{u}(\vec{x}, t) + o(1)$$

where $\bar{u}$ is computed via (8.16).

Next, I summarize the structure of these equations for a general family of weakly nonlinear oscillators. Then I turn to some geophysical examples including RSWE.

EXAMPLE 8.1. **Finite-Dimensional Family of Weakly Nonlinear Oscillators.** We will work here with complex quantities. We are interested in a situation where a system has both very fast frequencies and some zero frequencies coupled through nonlinearity. For $\vec{u} \in \mathbb{C}^n$, we consider

$$(8.17) \qquad \frac{\partial \vec{u}}{\partial t} + \epsilon^{-1} \mathcal{L} \vec{u} + B(\vec{u}, \vec{u}) = 0 \,.$$

The nonlinear term is a matrix-valued bilinear operator. Let the system have m fast frequencies, and $n - m$ zero frequencies, and let the skew-symmetric linear operator be already diagonalized,

$$(8.18) \qquad \mathcal{L} = \begin{pmatrix} \mathcal{L}_1 & 0 \\ 0 & 0 \end{pmatrix} \quad \text{where} \quad \mathcal{L}_1 = \begin{pmatrix} \iota\omega_1 & 0 & \cdots & 0 \\ 0 & \omega_2 & \ddots & \vdots \\ \vdots & \ddots & \ddots & 0 \\ 0 & \cdots & 0 & \iota\omega_m \end{pmatrix} \,.$$

It is natural to divide the wave amplitudes into fast modes $\vec{u}_{\mathrm{I}}$ and slow modes $\vec{u}_{\mathrm{II}}$,

$$(8.19) \qquad \vec{u}_{\mathrm{I}} = \begin{pmatrix} u_1 \\ \vdots \\ u_m \\ 0 \\ \vdots \\ 0 \end{pmatrix}, \quad \vec{u}_{\mathrm{II}} = \begin{pmatrix} 0 \\ \vdots \\ 0 \\ u_{m+1} \\ \vdots \\ u_n \end{pmatrix} \,.$$

The general multiple-scale procedure described above yields the
AVERAGED LEADING-ORDER EQUATION:

$$(8.20) \qquad \frac{d\vec{u}}{dt} + \lim_{\tau \to \infty} \frac{1}{\tau} \int_0^\tau e^{\mathcal{L}s} B(e^{-\mathcal{L}s}\vec{u}, e^{-\mathcal{L}s}\vec{u}) ds = 0$$

where

$$\vec{u}^\epsilon = e^{-\frac{t}{\epsilon}\mathcal{L}}\vec{u}(t) + o(1) \,.$$

The next step is to see that the averaged equation has various resonance mechanisms in it.

Resonances in the Averaged Equation. We begin by breaking up the nonlinear term into components as it hits various combinations of the slow and fast waves. Let

$$B(\vec{u}, \vec{u}) = \begin{pmatrix} B_{11}^1(\vec{u}_{\mathrm{I}}, \vec{u}_{\mathrm{I}}) + B_{12}^1(\vec{u}_{\mathrm{I}}, \vec{u}_{\mathrm{II}}) + B_{22}^1(\vec{u}_{\mathrm{II}}, \vec{u}_{\mathrm{II}}) \\ B_{11}^2(\vec{u}_{\mathrm{I}}, \vec{u}_{\mathrm{I}}) + B_{12}^2(\vec{u}_{\mathrm{I}}, \vec{u}_{\mathrm{II}}) + B_{22}^2(\vec{u}_{\mathrm{II}}, \vec{u}_{\mathrm{II}}) \end{pmatrix} \,.$$

Thus the superscript B^1 denotes the projection of the quadratic nonlinear terms on the fast waves while B^2 denotes the projection on the slow waves. In B_{12}^1 and B_{12}^2 I have grouped together all quadratic cross terms involving $\vec{u}_{\mathrm{I}}$ and $\vec{u}_{\mathrm{II}}$. With the linear operator given in diagonalized form (8.18), let $\vec{u}^j$ be the jth component of

$\vec{u}$ and let $\vec{e}_j$ be the m-vector with 1 in the jth component and 0s elsewhere. The averaged leading-order equation looks like

$$
\frac{d\bar{u}^j}{dt} + \lim_{\tau \to \infty} \frac{1}{\tau} \int_0^\tau ds\, e^{\iota \omega_j s} \bigg\{ \vec{e}_j \cdot B_{11}^1 \bigg(\sum_{k=1}^m e^{-\iota \omega_k s} \bar{u}^k \vec{e}_k, \sum_{l=1}^m e^{-\iota \omega_l s} \bar{u}^l \vec{e}_m \bigg)
$$

$$
+ \vec{e}_j \cdot B_{12}^1 \bigg(\sum_{k=1}^m e^{-\iota \omega_k s} \bar{u}^k \vec{e}_l, \bar{u}_{\text{II}} \bigg) + \vec{e}_j \cdot B_{22}^1 (\bar{u}_{\text{II}}, \bar{u}_{\text{II}}) \bigg\} = 0 \,,
$$

(8.21)

$$
\frac{d\bar{u}_{11}}{dt} + \lim_{\tau \to \infty} \frac{1}{\tau} \int_0^\tau ds \bigg\{ B_{11}^2 \bigg(\sum_{k=1}^m e^{-\iota \omega_k s} \bar{u}^k \vec{e}_k, \sum_{l=1}^m e^{-\iota \omega_l s} \bar{u}^l \vec{e}_l \bigg)
$$

$$
+ \cdot B_{12}^2 \bigg(\sum_{k=1}^m e^{-\iota \omega_k s} \bar{u}^k \vec{e}_k, \bar{u}_{11} + \cdot B_{22}^2 (\bar{u}_{\text{II}}, \bar{u}_{\text{II}}) \bigg) \bigg\} = 0 \,.
$$

Expand the bilinear terms and let $B^{1j} = \vec{e}_j \cdot B^1$. The slow-scale terms can be taken outside the integral and the limit evaluated as

$$
\frac{d\bar{u}^j}{dt} + \sum_{k,l=1}^m B_{11}^{1j} (\bar{u}^k \vec{e}_k, \bar{u}^l \vec{e}_l) f(\omega_j - \omega_k - \omega_l)
$$

$$
+ \sum_{k=1}^m B_{12}^{1j} (\bar{u}^k \vec{e}_k, \bar{u}_{\text{II}}) f(\omega_j - \omega_k) + B_{22}^{1j} (\bar{u}_{\text{II}}, \bar{u}_{\text{II}}) f(\omega_j) = 0 \,,
$$

(8.22)

$$
\frac{d\bar{u}_{\text{II}}}{dt} + \sum_{k,l=1}^m B_{11}^2 (\bar{u}^k \vec{e}_k, \bar{u}^l \vec{e}_l) f(-\omega_k - \omega_l)
$$

$$
+ \sum_{k=1}^m B_{12}^2 (\bar{u}^k \vec{e}_k, \bar{u}_{\text{II}}) f(-\omega_k) + B_{22}^2 (\bar{u}_{\text{II}}, \bar{u}_{\text{II}}) = 0 \,.
$$

Here the function f contains the limit of the fast-time integral and is given by

(8.23)
$$
f(\Delta) = \lim_{\tau \to \infty} \frac{1}{\tau} \int_0^\tau ds\, e^{\iota \Delta s} = \begin{cases} 1 & \text{for } \Delta = 0 \\ 0 & \text{for } \Delta \neq 0 \,. \end{cases}
$$

In this limiting process, only those waves whose fast frequencies exactly cancel contribute on the slow time. This is the resonance process. Since each ω is nonzero, the terms B_{22}^1 and B_{12}^2 contribute zero. The resulting equations have the general form

$$
\text{fast waves:} \frac{d\bar{u}^j}{dt} + \sum_{\substack{k,l=1 \\ \omega_k + \omega_l = \omega_j}}^m \bar{u}^k \bar{u}^l B_{11}^{1j} (\vec{e}_k, \vec{e}_l) + \sum_{\substack{k=1 \\ \omega_k = \omega_j}}^m \bar{u}^k B_{12}^{1j} (\vec{e}_k, \bar{u}_{11}) = 0 \,,
$$

$$
j = 1, \ldots, m \,,
$$

(8.24)

$$
\text{slow waves:} \frac{d\bar{u}_{\text{II}}}{dt} + \sum_{\substack{k,l=1 \\ \omega_k + \omega_l = 0}}^m \bar{u}^k \bar{u}^l B_{11}^2 (\vec{e}_k, \vec{e}_l) + B_{22}^2 (\bar{u}_{\text{II}}, \bar{u}_{\text{II}}) = 0 \,.
$$

Note that similar resonance conditions appear as were discussed in the last two sections of Chapter 5. The B_{11}^1 term involves resonances among three fast waves where $\omega_k + \omega_l = \omega_j$. The B_{11}^2 and B_{12}^1 terms involve wave–mean flow resonance where either $\omega_k = \omega_j$ or $\omega_k + \omega_l = 0$.

When does the slow dynamics evolve independently from the fast dynamics for the ODE system considered here? A necessary and sufficient condition for this to occur is that the following interaction coefficients vanish:

$$(8.25) \qquad B_{11}^2(\vec{e}_k, \vec{e}_l) \equiv 0 \quad \text{whenever} \quad \omega_k + \omega_l = 0.$$

For physical reasons related to the conservation of potential vorticity, we show below that the geophysical examples from Chapters 4 and 7 satisfy (8.25). Next we give an example of applying this ODE averaging theorem directly to the elementary solutions of the stratified Boussinesq equations from Chapter 2. This example is motivated by an interesting physical issue related to the nature of the low Froude limit described in Chapter 6.

8.4. Elementary Analytic Models for Comparing Instabilities at Low Froude Numbers with the Low Froude Number Limit Dynamics

In Chapter 6 we discussed the limiting dynamics with strong stratification for the Boussinesq equations and found some remarkable behavior in elementary solutions of these limit equations. Here we utilize the elementary solutions developed in Chapter 2 together with the averaging principle developed above to address the following basic physical question: For laminar solutions of the equations for the low Froude number limiting dynamics, are there instabilities in these laminar solutions for the Boussinesq equations with $\varepsilon \ll 1$ that cannot be captured by the low Froude number limiting dynamics?

8.4.1. Elementary Exact Solutions of the Boussinesq Equations.
From Chapter 2, after a trivial Galilean transformation, standard kinematic formulas establish that every stratified fluid motion in the vicinity of the origin locally has the form

$$(8.26) \qquad \vec{v} = \mathcal{D}\vec{x} + \frac{1}{2}\vec{\omega} \times \vec{x} + O(|\vec{x}|^2), \quad \rho = \vec{b} \cdot \vec{x} + O(|\vec{x}|^2),$$

where $\omega = \text{curl } v$ is the vorticity, $\vec{b} = \nabla\rho$ is the density gradient, and $\mathcal{D}$ is the 3×3 symmetric deformation matrix, $\mathcal{D} = \frac{1}{2}(\nabla\vec{v} + {}^{\mathsf{T}}\nabla\vec{v})$, with vanishing trace, $\text{tr } \mathcal{D} = 0$. Thus, $\mathcal{D}$, $\vec{\omega}$, and $\vec{b}$ characterize the local structure of a stratified fluid flow. In Chapter 2 we established that there are elementary nonlinear exact solutions of the Boussinesq equations in (2.1) with the form described in (8.26) that also evolve self-consistently according to the Boussinesq equations. These exact solutions of (2.1) are given by

$$(8.27) \qquad \vec{v}(\vec{x}, t) = \mathcal{D}\vec{x} + \frac{1}{2}\vec{\omega}(t) \times \vec{x} \equiv V(t)\vec{x}, \quad \rho(\vec{x}, t) = \vec{b}(t) \cdot \vec{x},$$

for a prescribed constant, symmetric, traceless deformation matrix $\mathcal{D}$, where $\vec{\omega}(t)$ and $\vec{b}(t)$ satisfy the coupled 6×6 system of ODEs

(8.28)
$$\frac{d\vec{\omega}}{dt} = \mathcal{D}\vec{\omega} + \varepsilon^{-1} \begin{pmatrix} -b_2 \\ b_1 \\ 0 \end{pmatrix},$$

$$\frac{d\vec{b}}{dt} = -\mathcal{D}\vec{b} + \frac{1}{2}\vec{\omega} \times \vec{b} + \varepsilon^{-1}\left(\mathcal{D}\vec{e}_3 - \frac{1}{2}\vec{\omega} \times \vec{e}_3 \right),$$

with ε defined by the Froude number. In (8.28) we have used the low Froude number scaling discussed in Section 6.1 for these solutions. Here for simplicity in exposition, both viscosity and the effects of finite rotation are ignored. Besides these elementary laminar solutions as developed in Chapter 2, one can build a more general family of exact solutions for the Boussinesq equations by superimposing nonlinear plane waves on the laminar solutions in (8.27),

(8.29)
$$\vec{v}(\vec{x}, t) = \mathcal{D}\vec{x} + \frac{1}{2}\vec{\omega}(t) \times \vec{x} + \vec{A}(t)F(\vec{\alpha}(t) \cdot \vec{x}),$$

$$\rho(\vec{x}, t) = \vec{b}(t) \cdot \vec{x} + B(t)F(\vec{\alpha}(t) \cdot \vec{x}),$$

with $F(s)$ an arbitrary wave form. Solutions with the general structure in (8.29) satisfy the Boussinesq equations if, in addition to the equations for $\vec{\omega}(t)$ and $\vec{b}(t)$ in (8.28), the wave vector $\vec{\alpha}(t)$ and the amplitudes $\vec{A}(t)$ and $B(t)$ satisfy the ODEs

(8.30)
$$\frac{d\vec{\alpha}}{dt} = -{}^{\mathsf{T}}V(t)\vec{\alpha}(t),$$

$$\frac{d\vec{A}}{dt} = -V(t)\vec{A}(t) + \vec{\alpha}\frac{2({}^{\mathsf{T}}V(t)\vec{\alpha}(t) \cdot \vec{A}(t))}{|\vec{\alpha}|^2} + \varepsilon^{-1}\left(\frac{\alpha_3}{|\vec{\alpha}|^2}\vec{\alpha} - \vec{e}_3 \right)B(t),$$

$$\frac{dB}{dt} = -\vec{A}(t) \cdot (\vec{b} - \varepsilon^{-1}\vec{e}_3).$$

The detailed verification of all of these exact solutions can be found in Sections 2.5 and 2.6.

We regard the laminar solutions of the Boussinesq equations in (8.27) and (8.28) and their plane wave perturbations in (8.29) and (8.30) as a robust elementary context for quantitative comparison of laminar solutions for the low Froude number limiting dynamics and their instabilities compared with those in the Boussinesq solutions at finite small Froude numbers ε, with $\varepsilon \ll 1$. We shall see below that an interesting geometric criterion emerges in characterizing the instabilities that emerge in this process.

8.4.2. Low Froude Number Limiting Dynamics for the Laminar Elementary Solutions.
As we discussed in much more generality in Chapter 6, the low Froude number limiting dynamics suppresses all fast-wave motions to leading order in ε. For the special solutions in (8.27) and (8.28), this condition is automatically satisfied provided that

(8.31)
$$\mathcal{D}\vec{e}_3 - \frac{1}{2}\vec{\omega} \times \vec{e}_3 = 0, \quad b_1, b_2 = 0.$$

The first equation in (8.31) constrains both the horizontal vorticity and the deformation matrix $\mathcal{D} = (d_{ij})$ so that $d_{33} = 0$, where without loss of generality, $\mathcal{D}$ has the form

$$(8.32) \qquad \mathcal{D} = \begin{pmatrix} \gamma & 0 & d_{13} \\ 0 & -\gamma & d_{23} \\ d_{13} & d_{23} & 0 \end{pmatrix}.$$

Here the velocity field is purely horizontal and given by

$$(8.33) \qquad \vec{V}_H = \tilde{\gamma} z \vec{e}_H + \begin{pmatrix} \gamma(t) & -\frac{\bar{\omega}_0}{2} \\ \frac{\bar{\omega}_0}{2} & -\gamma(t) \end{pmatrix} \begin{pmatrix} x \\ y \end{pmatrix}$$

with $\tilde{\gamma}\vec{e}_H = 2^{\top}(d_{13}, d_{23})$ and the constant vertical vorticity ω_0. *The steady flows in* (8.33) *are the special exact solutions for the laminar low Froude number limiting dynamics* from Chapter 6. As an independent confirmation, we invite the reader to verify that the flows in (8.33) automatically satisfy the vorticity-stream equations for the low Froude number limit presented in Chapter 6.

8.4.3. Instabilities for Finite Small Froude Numbers for the Laminar Elementary Flows.

Here we take the initial data in (8.33) and perturb it within the class of laminar flow solutions of the Boussinesq equations defined by (8.28). Below we develop a rigorous analytic criterion involving the vorticity and strain, ω_0 and γ, of the low Froude number limit flow in (8.33), so that stability or instability occurs in the family of laminar solutions of the Boussinesq equations from (8.27) and (8.28) for arbitrarily small values of the Froude number ε. We will confirm the predictions of this criterion with some elementary numerical solutions.

To develop this asymptotic criterion, we apply the general fast-wave averaging developed in Section 8.3 to the explicit ODEs in (8.28) that clearly display a separation of time scales in the limit $\varepsilon \to 0$. Thus, we set $\tau = \frac{t}{\varepsilon}$ and have leading-order asymptotic solutions as $\varepsilon \to 0$ with the form

$$
\begin{aligned}
\omega_1^0 &= -2d_{23} + C_1 \sin\left(\frac{\tau}{\sqrt{2}}\right) + C_2 \cos\left(\frac{\tau}{\sqrt{2}}\right) + O(\varepsilon), \\
\omega_2^0 &= 2d_{13} + C_3 \sin\left(\frac{\tau}{\sqrt{2}}\right) + C_4 \cos\left(\frac{\tau}{\sqrt{2}}\right) + O(\varepsilon), \\
\omega_3^0 &= \bar{\omega}_0 + O(\varepsilon), \\
b_1^0 &= \frac{C_3}{\sqrt{2}} \cos\left(\frac{\tau}{\sqrt{2}}\right) - \frac{C_4}{\sqrt{2}} \sin\left(\frac{\tau}{\sqrt{2}}\right) + O(\varepsilon), \\
b_2^0 &= -\frac{C_1}{\sqrt{2}} \cos\left(\frac{\tau}{\sqrt{2}}\right) + \frac{C_2}{\sqrt{2}} \sin\left(\frac{\tau}{\sqrt{2}}\right) + O(\varepsilon), \\
b_3^0 &= \bar{b}_3^0 + O(\varepsilon).
\end{aligned}
$$

(8.34)

Here the coefficients $C_j(t)$ depend on the slow time t. The averaging procedure shows that nonsecular terms arise at leading order for $\varepsilon \ll 1$ provided that the fast-wave amplitude vector $\vec{C}$ satisfies the ODEs

$$\frac{dC_1}{dt} = \gamma C_1 - \frac{\bar{\omega}_0}{4} C_3 + \frac{\sqrt{2b_3^0}}{4} C_2 \,,$$

$$\frac{dC_2}{dt} = \gamma C_2 - \frac{\bar{\omega}_0}{4} C_4 - \frac{\sqrt{2b_3^0}}{4} C_1 \,,$$

(8.35)

$$\frac{dC_3}{dt} = -\gamma C_3 + \frac{\bar{\omega}_0}{4} C_1 + \frac{\sqrt{2b_3^0}}{4} C_4 \,,$$

$$\frac{dC_4}{dT} = -\gamma C_4 + \frac{\bar{\omega}_0}{4} C_2 - \frac{\sqrt{2b_3^0}}{4} C_3 \,.$$

We omit the lengthy details here for this standard asymptotic evaluation and leave them as an exercise for the interested reader.

The equations in (8.35) for the vector $\vec{C}(t)$ govern the growth of the fast waves on the eddy turnover time scale. The constant-coefficient matrix on the right-hand side of (8.35) has the four explicit eigenvalues

(8.36)
$$\lambda_{1,2,3,4} = \pm \frac{1}{2}\sqrt{16\gamma^2 - (\bar{\omega}_0)^2} \pm \frac{i\sqrt{2}}{4} \bar{b}_3^0 \,.$$

From (8.36), we immediately have the following *analytic criterion for instability of the low Froude number limiting process* on eddy turnover times:

Criterion for Stability/Instability. The elementary laminar solutions of the low Froude number limit dynamics in (8.33) are stable against the laminar solutions of the Boussinesq equations in (8.28) at finite small Froude numbers $\varepsilon \ll 1$, provided the vorticity and strain in (8.33) satisfy

(8.37)
$$|\gamma| < \frac{|\bar{\omega}_0|}{4} \,.$$

Instability occurs provided the vorticity and strain satisfy

(8.38)
$$|\gamma| > \frac{|\bar{\omega}_0|}{4} \,.$$

We note the surprising fact that this criterion does not involve the vertical shear $\bar{\gamma}$ in (8.33).

We numerically integrated the ODEs in (8.28) for the Froude numbers $\varepsilon = 0.01$ and $\varepsilon = 0.02$ with initial data given by the low Froude number limiting solutions in (8.33) to check whether the predictions of the analytic criteria in (8.37) and (8.38) are realized at these finite Froude numbers. In Figure 8.1 we plot the physically interesting vertical component of the density gradient b_3 for the numerical solution at these two Froude numbers for the two values of initial vorticity and strain, $\bar{\omega}_0$ and γ, given by $\bar{\omega}_0 = 5$, $\gamma = 1$ (Figure 8.1(a)) with $\bar{\omega}_0 = 3$, $\gamma = 1$ (Figure 8.1(b)). As predicted by the analytic criteria in (8.37) and (8.38), we clearly have nonlinear stability in the case from Figure 8.1(a) and instability in the case from Figure 8.1(b) although for the larger Froude number $\varepsilon = 0.1$ the oscillations

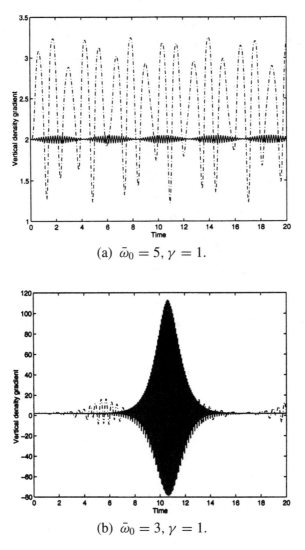

(a) $\bar{\omega}_0 = 5, \gamma = 1$.

(b) $\bar{\omega}_0 = 3, \gamma = 1$.

FIGURE 8.1. The vertical component of the density gradient b_3 of perturbations with the linear spatial structure to the low Froude number limiting dynamics laminar solutions at small Froude numbers $\varepsilon = 0.1$ (dashed line) and $\varepsilon = 0.02$ (solid line). Note the bounded transient growth in the stable situation in (a) with smaller amplitude as ε decreases compared with the very large transient amplitudes in case (b), where the analytic criterion predicts instabilities.

are of large amplitude in the stable case. While we don't discuss this in detail here, all of our elementary numerical integrations of the system in (8.28) confirm the stability criteria in (8.37) and (8.38) for sufficiently small values of the Froude number, and the other physical variables behave in a similar fashion to the vertical density gradient.

8.4.4. The Effect of Small-Scale Perturbations in Boussinesq Flows at Low Froude Numbers. It is obviously an interesting issue whether the explicit criterion for stability-instability in (8.37) and (8.38) for laminar elementary solutions of the Boussinesq equations for small Froude numbers remains valid for much more general solutions. The general plane wave perturbation solutions of the laminar elementary flows defined in (8.29) and (8.30) provide an accessible family of basic solutions that are useful for addressing this question.

We numerically integrated the equations in (8.30) for finite small Froude numbers utilizing the laminar solutions in (8.33) of the low Froude number limiting dynamics as basic states in (8.30) for a wide variety of strain and vertical vorticity parameters γ and $\bar{\omega}_0$. We used initial data for (8.30) compatible with the low Froude number limiting dynamics for the small-scale flow with

$$(8.39) \qquad \vec{\alpha}(0) = \vec{\alpha}_H, \quad \vec{A}_H(0) \cdot \vec{\alpha}_H(0) = 0, \quad A_3(0) = 0, \quad B(0) = 0.$$

We utilized the two small Froude numbers, $\varepsilon = 0.1$ and $\varepsilon = 0.01$, in this study.

Conclusions. For the Froude number $\varepsilon = 0.01$, *all of the laminar solutions of the low Froude number limit equations are stable to general plane wave perturbations of this type for finite eddy turnover times*. Thus, the *laminar criterion for stability-instability in (8.37) and (8.38) remains valid for the small-scale plane wave perturbations in* (8.30). For the larger Froude number $\varepsilon = 0.1$, there can be mild transient growth.

We illustrate these results in Figure 8.2, where we plot the amplitude of the vertical density gradient perturbation $b_3(t)$ for the Froude numbers $\varepsilon = 0.1$ (Figure 8.2(a)) and $\varepsilon = 0.01$ (Figure 8.2(b)) for the stable laminar low Froude number limit flow with $\bar{\omega}_0 = 5$ and $\gamma = 1$ from Figure 8.1(a). Mild transient growth is evident for $\varepsilon = 0.1$ with stability for $\varepsilon = 0.01$. We remark that a similar numerical integration for the Froude number $\varepsilon = 0.001$ (not depicted here) has basically the same pattern as in Figure 8.2(b) with a much smaller amplitude.

It would be interesting to study whether the general criterion in (8.37) and (8.38) for instability/stability in the low Froude number limit applies for more general solutions of the Boussinesq equations with strong stratification. The same type of analysis can be applied to the rotating Boussinesq equations with strong rotation and strong stratification as described in Chapter 6. The result is that the quasi-geostrophic states are *always stable* in the limit on eddy turnover time scales for ε small enough. I leave the details as an amusing exercise for the interested reader.

8.5. The Rapidly Rotating Shallow Water Equations with Unbalanced Initial Data in the Quasi-Geostrophic Limit

Here we apply the abstract fast-wave averaging principle developed in Section 8.3 to the main motivating example of this chapter, the rapidly rotating shallow water equations (RSWE) from (8.17) in the formal geostrophic limit $\varepsilon \ll 1$ with general unbalanced initial data. With $\vec{u} = {}^T(\vec{v}, h)$ and the operators $\mathcal{L}(\vec{u})$ and $B(\vec{u}, \vec{u})$ defined in (8.19), RSWE exactly fits the abstract framework of Section 8.3.

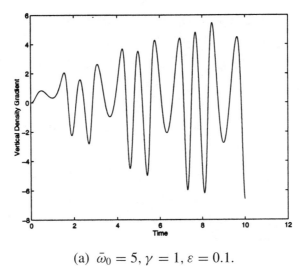

(a) $\bar{\omega}_0 = 5, \gamma = 1, \varepsilon = 0.1.$

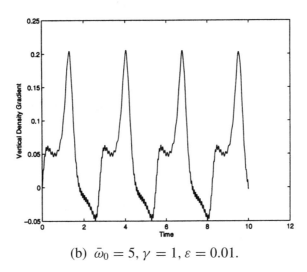

(b) $\bar{\omega}_0 = 5, \gamma = 1, \varepsilon = 0.01.$

FIGURE 8.2. The vertical component of the density gradient b_3 of perturbations with the plane-wave spatial structure of the low Froude number limiting dynamics laminar solutions. Note the small amplitude of order 2 of the evolving perturbations in (b) with $\varepsilon = .01$.

Below we give formal calculations that show that dispersive effects drastically deplete the nonlinear interactions in these equations. In fact, we have the following:

Reduced Dynamical Equations for Rapidly Rotating Shallow Water. The solution $\vec{u}^\epsilon(\vec{x}, t)$ of the rotating shallow water equations from (8.17) has the form

$$(8.40) \qquad \vec{u}^\epsilon(\vec{x}, t) = \bar{u}_I(\vec{x}, t) + e^{-\frac{t}{\epsilon}\mathcal{L}} \bar{u}_{II}(\vec{x}, t) + o(I) \quad \text{as } \epsilon \to 0$$

where the slow vortical component $\vec{u}_I(\vec{x}, t)$ is determined by the quasi-geostrophic equations

$$\vec{v}_I = F^{-\frac{1}{2}}\nabla^\perp h_I, \quad q_I = \omega_I - F^{\frac{1}{2}}h_I = F^{-\frac{1}{2}}\Delta h_I - F^{\frac{1}{2}}h_I,$$

(8.41)

$$\frac{Dq_I}{Dt} = \frac{\partial q_I}{\partial t} + \vec{v}_I \cdot \nabla q_I = 0.$$

On the other hand, the fast component $\vec{u}_{II}(\vec{x}, t)$ is given by Fourier modes in energy shells

(8.42)
$$\bar{u}_{II}(\vec{x}, t) = \sum_{l=0}^{\infty} \sum_{|\vec{k}|=\Lambda_l} e^{i\vec{k}\cdot\vec{x}} \left(\sum_{\alpha=\pm 1} \sigma_{(\vec{k})}^{(\alpha)} \vec{r}_{(\vec{k})}^{(\alpha)} \right)$$

where the dynamics of $\vec{u}_{II}(\vec{x}, t)$ is given at each energy level $|\vec{k}| = \Lambda_l$ by the reduced dynamics equations

(8.43)
$$\frac{\partial \sigma_{(\vec{k})}^{(\alpha)}}{\partial t} = \sum_{|\vec{k}_1|=\Lambda_l} \mathcal{C}_{(\vec{k}_1,\vec{k})}^{(\alpha)}(\vec{u}_I)\sigma_{(\vec{k}_1)}^{(\alpha)},$$

where $(\mathcal{C}_{(\vec{k}_1,\vec{k})}^{(\alpha)}(\vec{u}_I))$ is an explicit Hermitian matrix depending on the quasi-geostrophic component $\vec{u}_I$. Furthermore, with $\omega(\vec{k}) = (1 + F^{-1}|\vec{k}|^2)^{1/2}$,

$$e^{-\frac{t}{\epsilon}\mathcal{L}}\bar{u}_{II} = \sum_{l=0}^{\infty} \sum_{|\vec{k}|=\Lambda_l} e^{i(\vec{k}\cdot\vec{x})} \left(\sum_{\alpha=\pm 1} \right) e^{i\mp\frac{\omega(\vec{k})t}{\epsilon}} \sigma_{(\vec{k})}^{(\alpha)} \vec{r}_{(\vec{k})}^{(\alpha)}.$$

Here $\vec{r}_{(\vec{k})}^{(\alpha)}$, $\alpha = \pm 1$ are the right eigenvectors for the inertio-gravity waves given below.

Nonlinear Rossby Adjustment. With the explicit form of the limiting dynamics described above, it is a simple matter to prove that nonlinear Rossby adjustment occurs even for general unbalanced initial data. The key underlying mathematical fact is the Riemann-Lebesgue lemma (Folland [**10**]) that guarantees that

$$\int \phi(t)e^{\mp i\frac{\omega(\vec{k})t}{\epsilon}} dt \to 0$$

for any integrable function $\phi(t)$ since $|\omega(\vec{k})| > 0$ for any $\vec{k}$. Thus, if one applies any smooth space-time filter $\vec{\phi}(\vec{x}, t)$ to the RSWE for $\varepsilon \ll 1$, the asymptotic description in (8.40) and the above remark guarantee that

$$\lim_{\varepsilon \to 0} \int\int \vec{\phi}(\vec{x}, t) \cdot \vec{u}^\epsilon(\vec{x}, t)dx\, dt = \int\int \vec{\phi}(\vec{x}, t) \cdot \vec{u}_I(\vec{x}, t)dx\, dt.$$

Thus, even though there are very large oscillations in the gravity waves for unbalanced initial data, these oscillations cancel out for $\varepsilon \ll 1$ and only the quasi-geostrophic slow component u_I is observed in the limit after space-time filtering. This is a refined version of nonlinear Rossby adjustment for $\varepsilon \ll 1$.

Next, we set up the material to utilize the fast-wave averaging principle in (8.16) and formally deduce the structure in (8.40)–(8.43).

Fourier Decomposition of the Linear Operator $\mathcal{L}(\vec{u})$. Consider the linear operator $\mathcal{L}(\vec{u})$ from (8.19). Recall that $\mathcal{L}(\vec{u})$ has the form

$$(8.44) \qquad \mathcal{L}(\vec{u}) = \begin{pmatrix} \vec{v}^\perp + F^{-1/2}\nabla h \\ F^{-1/2}\,\mathrm{div}\,\vec{v} \end{pmatrix}.$$

We seek spatially periodic eigenfunctions of $\mathcal{L}(\vec{u})$ of the form

$$\vec{u} = e^{i\vec{k}\cdot\vec{x}}\vec{r}.$$

Clearly the function $\vec{u}$ satisfies

$$\mathcal{L}(e^{i\vec{k}\cdot\vec{x}}) = e^{i\vec{k}\cdot\vec{x}}\mathcal{L}(i\vec{k})\vec{r}$$

where the matrix symbol $\mathcal{L}(i\vec{k})$ is given by

$$(8.45) \qquad \mathcal{L}(i\vec{k}) = \begin{pmatrix} 0 & -1 & iF^{-1/2}k_1 \\ 1 & 0 & iF^{-1/2}k_2 \\ iF^{-1/2}k_1 & iF^{-1/2}k_2 & 0 \end{pmatrix}.$$

From (8.45) $\mathcal{L}(i\vec{k})$ is skew-Hermitian, $(\mathcal{L}^T)^*(i\vec{k}) = -\mathcal{L}(i\vec{k})$, and therefore it has an orthonormal basis of eigenvectors with purely imaginary eigenvalues. The eigenvalue problem for the operator $\mathcal{L}(\vec{u})$ reduces to the algebraic eigenvalue problem for the symbol $\mathcal{L}(i\vec{k})$ that we summarize below:

Algebraic Eigenvalue Problem for $\mathcal{L}(i\vec{k})$.

EIGENVALUE PROBLEM:

$$(8.46) \qquad (-i\omega I + \mathcal{L}(i\vec{k}))\vec{r} = 0.$$

CHARACTERISTIC DETERMINANT:

$$\det(-i\omega I + \mathcal{L}(i\vec{k})) = i\omega(\omega^2 - (1 + F^{-1}|\vec{k}|^2)).$$

The eigenvalues and eigenvectors are given by

EIGENVALUES:

$$(8.47) \qquad \begin{array}{ccc} \text{gravity mode} & \text{vortical mode} & \text{gravity mode} \\ \omega_{(\vec{k})}^{(-1)} = -\omega(\vec{k}), & \omega_{(\vec{k})}^{(0)} = 0, & \omega_{(\vec{k})}^{(1)} = \omega(\vec{k}), \end{array}$$

where the frequency $\omega(\vec{k})$ is given by

$$(8.48) \qquad \omega(\vec{k}) = (1 + F^{-1}|\vec{k}|^2)^{1/2}.$$

RIGHT EIGENVECTORS (CASE $\vec{k} \neq 0$):

$$(8.49) \qquad \begin{array}{ccc} \text{gravity mode} & \text{vortical mode} & \text{gravity mode} \\ \vec{r}_{(\vec{k})}^{(1)} = \begin{pmatrix} \dfrac{-\omega(\vec{k})k_1 + ik_2}{\sqrt{2}\,\omega(\vec{k})|\vec{k}|} \\ \dfrac{-\omega(\vec{k})k_2 - ik_1}{\sqrt{2}\,\omega(\vec{k})|\vec{k}|} \\ \dfrac{F^{-1}|\vec{k}|}{\sqrt{2}\,\omega(\vec{k})} \end{pmatrix}, & \vec{r}_{(\vec{k})}^{(0)} = \begin{pmatrix} \dfrac{-iF^{-1/2}k_2}{\omega(\vec{k})} \\ \dfrac{iF^{-1/2}k_1}{\omega(\vec{k})} \\ \dfrac{1}{\omega(\vec{k})} \end{pmatrix}, & \vec{r}_{(\vec{k})}^{(1)} = \begin{pmatrix} \dfrac{\omega(\vec{k})k_1 + ik_2}{\sqrt{2}\,\omega(\vec{k})|\vec{k}|} \\ \dfrac{\omega(\vec{k})k_2 - ik_1}{\sqrt{2}\,\omega(\vec{k})|\vec{k}|} \\ \dfrac{F^{-1/2}|\vec{k}|}{\sqrt{2}\,\omega(\vec{k})} \end{pmatrix}. \end{array}$$

The formulas for the right eigenvectors in (8.49) are valid for $\vec{k} \neq 0$. If $\vec{k} = 0$ the formulas for the gravity-wave eigenvectors are given by

GRAVITY-WAVE RIGHT EIGENVECTORS (CASE $\vec{k} = 0$):

$$(8.50) \qquad \vec{r}^{(-1)}_{(\vec{k})} = \begin{pmatrix} \frac{-i}{\sqrt{2}} \\ \frac{1}{\sqrt{2}} \\ 0 \end{pmatrix}, \quad \vec{r}^{(0)}_{(\vec{k})} = \begin{pmatrix} 0 \\ 0 \\ 1 \end{pmatrix}, \quad \vec{r}^{(1)}_{(\vec{k})} = \begin{pmatrix} \frac{i}{\sqrt{2}} \\ \frac{1}{\sqrt{2}} \\ 0 \end{pmatrix}.$$

The eigenvectors given in (8.49) and (8.50) are an orthonormal basis and satisfy the

SYMMETRY RELATIONS FOR THE EIGENVECTORS:

$$(8.51) \qquad (\vec{r}^{(\alpha)}_{(\vec{k})})^* = \vec{r}^{(-\alpha)}_{(-\vec{k})} \quad \text{for } \alpha = -1, 0, 1.$$

To compute the explicit averaged equations in (8.16) for the singular limit of the rotating shallow water equations, we expand the solution $\bar{u}(\vec{x}, t)$ into eigenmodes so that

$$(8.52) \qquad \bar{u}(\vec{x}, t) = \sum_{\vec{k} \in \mathbb{Z}^2} \sum_{\alpha=-1}^{1} e^{i\vec{k}\cdot\vec{x}} \sigma^{(\alpha)}_{(\vec{k})}(t) \vec{r}^{(\alpha)}_{(\vec{k})}$$

where the $\vec{r}^{(\alpha)}_{(\vec{k})}$ are the explicit right eigenvectors in (8.45) and (8.46) with $\alpha = 1, -1$, corresponding to the two gravity modes at a fixed wavenumber and $\alpha = 0$ corresponding to the vortical mode. Since $\bar{u} = \binom{\bar{v}}{\bar{h}}$ is a real-valued function with the symmetry relation in (8.47), we require the coefficients to satisfy

$$(8.53) \qquad (\sigma^{(\alpha)}_{(\vec{k})}(t))^* = \sigma^{(-\alpha)}_{(-\vec{k})}(t).$$

With this Fourier representation, the exponential $e^{-\tau\mathcal{L}}\bar{u}(\vec{x}, t)$ is given by

$$(8.54) \qquad e^{-\mathcal{L}\tau}\bar{u}(\vec{x}, t) = \sum_{\vec{k} \in \mathbb{Z}^2} \sum_{\alpha=-1}^{1} e^{i(\vec{k}\cdot\vec{x} - \omega^{(\alpha)}_{(\vec{k})}\tau)} \sigma^{(\alpha)}_{(\vec{k})} \vec{r}^{(\alpha)}_{(\vec{k})}.$$

With the explicit bilinear term $\mathcal{B}(\vec{u}, \vec{u})$ for the rotating shallow water equations given by $\mathcal{B}(\vec{u}, \vec{u}) = \binom{\bar{v}\cdot\nabla\bar{v}}{\text{div}(h\bar{v})}$, we utilize (8.54) and calculate that

$$(8.55) \quad \mathcal{B}(e^{-\mathcal{L}\tau}\bar{u}, e^{-\mathcal{L}\tau}\bar{u}) =$$

$$\sum_{\vec{k} \in \mathbb{Z}^2} \sum_{\alpha=-1}^{1} \left\{ \sum_{\vec{k}_1 + \vec{k}_2 = \vec{k}} \sum_{\alpha_1, \alpha_2 = 1}^{1} e^{i(\vec{k}\cdot\vec{x} - (\omega^{(\alpha_1)}_{(\vec{k}_1)} + \omega^{(\alpha_2)}_{(\vec{k})})\tau)} C^{(\alpha_1, \alpha_2, \alpha)}_{(\vec{k}_1, \vec{k}_2, \vec{k})} \sigma^{(\alpha_1)}_{(\vec{k}_1)}(t) \sigma^{(\alpha_2)}_{(\vec{k}_2)}(t) \right\} \vec{r}^{(\alpha)}_{(\vec{k})},$$

where in equation (8.55) we have already expanded the vector $\vec{r}^{(\alpha_2)}_{(\vec{k}_2)}$ in terms of the basis $\{\vec{r}^{(\alpha)}_{(\vec{k})}\}$. The interaction coefficient $C^{(\alpha_1, \alpha_2, \alpha)}_{(\vec{k}_1, \vec{k}_2, \vec{k})}$ is given by

$$(8.56) \qquad C^{(\alpha_1, \alpha_2, \alpha)}_{(\vec{k}_1, \vec{k}_2, \vec{k})} = \frac{i}{2}\Big[\big(\vec{v}^{(\alpha_1)}_{(\vec{k}_1)} \cdot \vec{k}_2\big)\big\langle\vec{r}^{(\alpha_2)}_{(\vec{k}_2)}, \vec{r}^{(\alpha)}_{(\vec{k})}\big\rangle + \big(\vec{v}^{(\alpha_2)}_{(\vec{k}_2)} \cdot \vec{k}_1\big)\big\langle\vec{r}^{(\alpha_1)}_{(\vec{k}_1)}, \vec{r}^{(\alpha)}_{(\vec{k})}\big\rangle$$
$$+ \big(\vec{v}^{(\alpha_1)}_{(\vec{k}_1)} \cdot \vec{k}_1\big)h^{(\alpha_2)}_{(\vec{k}_2)}h^{(\alpha)}_{(\vec{k})} + \big(\vec{v}^{(\alpha_2)}_{(\vec{k}_2)} \cdot \vec{k}_2\big)h^{(\alpha_1)}_{(\vec{k}_1)}h^{(\alpha_1)}_{(\vec{k}_1)}h^{(\alpha)}_{(\vec{k})}\Big].$$

where $C^{(\alpha_1,\alpha_2,\alpha)}_{(\vec{k}_1,\vec{k}_2,\vec{k})}$ in (8.56) has been written so that the sum inside the braces in (8.55) is symmetric in the indices $\binom{\alpha_1}{k_1}$ and $\binom{\alpha_1}{k_1}$, i.e., $C^{(\alpha_1,\alpha_2,\alpha)}_{(\vec{k}_1,\vec{k}_2,\vec{k})} = C^{(\alpha_2,\alpha_1,\alpha)}_{(\vec{k}_2,\vec{k}_1,\vec{k})}$. From (8.55) it now follows that

$$(8.57) \quad e^{\tau \mathcal{L}} \mathcal{B}(e^{-\mathcal{L}\tau}\bar{u}, e^{-\mathcal{L}\tau}\bar{u}) =$$

$$\sum_{\vec{k}\in\mathbb{Z}^2}\sum_{\alpha=-1}^{1}\left\{\sum_{\vec{k}_1+\vec{k}_2=\vec{k}}\sum_{\alpha_1,\alpha_2=-1}^{1} e^{i(\vec{k}\cdot\vec{x}-(\omega^{(\alpha_1)}_{(\vec{k}_1)}+\omega^{(\alpha_2)}_{(\vec{k}_2)}-\omega^{(\alpha)}_{(\vec{k})})\tau)} C^{(\alpha_1,\alpha_2,\alpha)}_{(\vec{k}_1,\vec{k}_2,\vec{k})}\sigma^{(\alpha_1)}_{(\vec{k}_1)}(t)\sigma^{(\alpha_2)}_{(\vec{k}_2)}(t)\right\}\vec{r}^{(\alpha)}_{(\vec{k})}.$$

Next, we substitute (8.57) back into the averaged equations from (8.16) and compute that

$$\frac{\partial \bar{u}}{\partial t}(\vec{x},t)$$

$$= -\lim_{\tau\to\infty}\frac{1}{\tau}\int_0^\tau e^{s\mathcal{L}}\mathcal{B}(\vec{x},t,e^{-s\mathcal{L}}\vec{u}(\vec{x},t),e^{-s\mathcal{L}}\vec{u}(\vec{x},t))ds$$

$$(8.58)$$

$$= -\lim_{\tau\to\infty}\frac{1}{\tau}\int_0^\tau \sum_{\vec{k}\in\mathbb{Z}^2}\sum_{\alpha=-1}^{1}\left\{\sum_{\vec{k}_1+\vec{k}_2=\vec{k}}\sum_{\alpha_1,\alpha_2=-1}^{1} e^{i(\vec{k}\cdot\vec{x}-(\omega^{(\alpha_1)}_{(\vec{k}_1)}+\omega^{(\alpha_2)}_{(\vec{k}_2)}-\omega^{(\alpha)}_{(\vec{k})})s)}\right.$$

$$\left.\times C^{(\alpha_1,\alpha_2,\alpha)}_{(\vec{k}_1,\vec{k}_2,\vec{k})}\sigma^{(\alpha_1)}_{(\vec{k}_1)}(t)\sigma^{(\alpha_2)}_{(\vec{k}_2)}(t)\right\}\vec{r}^{(\alpha)}_{(\vec{k})}ds.$$

With the explicit form of the averaged equations, it is straightforward to evaluate the average in (8.58). Whenever $\vec{k}_1 + \vec{k}_2 = \vec{k}$ but

$$\omega^{(\alpha_1)}_{(\vec{k}_1)} + \omega^{(\alpha_2)}_{(\vec{k}_2)} - \omega^{(\alpha)}_{(\vec{k})} \neq 0,$$

the corresponding term in (8.58) is oscillatory and the averaging procedure in (8.58) will give a zero contribution as described in Section 8.3 for the simple example of a nonlinear oscillator. Consequently, the only nonzero contributions that survive the averaging process are the direct three-wave resonances with

$$(8.59) \qquad \vec{k}_1 + \vec{k}_2 = \vec{k}, \qquad \omega^{(\alpha_1)}_{(\vec{k}_1)} + \omega^{(\alpha_2)}_{(\vec{k}_2)} = \omega^{(\alpha)}_{(\vec{k})}.$$

After taking into account the cancellation except when (8.59) is satisfied and taking the inner product of (8.58) with the eigenvector $\vec{r}^{(\alpha)}_{(\vec{k})}$, we determine that the averaged equations for the Fourier amplitudes $\sigma^{(\alpha)}_{(\vec{k})}(t)$ of $\bar{u}(\vec{x},t)$ from (8.52) are given by

$$(8.60) \qquad \frac{\partial \sigma^{(\alpha)}_{(\vec{k})}}{\partial t} + \sum_{(\vec{k}_1,\vec{k}_2,\alpha_1,\alpha_2)\in S_{\alpha,\vec{k}}} C^{(\alpha_1,\alpha_2,\alpha)}_{(\vec{k}_1,\vec{k}_2,\vec{k})}\sigma^{(\alpha_1)}_{(\vec{k}_1)}\sigma^{(\alpha_2)}_{(\vec{k}_2)} = 0$$

where $C^{(\alpha_1,\alpha_2,\alpha)}_{(\vec{k}_1,\vec{k}_2,\vec{k})}$ is given in (8.56) and $S_{\alpha,\vec{k}}$ is the set associated with all three-wave resonances from (8.58) and is defined by

$$(8.61) \qquad S_{\alpha,\vec{k}} = \left\{(\vec{k}_1,\vec{k}_2,\alpha_1,\alpha_2): \vec{k}_1+\vec{k}_2=\vec{k}, \omega^{(\alpha_1)}_{(\vec{k}_1)}+\omega^{(\alpha_2)}_{(\vec{k}_2)}=\omega^{(\alpha)}_{(\vec{k})}\right\}.$$

Clearly, the set of resonances includes all possible three-wave interactions of the vortical modes since $\omega^{(0)}_{(\vec{k})} = 0$; i.e., all vortical modes resonate with each other.

This is completely expected since we already know from Chapter 4 that the quasi-geostrophic equations in (8.41) are valid in the limit of vanishing Rossby number when the initial gravity waves are completely absent. What is especially interesting for the geophysical application developed here is that the remainder of the set $S_{\alpha,\vec{k}}$ is extremely small and involves only gravity mode/vortical mode resonances.

Gravity Mode Three-Wave Resonances. Here we will show that no three-wave interactions occur solely among the gravity waves in the averaged equations in (8.59). Three-wave interactions that involve gravity waves exclusively necessarily satisfy $\vec{k}_1 + \vec{k}_2 = \vec{k}$ and either

$$(8.62) \qquad\qquad \omega(\vec{k}_1) \pm \omega(\vec{k}_2) = \omega(\vec{k})$$

where $\omega(\vec{k}) = (1 + F^{-1}|\vec{k}|^2)^{1/2}$. Since $\omega(\vec{k})$ is a radial function, we reduce the second equation in (8.62) to the first case in (8.62) by interchanging the roles of $\vec{k}$ and $\vec{k}_2$. Furthermore, there are never any solutions of the first equality in (8.62) because, in fact,

$$(8.63) \qquad\qquad \omega(\vec{k}_1 + \vec{k}_2) < \omega(\vec{k}_1) + \omega(\vec{k}_2) .$$

Since $\omega(\vec{k}) = (1 + F^{-1}|\vec{k}|^2)^{1/2}$, the inequality in (8.63) is a direct consequence of the elementary inequality

$$(1 + (x + y)^2)^{1/2} < (1 + x^2)^{1/2} + (1 + y^2)^{1/2} ,$$

which is readily verified by squaring each side. Thus, three-wave resonances involving gravity modes exclusively are impossible. This fact also supplies the needed information postponed in the discussion from Section 5.7.

Gravity Mode/Vortical Mode Resonances. The final nontrivial possibility for resonant interactions to occur and satisfy (8.59) is the situation when two gravity modes and a vortical mode resonate. Since the dispersion relation $\omega(\vec{k})$ is a radial function, such vortical mode/gravity mode three-wave resonances are possible only if the two gravity modes with wavenumbers $\vec{k}_1$ and $\vec{k}_2$ satisfy

$$(8.64) \qquad\qquad \vec{k} = \vec{k}_1 + \vec{k}_2 \quad \text{and} \quad |\vec{k}_1| = |\vec{k}_2| ,$$

where $\vec{k}$ is the wavenumber of the shear mode. Thus, three-wave interactions of two gravity modes and a vortical mode are possible only under the strong geometric constraint when the magnitude of the wavenumbers for the two gravity modes are equal.

Derivation of the Reduced Dynamical Equations for Rotating Shallow Water in the Singular Limit. We decompose the function $\bar{u}(x, t)$ satisfying the averaged equation in (8.58) into a (slow) quasi-geostrophic field $\bar{u}_{\mathrm{I}}$ defined by

$$(8.65) \qquad\qquad \bar{u}_{\mathrm{I}}(\vec{x}, t) = \sum_{\vec{k} \in \mathbb{Z}^2} e^{i\vec{k} \cdot \vec{x}} \sigma_{(\vec{k})}^{(0)}(t) \vec{r}_{(\vec{k})}^{(0)}$$

and a (fast) gravity wave field $\bar{u}_{\mathrm{II}}$ defined by

$$(8.66) \qquad\qquad \bar{u}_{\mathrm{II}}(\vec{x}, t) = \sum_{\vec{k} \in \mathbb{Z}^2} \sum_{\alpha = \pm 1} e^{i\vec{k} \cdot \vec{x}} \sigma_{(\vec{k})}^{(\alpha)}(t) \vec{r}_{(\vec{k})}^{(\alpha)} .$$

Next we utilize the information contained in (8.60)–(8.64) to derive from (8.60) the averaged equations for the quasi-geostrophic field $\bar{u}_I$ and the gravity wave field $\bar{u}_{II}$ for the rotating shallow water equations. First we consider the quasi-geostrophic field $\bar{u}_I$. From (8.61)–(8.64) we know that given vortical mode $\sigma^{(0)}_{(\vec{k})}(t)$, the three-wave resonant set $S_{\alpha,\vec{k}}$ in (8.60) involves either a pair of vortical modes $\sigma^{(0)}_{(\vec{k}_1)}$ and $\sigma^{(0)}_{(\vec{k}_2)}$ with $\vec{k}_1 + \vec{k}_2 = \vec{k}$, or a pair of gravity modes $\sigma^{(\alpha_1)}_{(\vec{k}_1)}$ and $\sigma^{(\alpha_2)}_{(\vec{k}_2)}$ with $\vec{k}_1 + \vec{k}_2 = \vec{k}$, $|\vec{k}_1| = |\vec{k}_2|$, and $\alpha_1 + \alpha_2 = 0$. Therefore (8.60) reduces to

$$(8.67) \qquad \frac{\partial \sigma^{(0)}_{(\vec{k})}}{\partial t} = \sum_{\vec{k}_1 + \vec{k}_2 = \vec{k}} C^{(0,0,0)}_{(\vec{k}_1,\vec{k}_2,\vec{k})} \sigma^{(0)}_{(\vec{k}_1)} \sigma^{(0)}_{(\vec{k}_2)} + \sum_{\substack{\vec{k}_1 + \vec{k}_2 = \vec{k} \\ |\vec{k}_1| = |\vec{k}_2|}} \sum_{\substack{\alpha_1 + \alpha_2 = 0 \\ \alpha_1 = \pm 1}} C^{(\alpha_1,\alpha_2,0)}_{(\vec{k}_1,\vec{k}_2,\vec{k})} \sigma^{(\alpha_1)}_{(\vec{k}_1)} \sigma^{(\alpha_2)}_{(\vec{k}_2)}.$$

The straightforward calculation of the vortical/gravity wave resonant interaction coefficient $C^{(\alpha_1,\alpha_2,0)}_{(\vec{k}_1,\vec{k}_2,\vec{k})}$ in the second sum in (8.67) utilizing the formula for the interaction coefficient in (8.56) together with the explicit formulas for the right eigenvectors in (8.49)–(8.50) yields the remarkable fact that

$$(8.68) \qquad C^{(1,-1,0)}_{(\vec{k}_1,\vec{k}_2,\vec{k}_1+\vec{k}_2)} = C^{(-1,1,0)}_{(\vec{k}_1,\vec{k}_2,\vec{k}_1+\vec{k}_2)} = 0 \quad \text{for } \vec{k}_1 + \vec{k}_2 = \vec{k} \text{ and } |\vec{k}_1| = |\vec{k}_2|.$$

Therefore the vortical modes only interact among themselves and the equation in (8.67) reduces to

$$(8.69) \qquad \frac{\partial \sigma^{(0)}_{(\vec{k})}}{\partial t} = \sum_{\vec{k}_1 + \vec{k}_2 = \vec{k}} C^{(0,0,0)}_{(\vec{k}_1,\vec{k}_2,\vec{k})} \sigma^{(0)}_{(\vec{k}_1)} \sigma^{(0)}_{(\vec{k}_2)},$$

where the interaction coefficients $C^{(0,0,0)}_{(\vec{k}_1,\vec{k}_2,\vec{k})}$ are computed from (8.56) and (8.49)–(8.50), and are given explicitly by

$$(8.70) \qquad C^{(0,0,0)}_{(\vec{k}_1,\vec{k}_2,\vec{k}_1+\vec{k}_2)} = \frac{F^{-3/2}(\vec{k}_1 \cdot \vec{k}_2^\perp)(|\vec{k}_2|^2 - |\vec{k}_1|^2)}{2\omega(\vec{k}_1)\omega(\vec{k}_2)\omega(\vec{k}_1 + \vec{k}_2)}.$$

We claim that the equation in (8.69) for the quasi-geostrophic component $\vec{u}_I$ is equivalent to the quasi-geostrophic equations in (8.41). Let $\vec{u}_I = (\vec{v}_I, h_I) = (v_{I1}, v_{I2}, h_I)$ be the quasi-geostrophic field in (8.65). Since every vortical eigenmode in (8.65) satisfies the geostrophic balance condition in (8.41), so does $\vec{u}_I$

$$(8.71) \qquad \vec{v}_I = F^{-1/2} \nabla^\perp h_I.$$

Next we show that the potential vorticity $q_I = \omega_I - F^{1/2} h_I$ satisfies the potential vorticity equation in (8.41) with $\vec{v} = \vec{v}_I$. Since $\vec{u}_I$ satisfies (8.71) and $\omega_I = \partial \vec{v}_{I2}/\partial x_1 - \partial \vec{v}_{I1}/\partial x_2$, it follows that $q_I = F^{-1/2}\Delta h_I - F^{1/2} h_I$, and that q_I has the Fourier series expansion

$$(8.72) \qquad q_I = \sum_{\vec{k} \in \mathbb{Z}^2} e^{i\vec{k}\cdot\vec{x}} \sigma^{(0)}_{(\vec{k})} q_{(\vec{k})}$$

where $q_{(\vec{k})}$ is given by

$$(8.73) \qquad q_{(\vec{k})} = -(|\vec{k}|^2 F^{-1/2} + F^{1/2}) h^{(0)}_{(\vec{k})} = -F^{1/2}\omega(\vec{k}),$$

and $h_{(\vec{k})}^{(0)}$ is given by (8.49). Next we compute the Fourier expansions for $\frac{Dq_I}{Dt}$:

$$(8.74) \quad \frac{Dq_I}{Dt} = \frac{\partial q_I}{\partial t} + \vec{v}_I \cdot \nabla q_I = \sum_{\vec{k} \in \mathbb{Z}^2} e^{i\vec{k} \cdot \vec{x}} \left(q_{(\vec{k})} \frac{\partial \sigma_{(\vec{k})}^{(0)}}{\partial t} + \sum_{\vec{k}_1 + \vec{k}_2 = \vec{k}} D_{(\vec{k}_1, \vec{k}_2, \vec{k})} \sigma_{(\vec{k}_1)}^{(0)} \sigma_{(\vec{k}_2)}^{(0)} \right)$$

where $D_{(\vec{k}_1, \vec{k}_2, \vec{k})}$ is given by

$$(8.75) \qquad D_{(\vec{k}_1, \vec{k}_2, \vec{k})} = \frac{i}{2} \left[(\vec{v}_{(\vec{k}_1)}^{(0)} \cdot \vec{k}_2) q_{(\vec{k}_2)} + (\vec{v}_{(\vec{k}_2)}^{(0)} \cdot \vec{k}_1) q_{(\vec{k}_1)} \right].$$

The direct calculation of $D_{(\vec{k}_1, \vec{k}_2, \vec{k})}$ utilizing (8.74) together with (8.47)–(8.49) shows that

$$(8.76) \qquad D_{(\vec{k}_1, \vec{k}_2, \vec{k})} = \frac{F^{-1/2}(\vec{k}_1 \cdot \vec{k}_2^{\perp})}{2\omega_{\vec{k}_1} \omega_{\vec{k}_2}} [\omega^2(\vec{k}_1) - \omega^2(\vec{k}_2)] = q_{(\vec{k})} C_{(\vec{k}_1, \vec{k}_2, \vec{k}_1 + \vec{k}_2)}^{(0,0,0)}.$$

Introducing (8.76) back into (8.74) and utilizing the equation in (8.69) for $\vec{u}_I$, we obtain

$$(8.77) \quad \begin{aligned} \frac{Dq_I}{Dt} &= \frac{\partial q_I}{\partial t} + \vec{v}_I \cdot \nabla q_I \\ &= \sum_{\vec{k} \in \mathbb{Z}^2} e^{i\vec{k} \cdot \vec{x}} q_{(\vec{k})} \left(\frac{\partial \sigma_{(\vec{k})}^{(0)}}{\partial t} + \sum_{\vec{k}_1 + \vec{k}_2 = \vec{k}} C_{(\vec{k}_1, \vec{k}_2, \vec{k}_1 + \vec{k}_2)}^{(0,0,0)} \sigma_{(\vec{k}_1)}^{(0)} \sigma_{(\vec{k}_2)}^{(0)} \right) = 0, \end{aligned}$$

which proves that $\vec{u}_I$ solves the potential vorticity equation in (8.41). Collecting the results from (8.71) and (8.77), we conclude that the evolution of $\vec{u}_I(\vec{x}, t)$ is determined by the quasi-geostrophic equations in (8.41).

Next, we consider the averaged equations for the gravity wave field $\vec{u}_{II}$. From (8.64) we know that the set of three-wave resonances $S_{\alpha, \vec{k}}$ in (8.61) for the gravity mode $\sigma_{(\vec{k})}^{(\alpha)}$, $\alpha = \pm 1$, in (8.60) is equivalent to the restrictions

$$(8.78) \qquad \begin{aligned} \alpha_1 &= \alpha, & \alpha_2 &= 0, & \vec{k}_1 + \vec{k}_2 &= \vec{k}, & |\vec{k}_1| &= |\vec{k}|, & \text{or} \\ \alpha_2 &= \alpha, & \alpha_1 &= 0, & \vec{k}_1 + \vec{k}_2 &= \vec{k}, & |\vec{k}_2| &= |\vec{k}|. \end{aligned}$$

Since the second condition in (8.78) is equivalent to the first one by interchanging the variables $\vec{k}_1$ and $\vec{k}_2$, it follows that equation (8.60) for the gravity mode $\sigma_{(\vec{k})}^{(\alpha)}$ reduces to

$$(8.79) \qquad \frac{\partial \sigma_{(\vec{k})}^{(\alpha)}}{\partial t} = - \sum_{|\vec{k}_1| = |\vec{k}|} 2 C_{(\vec{k}_1, \vec{k} - \vec{k}_1, \vec{k})}^{(\alpha, 0, \alpha)} \sigma_{(\vec{k} - \vec{k}_1)}^{(0)} \sigma_{(\vec{k}_1)}^{(\alpha)},$$

and where the interaction coefficients $C^{(\alpha,0,\alpha)}_{(\vec{k}_1,\vec{k}_2,\vec{k})}$, $\alpha = \pm 1$, from (8.56) are given explicitly by

$$(8.80) \quad C^{(1,0,1)}_{(\vec{k}_1,\vec{k}-\vec{k}_1,\vec{k})} = \left(C^{(-1,0,-1)}_{(\vec{k}_1,\vec{k}-\vec{k}_1,\vec{k})} \right)^* =$$

$$-\frac{F^{-1/2}}{4\omega(\vec{k}_2)\omega(\vec{k})^2|\vec{k}|^2}\{[|\vec{k}|^2(\vec{k}_1 \cdot \vec{k}^\perp) - (\vec{k}_1 \cdot \vec{k})(\vec{k}_1 \cdot \vec{k}^\perp)]$$

$$+ i\omega(\vec{k})[2|\vec{k}|^4 - |\vec{k}|^2(\vec{k}_1 \cdot \vec{k}) - (\vec{k}_1 \cdot \vec{k}^\perp)^2]\}\,.$$

In particular, the dynamical equation (8.79) shows that the resonant interaction with a given gravity wave with $|\vec{k}| = \Lambda$ is strongly constrained to gravity modes with wavenumber $\vec{k}_1$ confined to the same shell $|\vec{k}_1| = \Lambda$. The dynamical equations in (8.79) are a finite-dimensional system of linear equations for the amplitudes $\sigma^{(\alpha)}_{(\vec{k})}$ of the form

$$(8.81) \qquad \frac{\partial \sigma^{(\alpha)}_{\vec{k}}}{\partial t} = \sum_{|\vec{k}_1|=\Lambda} \mathcal{C}^{(\alpha)}_{(\vec{k}_1,\vec{k})}(\vec{u}_{\mathrm{I}})\sigma^{(\alpha)}_{(\vec{k}_1)}$$

where the coefficients $\mathcal{C}^{(\alpha)}_{(\vec{k}_1,\vec{k})}(\vec{u}_{\mathrm{I}}) = \mathcal{C}^{(\alpha,0,\alpha)}_{(\vec{k}_1,\vec{k}-\vec{k}_1,\vec{k})}\sigma^{(0)}_{(\vec{k}-\vec{k}_1)}$ depend on the quasi-geostrophic component $\vec{u}_{\mathrm{I}}$ and follow directly from (8.80) and the symmetry conditions in (8.53),

$$(8.82) \quad \mathcal{C}^{(1)}_{(\vec{k}_1,\vec{k})}(\vec{u}_{\mathrm{I}}) = \left(\mathcal{C}^{(-1)}_{(\vec{k}_1,\vec{k})}(\vec{u}_{\mathrm{I}}) \right)^* =$$

$$-\frac{F^{-1/2}}{4\omega(\vec{k}_2)\omega(\vec{k})^2|\vec{k}|^2}\{[|\vec{k}|^2(\vec{k}_1 \cdot \vec{k}^\perp) - (\vec{k}_1 \cdot \vec{k})(\vec{k}_1 \cdot \vec{k}^\perp)]$$

$$+ i\omega(\vec{k})[2|\vec{k}|^4 - |\vec{k}|^2(\vec{k}_1 \cdot \vec{k}) - (\vec{k}_1 \cdot \vec{k}^\perp)^2]\}\sigma^{(0)}_{(\vec{k}-\vec{k}_1)}\,.$$

Moreover, a direct calculation with the formula in (8.82) shows that the coefficient matrix $(\mathcal{C}^{(\alpha)}_{(\vec{k}_1,\vec{k})}(\vec{u}_{\mathrm{I}}))$ is skew-Hermitian, i.e.,

$$(8.83) \qquad \left(\mathcal{C}^{(\alpha)}_{(\vec{k}_1,\vec{k})}(\vec{u}_{\mathrm{I}}) \right)^* = -\mathcal{C}^{(\alpha)}_{(\vec{k},\vec{k}_1)}(\vec{u}_{\mathrm{I}})\,.$$

This completes our discussion of the decomposition for the reduced dynamics that we claimed at the beginning of this section.

8.6. The Interaction of Fast Waves and Slow Dynamics in the Rotating Stratified Boussinesq Equations

In this final section of the chapter, I briefly discuss the interaction of fast and slow dynamics in the rotating stratified Boussinesq equations. As we discussed in Chapters 6 and 7, there are at least two regimes of interest with slow-wave limiting dynamics having a very different structure. How do the slow waves interact with the fast waves in these regimes? Below, we will see that nonlinear Rossby adjustment occurs in these regimes, too, and I provide a conceptual mathematical proof here to contrast with the argument presented in the last section for RSWE that involves detailed algebraic manipulations. Here I will discuss only the limiting behavior of the rapidly rotating strongly stratified Boussinesq equation (RSSBE) from Chapter 7 for simplicity in exposition. The two different limits from Chapters 6 and 7 and the interaction of fast and slow dynamics in these two regimes is compared and contrasted in the paper by Embid and Majda [**8**].

First, I recall the quasi-geostrophic scaling for the Boussinesq equations from Chapter 7:

$$\text{Ro} = \epsilon \ \text{ with } \epsilon \ll 1 , \quad \overline{P} = \epsilon^{-1} , \quad \text{Fr} = \epsilon F \ \text{ with } F \neq 0 ,$$

(8.84)

$$\Gamma = (\text{Fr})^{-1} \quad \text{or equivalently} \quad B = \frac{NU}{g} .$$

With these distinguished values for the parameters from (8.84), the Boussinesq equations assume the form

(8.85)
$$\frac{D^H \vec{v}_H}{Dt} + w \frac{\partial \vec{v}_H}{\partial x_3} = -\epsilon^{-1} (\vec{v}_H^\perp + \nabla_H \phi) ,$$

$$\frac{D^H w}{Dt} + w \frac{\partial w}{\partial x_3} = -\epsilon^{-1} \frac{\partial \phi}{\partial x_3} - \epsilon^{-1} F^{-1} \rho ,$$

$$\frac{D\rho}{Dt} = \epsilon^{-1} F^{-1} w , \quad \text{div}_H \vec{v}_H + \frac{\partial w}{\partial x_3} = 0 .$$

We utilize the nonlocal form of the Boussinesq equations to remove the pressure ϕ as discussed in Chapter 7. Then the equations in (8.85) assume the abstract form

(8.86)
$$\frac{\partial \vec{u}}{\partial t} + \epsilon^{-1} \mathcal{L}(\vec{u}) + \mathcal{B}(\vec{u}, \vec{u}) = 0 , \quad \vec{u}\big|_{t=0} = \vec{u}_0(w) .$$

Here $\vec{u} = {}^T(\vec{v}_H, w, \rho)$. The linear operator $\mathcal{L}(\vec{u})$ is given by

(8.87)
$$\mathcal{L}(\vec{u}) = \begin{pmatrix} \vec{v}_H^\perp + \nabla_H \Delta^{-1}\left(\omega_3 - F^{-1}\frac{\partial \rho}{\partial x_3}\right) \\ F^{-1}\rho + \frac{\partial}{\partial x_3}\Delta^{-1}\left(\omega_3 - F^{-1}\frac{\partial \rho}{\partial x_3}\right) \\ -F^{-1}\omega \end{pmatrix} .$$

The quadratic operator $\mathcal{B}(\vec{u}, \vec{u})$ is given by

(8.88)
$$\mathcal{B}(\vec{u}, \vec{u}) = \begin{pmatrix} \vec{v} \cdot \nabla \vec{v}_H - \nabla_H \Delta^{-1}(\text{div}(\vec{v} \cdot \nabla \vec{v})) \\ \vec{v} \cdot \nabla w - \frac{\partial}{\partial x_3}\Delta^{-1}(\text{div}(\vec{v} \cdot \nabla \vec{v})) \\ \vec{v} \cdot \nabla \rho \end{pmatrix} .$$

With the distinguished scaling in (8.84), as discussed in Chapter 7, the conservation of potential vorticity becomes

(8.89)
$$\frac{D}{Dt}\left\{\omega_3 - F\frac{\partial \rho}{\partial x_3} - \epsilon F\left(\vec{\omega}_H \cdot \nabla_H \rho + \omega_3 \frac{\partial \rho}{\partial x_3}\right)\right\} = 0 .$$

As explained in Chapter 7, the quasi-geostrophic equations arise as a formal limit of the solutions of (8.85) as $\epsilon \rightarrow 0$ since the terms of order ϵ^{-1} in (8.85)

imply formally that

$$\vec{v}_H = \nabla_H^\perp \phi \qquad \text{geostrophic balance,}$$

(8.90)
$$\frac{\partial \phi}{\partial x_3} = -F^{-1}\rho \qquad \text{hydrostatic balance,}$$

$$w = 0\,,$$

and (8.90) combined with the conservation of potential vorticity in (8.89) yields the quasi-geostrophic equations. The rigorous proof of this formal limit presented in Chapter 7 requires that the initial data is in geostrophic balance so that all of the conditions in (8.90), to leading order in ϵ, are satisfied at time $t = 0$. Here we study the limiting behavior when the initial data are unbalanced so that the conditions in (8.90) are not satisfied.

For completeness, we record the

STRATIFIED QUASI-GEOSTROPHIC EQUATIONS:

$$(8.91) \quad \frac{\partial q}{\partial t} + \vec{v}_H \cdot \nabla q = 0\,, \quad \Delta_H \phi + F^2 \frac{\partial^2 \phi}{\partial x_3^2} = q\,, \quad \vec{v}_H = \nabla^\perp \phi = \begin{pmatrix} -\frac{\partial \phi}{\partial x_2} \\ \frac{\partial \phi}{\partial x_1} \end{pmatrix},$$

provided that the initial data is in geostrophic and hydrostatic balance. In equations (8.91), $q(x, t)$ is the potential vorticity,

$$x = (x_1, x_2, x_3)\,, \qquad \Delta_H = \frac{\partial^2}{\partial x_1^2} + \frac{\partial^2}{\partial x_2^2}\,,$$

and $F = \frac{f}{N}$ is the ratio of the buoyancy time to the rotation time.

In the rigorous analysis in part (i) of the theorem in Chapter 7, we utilized energy estimates to get the following behavior independent of ε for general (unbalanced) initial data:

PROPOSITION 8.1 *Assume that the initial data $\vec{u}_0(x)$ for the rotating Boussinesq equations in (8.85) or (8.86) belongs to X_s for integer s with $s \geq 3$, i.e., $\vec{u}_0(x) \in H^s(\mathbb{T}^3)$ and div $\vec{v}_0 = 0$. Then a classical solution $\vec{u}^\epsilon(x, t)$ of the nonlinear equations in (8.85) or (8.86) exists on a common time interval $[0, T]$, independent of ϵ for $0 < \epsilon \leq 1$, belongs to X_s for $0 \leq t \leq T$, and satisfies the a priori estimate*

$$(8.92) \qquad \max_{\substack{0 \leq t \leq T \\ 0 < \epsilon \leq 1}} \|\vec{u}^\epsilon(x, t)\|_s \leq C(T, \|\vec{u}_0\|_s)\,.$$

These key mathematical facts are essential when we discuss nonlinear Rossby adjustment for the Boussinesq equations. They also show that the abstract structure of Section 8.3 applies to these equations.

Energy Estimates. Consider the linear equation,

$$(8.93) \qquad \frac{\partial \vec{u}}{\partial \tau} + \mathcal{L}(\vec{u}) = 0\,, \quad \vec{u}\big|_{t=0} = \vec{u}_0(x)\,,$$

where $\mathcal{L}(\vec{u})$ is the operator in (8.87). As discussed in Chapter 7, the operator $\mathcal{L}(\vec{u})$ does not satisfy an energy principle for arbitrary initial data $\vec{u}_0(x)$. Nevertheless, an energy principle for solutions of (8.93) is valid provided that with $\vec{u}_0 = \begin{pmatrix} \vec{v}_0 \\ \rho_0 \end{pmatrix}$, the initial velocity field is incompressible, i.e., div $\vec{v}_0 = 0$. We know that if initially div $\vec{v}_0 = 0$, then the solution $\vec{u}(x, \tau)$ of (8.93) satisfies div $\vec{v}(x, \tau) = 0$ for all times τ, with $-\infty < \tau < \infty$. Thus, as discussed in Chapter 7, let $\mathcal{E} = \frac{1}{2}\vec{v} \cdot \vec{v} + \frac{1}{2}\rho^2$ and assume div $\vec{v} = 0$; then the solution $\vec{u}(x, t)$ of (8.93) satisfies

$$(8.94) \qquad \frac{\partial \mathcal{E}}{\partial \tau} = -\operatorname{div}((\mathcal{E} + \phi)\vec{v})$$

with ϕ the pressure for the linearized problem. This fact from (8.94) immediately implies that the operator $\mathcal{L}$ is skew-symmetric on the appropriate space.

PROPOSITION 8.2 *Let X_s be the closed subspace of the Sobolev space $H^s(\mathbb{T}^3)$ consisting of spatially periodic vectors $\vec{u}(x) = \binom{\vec{v}(x)}{\rho(x)}$ with div $\vec{v} = 0$. Denote the solution of (8.93) by $e^{-\tau\mathcal{L}}\vec{u}_0$. If $\vec{u}_0$ belongs to X_s, then $e^{-\tau\mathcal{L}}\vec{u}_0$ belongs to X_s and satisfies*

$$(8.95) \qquad \|e^{-\tau\mathcal{L}}\vec{u}_0\|_s = \|\vec{u}_0\|_s$$

for $-\infty < \tau < \infty$. In (8.94) $\|\cdot\|_s$ denotes the standard Sobolev norm for periodic functions.

The estimate in (8.95) is false if div $\vec{v} \neq 0$.

As established in Chapter 7, the nonlinear equations in either (8.86) or (8.87) admit a standard energy principle with the energy density $\mathcal{E} = \frac{1}{2}\vec{v} \cdot \vec{v} + \frac{1}{2}\rho^2$. Next, we explicitly diagonalize the linear operator $\mathcal{L}(\vec{u})$ restricted to the functions with div $\vec{v} = 0$.

Fourier Decomposition of $\mathcal{L}$. For simplicity we assume periodic boundary conditions for the spatial domain and look for the Fourier eigenfunctions of the operator $\mathcal{L}$. These periodic eigenfunctions $\vec{u}$ are of the form

$$(8.96) \qquad \vec{u} = \exp(i\vec{k} \cdot \vec{x} - i\omega(\vec{k})t)\vec{r}$$

where $\vec{u}(\vec{x}, t)$ must satisfy the incompressibility condition. There are three cases to consider depending on whether $\vec{k}_H \neq 0$, or $\vec{k}_H = 0$ but $\vec{k} \neq 0$, or else $\vec{k} = 0$. In the first case where $\vec{k}_H \neq 0$, the eigenfrequencies $\omega(\vec{k})$ associated to the wavenumber $\vec{k}$ are

$$(8.97) \qquad \begin{aligned} \omega_{(\vec{k})}^{(-1)} &= -\omega(\vec{k}) = -\frac{(|\vec{k}_H|^2 + F^2 k_3^2)^{1/2}}{|\vec{k}|}, \\[2mm] \omega_{(\vec{k})}^{(0)} &= 0 \text{ (double)}, \quad \omega_{(\vec{k})}^{(0)} = \omega(\vec{k}) = \frac{(|\vec{k}_H|^2 + F^2 k_3^2)^{1/2}}{|\vec{k}|}, \end{aligned}$$

and the corresponding right eigenvectors are given by

$$\vec{r}^{(-1)}_{(\vec{k})} \qquad\qquad \vec{r}^{(1)}_{(\vec{k})} \qquad\qquad \vec{r}^{(0)}_{(\vec{k})}$$

(8.98)
$$
\begin{pmatrix}
\dfrac{-Fk_2k_3 - i\omega(\vec{k})k_1k_3}{\sqrt{2}\,\omega(\vec{k})|\vec{k}_H||\vec{k}|} \\[2mm]
\dfrac{Fk_1k_3 - i\omega(\vec{k})k_2k_3}{\sqrt{2}\,\omega(\vec{k})|\vec{k}_H||\vec{k}|} \\[2mm]
i\dfrac{|\vec{k}_H|}{\sqrt{2}\,|\vec{k}|} \\[2mm]
\dfrac{|\vec{k}_H|}{\sqrt{2}\,\omega(\vec{k})|\vec{k}|}
\end{pmatrix}
\quad
\begin{pmatrix}
\dfrac{-Fk_2k_3 + i\omega(\vec{k})k_1k_3}{\sqrt{2}\,\omega(\vec{k})|\vec{k}_H||\vec{k}|} \\[2mm]
\dfrac{Fk_1k_3 + i\omega(\vec{k})k_2k_3}{\sqrt{2}\,\omega(\vec{k})|\vec{k}_H||\vec{k}|} \\[2mm]
-i\dfrac{|\vec{k}_H|}{\sqrt{2}\,|\vec{k}|} \\[2mm]
\dfrac{|\vec{k}_H|}{\sqrt{2}\,\omega(\vec{k})|\vec{k}|}
\end{pmatrix}
\quad
\begin{pmatrix}
-i\dfrac{k_2}{\omega(\vec{k})|\vec{k}|} \\[2mm]
i\dfrac{k_1}{\omega(\vec{k})|\vec{k}|} \\[2mm]
0 \\[2mm]
-i\dfrac{Fk_3}{\omega(\vec{k})|\vec{k}|}
\end{pmatrix}
$$

where the fourth eigenvector has been discarded because the corresponding eigensolution does not yield an incompressible velocity field. The eigenfunctions associated with $\vec{r}^{(\pm 1)}_{(\vec{k})}$ represent fast gravity waves, and the one associated with $\vec{r}^{(0)}_{(\vec{k})}$ represents a slow vortical mode.

In the second case when $\vec{k}_H = 0$ but $\vec{k} \neq 0$, the eigenfrequencies in equation (8.97) reduce to $\omega^{(\pm 1)}_{(\vec{k})} = \pm F$ and $\omega^{(0)}_{(\vec{k})} = 0$, and the corresponding eigenvectors are

$$\vec{r}^{(-1)}_{(\vec{k})} \qquad \vec{r}^{(0)}_{(\vec{k})} \qquad \vec{r}^{(1)}_{(\vec{k})}$$

(8.99)
$$
\begin{pmatrix}
-\dfrac{i}{\sqrt{2}} \\[2mm]
\dfrac{1}{\sqrt{2}} \\[2mm]
0 \\[2mm]
0
\end{pmatrix}
\quad
\begin{pmatrix}
0 \\[2mm]
0 \\[2mm]
0 \\[2mm]
1
\end{pmatrix}
\quad
\begin{pmatrix}
\dfrac{i}{\sqrt{2}} \\[2mm]
\dfrac{1}{\sqrt{2}} \\[2mm]
0 \\[2mm]
0
\end{pmatrix}.
$$

In this case the eigenvectors $\vec{r}^{(\pm 1)}_{(\vec{k})}$ correspond to fast rotation modes with the inertial rotation frequency $\epsilon^{-1}F$, whereas the remaining eigenvector $\vec{r}^{(0)}_{(\vec{k})}$ yields a slow mode. In the final case of mean flows with $\vec{k} = 0$ there are four admissible eigenfunctions, two with eigenfrequency $\omega^{(\pm 1)}_{(0)} = \pm 1$, and two others with eigenfrequency $\tilde{\omega}^{(\pm 1)}_{(0)} = \pm F$. The corresponding eigenfunctions are

$$\vec{r}^{(-1)}_{(0)} \qquad \vec{r}^{(1)}_{(0)} \qquad \tilde{\vec{r}}^{(-1)}_{(0)} \qquad \tilde{\vec{r}}^{(1)}_{(0)}$$

(8.100)
$$
\begin{pmatrix}
0 \\[2mm]
0 \\[2mm]
\dfrac{1}{\sqrt{2}} \\[2mm]
-\dfrac{i}{\sqrt{2}}
\end{pmatrix}
\quad
\begin{pmatrix}
0 \\[2mm]
0 \\[2mm]
\dfrac{1}{\sqrt{2}} \\[2mm]
\dfrac{i}{\sqrt{2}}
\end{pmatrix}
\quad
\begin{pmatrix}
-\dfrac{i}{\sqrt{2}} \\[2mm]
\dfrac{1}{\sqrt{2}} \\[2mm]
0 \\[2mm]
0
\end{pmatrix}
\quad
\begin{pmatrix}
\dfrac{i}{\sqrt{2}} \\[2mm]
\dfrac{1}{\sqrt{2}} \\[2mm]
0 \\[2mm]
0
\end{pmatrix}
$$

with $\vec{r}^{(\pm 1)}_{(0)}$ corresponding to fast gravity waves and $\tilde{\vec{r}}^{(\pm 1)}_{(0)}$ corresponding to fast rotation waves.

We remark that all the eigenvectors above have been normalized to give an orthonormal basis, and the corresponding Fourier eigenfunctions are an orthogonal family. In addition. these eigenvectors satisfy the symmetry conditions already discussed for RSWE needed to guarantee real-valued solutions. Thus, we have a concrete calculation showing that the operator $\mathcal{L}$ is skew-symmetric on the space of incompressible velocity fields as claimed earlier.

It is a natural question to ask whether the fast inertio-gravity waves defined from (8.96), (8.98), (8.99), and (8.100) as arbitrary superpositions of the Fourier eigenvectors associated with $\omega_{(k)}^{(\pm 1)}$ contribute to leading order to the potential vorticity equation in (8.88) for $\varepsilon \ll 1$. We have the following important fact:

LEMMA 8.3 *Consider the leading-order contribution to the potential vorticity,* $\omega_3 - F\frac{\partial \rho}{\partial x_3}$, *for* $\varepsilon \ll 1$. *Then* $\omega_3 - F\frac{\partial \rho}{\partial x_3}$ *vanishes identically for arbitrary contributions from the fast inertio-gravity waves.*

PROOF: We only need to show that $\omega_3 - F\frac{\partial \rho}{\partial x_3}$ vanishes for each of the explicit eigenmode solutions from (8.96) corresponding to the eigenvalues $\omega_{(k)}^{(\pm 1)}$ corresponding to the inertio-gravity waves, since every inertio-gravity is a time-dependent linear combination of these modes. This is a calculation utilizing the explicit form for the eigenvectors in (8.98), (8.99), and (8.100). For example, from (8.98), (8.96) for these eigenmodes individually, i.e., for $\vec{u} = e^{i\vec{k}\cdot\vec{x}}\vec{r}_{\vec{k}}^{\pm 1}$, we have

$$\omega_3 - F\frac{\partial \rho}{\partial x_3} = \frac{\partial}{\partial x_3}\left(e^{i\vec{k}\cdot\vec{x}}v_2^{\pm 1}(\vec{k})\right) - \frac{\partial}{\partial x_2}\left(e^{i\vec{k}\cdot\vec{x}}v_1^{\pm 1}(\vec{k})\right) - F\frac{\partial}{\partial x_3}\left(e^{i\vec{k}\cdot\vec{x}}\rho^{\pm 1}(\vec{k})\right)$$

$$= i\frac{\exp(i\vec{k}\cdot\vec{x})}{\sqrt{2}}\left(\frac{F(k_1^2 + k_2^2)k_3}{\omega^{\pm}(\vec{k})|\vec{k}_H||\vec{k}|} - F\frac{k_3|\vec{k}_H|}{\omega^{\pm}(\vec{k})|\vec{k}|}\right) = 0,$$

as needed in the lemma. The interested reader can check these straightforward algebraic calculations in more detail. We will see below that the above fact for geophysical flows is the special structure needed to guarantee nonlinear Rossby adjustment for $\varepsilon \ll 1$. $\square$

Nonlinear Rossby Adjustment for the Stratified Boussinesq Equations for $\varepsilon \ll 1$. Here I show that the stratified quasi-geostrophic dynamics from (8.91) emerge from the Boussinesq equations as $\varepsilon \to 0$ for general unbalanced initial data. I will also show that the slow dynamics for (8.85) as $\varepsilon \to 0$ is determined by the stratified quasi-geostrophic equations from (8.91). In contrast to the direct algebraic calculational proof for RSWE presented in the last section, the proof presented below will be conceptual and emphasizes the special structure of RSSBE as noted in the previous lemma.

The rigorous version of the averaging principle presented in Section 8.3 [**34**] guarantees that

$$(8.101) \qquad \vec{u}^\epsilon(\vec{x}, t) = e^{-\frac{t}{\epsilon}\mathcal{L}}\bar{u}(\vec{x}, t) + o(1)$$

for $\varepsilon \ll 1$ on the time interval $0 \leq t \leq T$. With the decomposition into slow and fast waves u_{I} and u_{II}, we have

$$(8.102) \qquad e^{-\frac{t}{\epsilon}\mathcal{L}}\bar{u}(\vec{x}, t) = \bar{u}_{\mathrm{I}}(\vec{x}, t) + e^{-\frac{t}{\epsilon}\mathcal{L}}\bar{u}_{\mathrm{II}}(\vec{x}, t) + o(1)$$

where

$$(8.103) \qquad \bar{u}_{\mathrm{I}}(\vec{x}, t) = \sum_{\vec{k}\in\mathbb{Z}^3\setminus\{0\}} e^{i\vec{k}\cdot\vec{x}}\sigma_{(\vec{k})}^{(0)}(t)\vec{r}_{(\vec{k})}^{(0)}$$

so that the slow waves belong to the nullspace of $\mathcal{L}$ and

$$\bar{u}_{\mathrm{II}}(\vec{x}, t) = \sum_{\vec{k} \in \mathbb{Z}^3 \setminus \{0\}} \sum_{\alpha = \pm 1} e^{i\vec{k}\cdot\vec{x}} \sigma_{(\vec{k})}^{(\alpha)}(t) \vec{r}_{(\vec{k})}^{(\alpha)} \tag{8.104}$$

with $\vec{r}_{(\vec{k})}^{(\alpha)}$ the explicit inertio-gravity wave eigenvectors for the rotating Boussinesq equations.

We combine (8.101) with (8.102)–(8.104) and write $\vec{u}^\epsilon(\vec{x}, t)$ in the form

$$\vec{u}^\epsilon(\vec{x}, t) = \bar{u}_{\mathrm{I}}(\vec{x}, t) + u'_\epsilon(\vec{x}, t) + o(1) \tag{8.105}$$

where

$$u'_\epsilon = \sum_{\vec{k} \in \mathbb{Z}^3 \setminus \{0\}} \sum_{\alpha = \pm 1} e^{i(\vec{k}\cdot\vec{x} - \omega_{(\vec{k})}^{(\alpha)} \frac{t}{\epsilon})} \sigma_{(\vec{k})}^{(\alpha)}(t) \vec{r}_{(\vec{k})}^{(\alpha)} \,. \tag{8.106}$$

The quantity u'_ϵ describes the rapidly oscillating gravity waves in the singular limit. Our objective is to verify that $\bar{u}_{\mathrm{I}}(\vec{x}, t)$ is determined by the solution of the quasi-geostrophic equations in (8.91) with the initial data for (8.103) determined from $\vec{u}_0(\vec{x})$ at $t = 0$ via the Fourier coefficients $\sigma_{(\vec{k})}^{(0)}(0)$.

First, we verify that $\bar{u}_{\mathrm{I}}(\vec{x}, t)$ with the form in (8.103) automatically satisfies the geostrophic and hydrostatic balance conditions in (8.90). To check this we define $\bar{\phi}$ by the formula

$$\bar{\phi}(\vec{x}, t) = \sum_{\vec{k} \in \mathbb{Z}^3 \setminus \{0\}} e^{i\vec{k}\cdot\vec{x}} \sigma_{(\vec{k})}^{(0)}(t) (\omega(\vec{k})|\vec{k}|)^{-1} \tag{8.107}$$

so that with the explicit formulas for the right eigenvectors, we verify all the conditions in (8.90). Next, we compute from (8.90) that

$$\bar{q} \stackrel{\text{def}}{=} \bar{\omega}_3 - F \frac{\partial \bar{\rho}}{\partial x_3} = \Delta_H \bar{\phi} + F^2 \frac{\partial^2 \bar{\phi}}{\partial x_3^2} \,. \tag{8.108}$$

We remark here that all of these manipulations as well as those below are justified because $\bar{u}(\vec{x}, t)$ belongs to $H^3(\mathbb{T}^3)$ uniformly for $0 \le t \le T$ so that the corresponding Fourier series of $\bar{u}_{\mathrm{I}}$ and its first derivatives converge absolutely. At this point in the argument, we have established that

$$\bar{v} = \nabla^\perp \bar{\phi} \,, \quad \bar{\rho} = -F^{-1} \frac{\partial \bar{\phi}}{\partial x_3} \,, \tag{8.109}$$

and (8.107).

What remains is to establish that

$$\frac{\partial \bar{q}}{\partial t} + \nabla^\perp \bar{\phi} \cdot \nabla \bar{q} = 0 \tag{8.110}$$

in order to complete the proof of weak convergence to the quasi-geostrophic solution for general unbalanced initial data and to verify nonlinear Rossby adjustment. To verify this, we utilize Ertel's theorem on conservation of potential vorticity in the weak sense. Thus, we multiply (8.89) by a smooth test function $\psi(\vec{x}, t)$ with compact support in space-time and integrate by parts to obtain

$$(8.111) \qquad \int_0^\infty \int_{\mathbb{T}^3} \psi_t q^\epsilon + \vec{v}^\epsilon q^\epsilon \cdot \nabla \psi \, dx \, dt = 0$$

where

$$(8.112) \qquad q^\epsilon = \omega_3^\epsilon - F\frac{\partial \rho^\epsilon}{\partial x_3} - \epsilon F\left(\vec{\omega}_H^\epsilon \cdot \nabla_H \rho^\epsilon + \omega_3^\epsilon \frac{\partial \rho^\epsilon}{\partial x_3}\right).$$

The special fact regarding the potential vorticity in (8.112) is that

$$(8.113) \qquad q^\epsilon = \bar{q} + o(1),$$

i.e., the potential vorticity has no leading-order gravity wave fluctuations for $\epsilon \ll 1$.

Here is the proof of (8.113). As a consequence of Proposition 8.1 yielding the uniform bound on $\vec{u}^\epsilon(\vec{x}, t)$ in $H^3(\mathbb{T}^3)$, $\vec{\omega}_H^\epsilon \cdot \nabla_H \rho^\epsilon + \omega_3^\epsilon \frac{\partial \rho^\epsilon}{\partial x_3}$ is uniformly bounded. Thus we have the fact that

$$(8.114) \qquad q^\epsilon = \bar{q} + q' + o(1)$$

where $q' = \omega_3' - F\frac{\partial \rho'}{\partial x_3}$ is computed from u'_ϵ in (8.106). The remarkable key algebraic fact presented above in Lemma 8.3 is that q' *vanishes identically for any arbitrary Fourier amplitudes* $\sigma_{(\vec{k})}^{(\pm 1)}(t)$, i.e., $q' \equiv 0$. All of these facts combine to yield (8.113).

With (8.113), it is easy to finish the proof from (8.105), (8.106), and (8.109),

$$(8.115) \qquad \vec{v}^\epsilon = \nabla^\perp \bar{\phi} + v'_\epsilon + o(1).$$

Furthermore, from (8.106), v'_ϵ converges weakly to zero as can be deduced from the same argument utilized for nonlinear Rossby adjustment for RSWE above (8.44). Thus (8.111) and (8.113) combine to yield the fact that

$$\int_0^\infty \int_{\mathbb{T}^3} \psi_t \bar{q} + \nabla^\perp \bar{\phi} \bar{q} \cdot \nabla \psi \, dx \, dt = 0$$

for all test functions ψ. Thus $\bar{u}_I$ defines a solution of the stratified quasi-geostrophic equations, and, furthermore, with space-time filtering, nonlinear Rossby adjustment occurs.

CHAPTER 9

Waves and PDEs for the Equatorial Atmosphere and Ocean

At the equator, the tangential projection of the Coriolis force from rotation vanishes identically. This has profound consequences physically that allows the tropics to behave as a waveguide with extremely warm surface temperatures. The resulting behavior profoundly influences our short-term midlatitude climate through monsoons, El Ninõ, and its resulting teleconnections in the midlatitude atmosphere. How this happens through detailed physical mechanisms is an important contemporary research problem in the AOS community. An introduction to the dynamics of El Ninõ can be found in the beautiful book by Philander [**30**], while the fashion in which tropical heating affects the atmospheric circulation through clouds, moisture, and convection is discussed in a series of contemporary papers in the volume edited by Smith [**35**]. This last topic has great importance with numerous unsolved AOS problems; in particular, contemporary numerical general circulation models (GCMs) used by nations around the world for weather and climate prediction, often fail to capture this interaction between tropical clouds, moisture, and the atmospheric circulation in an accurate fashion. Thus, the need for additional theory is very important for these disciplinary problems.

The goal in this final chapter is to introduce the reader to the fascinating new phenomena in both waves and PDEs that emerge as a consequence of the vanishing of the tangential projection of the Coriolis force terms at the equator. This will involve pulling together and applying many of the ideas used in earlier chapters of the book as well as geophysical issues that are unique to the equatorial region. Chapter 11 of Gill [**11**] is recommended as supplementary reading on the physical phenomena for the tropics. The rigorous mathematical theory of nonlinear PDEs for the equatorial regions is waiting to be developed with many novel issues emerging. Some of these new model nonlinear equations are derived through formal asymptotics in Section 9.3, but there are many more novel and important approximate equations emerging from contemporary research on the tropics; see the recent paper by Majda and Klein [**21**] for a discussion as well as additional references in the bibliography of that paper.

9.1. Introduction to Equatorial Waves for Rotating Shallow Water

We begin by recalling the shallow water equations

$$(9.1) \qquad \frac{D\vec{v}}{Dt} + f\vec{v}^\perp + g\nabla h = 0, \qquad \frac{Dh}{Dt} + (H+h)\operatorname{div}\vec{v} = 0,$$

where

$$\vec{v} = \begin{pmatrix} u \\ v \end{pmatrix}, \quad \vec{v}^\perp = \begin{pmatrix} -v \\ u \end{pmatrix}, \quad \frac{D}{Dt} = \frac{\partial}{\partial t} + \vec{v} \cdot \nabla.$$

In the β-plane approximation, the rotation parameter is

$$f = f_0 + \beta y.$$

In Chapter 4 we have studied the f-plane, i.e., $f_0 \neq 0$ and $\beta = 0$, and found in linear theory, the dispersion relations

$$\omega_\pm = \sqrt{f_0^2 + gH|\vec{k}|^2} \quad \text{gravity waves,}$$
$$\omega_0 = 0 \quad \text{geostrophic balance.}$$

We have also looked at the effects of finite β at midlatitudes in the midlatitude planetary equations in Section 5.1. We found in (5.7) the dispersion relation $\omega = \frac{-\beta k_1}{F + |\vec{v}|^2}$, with $F = \frac{f_0^2}{gH}$, midlatitude Rossby waves. A quick check shows that F has units inverse length squared. As discussed in Chapter 4, this length scale is called the Rossby deformation radius, or Rossby length for short:

$$(9.2) \qquad L_R = \frac{\sqrt{gH}}{f_0}, \qquad F = L_R^{-2}.$$

Beyond this length, the effects of rotation take over. For instance, in the potential vorticity of the midlatitude equations, $q = \Delta \psi - F \psi$, the latter term dominates at scales larger than the Rossby radius.

Equatorial waves comprise the remaining β-plane regime, $f_0 = 0$, but $\beta \neq 0$. But first we review a proof of Ertel's conservation theorem, which holds for the shallow water equations (9.1) for any f. To find the conserved quantity, we will use the vector identity

$$\vec{v} \cdot \nabla \vec{v} = \vec{\omega} \times \vec{v} + \nabla \left(\frac{1}{2} |\vec{v}|^2 \right).$$

This identity is true in three dimensions as well as in the reduction to two dimensions through $\vec{\omega} = (0, 0, \omega)^\mathsf{T}$ and $\omega = \frac{\partial v}{\partial x} - \frac{\partial u}{\partial y}$. The momentum equations in (9.1) can now be written component-wise as

$$(9.3) \qquad \begin{aligned} \frac{\partial u}{\partial t} - (f + \omega)v &= -\frac{\partial}{\partial x} \left(gh + \frac{1}{2}(u^2 + v^2) \right), \\ \frac{\partial v}{\partial t} + (f + \omega)u &= -\frac{\partial}{\partial y} \left(gh + \frac{1}{2}(u^2 + v^2) \right). \end{aligned}$$

Taking the curl kills the right-hand side, a gradient yielding

$$\frac{\partial \omega}{\partial t} + \frac{\partial}{\partial x}((f + \omega)u) + \frac{\partial}{\partial y}((f + \omega)v) = 0.$$

This equation implies conservation of vorticity: $\int \omega = \text{const}$. We rewrite it as

$$(9.4) \qquad \frac{1}{f + \omega} \frac{D\omega}{Dt} + \text{div } \vec{v} = 0.$$

The mass conservation equation in (9.1) can be written

$$\text{(9.5)} \qquad \frac{1}{H+h}\frac{D}{Dt}(H+h) + \text{div}\,\vec{v} = 0\,.$$

Combining (9.4) and (9.5) yields the Ertel theorem

$$\text{(9.6)} \qquad \frac{D}{Dt}\left(\frac{f+\omega}{H+h}\right) = 0\,.$$

That is, the quantity $Q = (f+\omega)/(H+h)$ is constant along characteristics carried by the flow. As remarked in Chapter 4, this gives us the infinite number of conserved quantities listed below (4.19).

There is also a conserved energy as discussed in Section 4.3. The energy density

$$\text{(9.7)} \qquad \mathcal{E} = \frac{1}{2}(H+h)|\vec{v}|^2 + \frac{1}{2}\frac{c^2}{H}h^2$$

satisfies the local conservation law

$$\text{(9.8)} \qquad \frac{\partial \mathcal{E}}{\partial T} + \text{div}\left(\left(\mathcal{E} + \frac{1}{2}\frac{c^2}{H}h^2 + c^2 h\right)\vec{v}\right) = 0$$

where $c = \sqrt{gH}$ is the gravity wave speed. So the total energy is conserved,

$$\frac{\partial}{\partial t}\int \mathcal{E}\,d\vec{x} = 0\,.$$

For the midlatitude equations, we found the conserved potential vorticity Q to be independent of the fast waves in the low Rossby number limit in Chapter 8.

Moving to the equatorial region, there are two interesting properties that should be mentioned. In the midlatitudes, there was a time-scale separation between the gravity waves and the geostrophic wave. The slowest gravity wave was much faster than the geostrophic waves in the low Rossby limit. This is not true at the equator, where many of the waves are on the same time scale. The second point is that the equator acts as a waveguide, as we will see shortly.

We will do linear theory with equatorial waves, as this is already an amazingly rich and subtle problem. We will linearize about the state $\vec{v} = 0$ and $h = 0$, which amounts to simply dropping the nonlinear terms.

LINEARIZED EQUATORIAL SHALLOW WATER EQUATIONS:

$$\text{(9.9)} \quad \frac{\partial u}{\partial t} - \beta y v = -g\frac{\partial h}{\partial x}\,, \quad \frac{\partial v}{\partial t} + \beta y u = -g\frac{\partial h}{\partial y}\,, \quad \frac{\partial h}{\partial t} + H\left(\frac{\partial u}{\partial x} + \frac{\partial v}{\partial y}\right) = 0\,.$$

9.1.1. Kelvin Waves. These waves will be in geostrophic balance in the meridional (north-south or y) direction, but will move at gravity wave speeds, so they have this mixture of the properties of what in the midlatitudes are separate classes of waves. They will be trapped in the equatorial waveguide. They are an important part of the observational record of the equatorial atmosphere and ocean. We seek flows parallel to the equator ($v = 0$), so that the v-component of the momentum equation yields the geostrophic balance,

$$(9.10) \qquad \beta y u = -g \frac{\partial h}{\partial y}.$$

The remaining evolution equations become

$$(9.11) \qquad \frac{\partial u}{\partial t} = -g \frac{\partial h}{\partial x}, \qquad \frac{\partial h}{\partial t} = -H \frac{\partial u}{\partial x}.$$

Introducing characteristic variables

$$(9.12) \qquad q = \left(\frac{g}{H} \right)^{1/2} h + u \quad \text{and} \quad r = \left(\frac{g}{H} \right)^{1/2} h - u$$

allows us to isolate the eastward and westward moving waves

$$(9.13) \qquad \frac{\partial q}{\partial t} + c \frac{\partial q}{\partial x} = 0 \;\; \text{eastward} \quad \text{and} \quad \frac{\partial r}{\partial t} - c \frac{\partial r}{\partial x} = 0 \;\; \text{westward}$$

where $c = \sqrt{gH}$ is the gravity wave speed. It looks like a general solution will be an arbitrary superposition of these two waves, but we need to inquire into their structure in y.

Eastward Group: $r \equiv 0$ (so $u = ch/H$). Solving (9.13) is naturally accomplished by separation of variables

$$q = G(y)q(x - ct)$$

where $G(y)$ is determined by the requirement of geostrophic balance (9.10), which leads to an equation for G

$$\frac{\partial G}{\partial y} + \frac{\beta}{c} y G(y) = 0,$$

which immediately integrates to gives

$$(9.14) \qquad G(y) = G_0 e^{-\frac{\beta}{c} \frac{y^2}{2}},$$

and we conclude that the equator acts as a waveguide, trapping the wave.

Westward Group: $q \equiv 0$. Again, separation of variables leads to an equation for the structure in y, but the sign is changed from the eastward group, so

$$\frac{\partial G}{\partial y} - \frac{\beta y}{c} G(y) = 0 \quad \text{gives} \quad G(y) = G_0 e^{\frac{\beta}{c} \frac{y^2}{2}}.$$

This solution violates finite energy and is clearly unphysical with large growth as y increases.

In summary, Kelvin waves only propagate in the eastward direction, do not disperse, and travel with the gravity wave speed $c = \sqrt{gH}$. They are in geostrophic balance and are trapped in the equatorial waveguide.

Inquiring into units, we see that there is a length scale that determines the trapping distance in y,

$$L_e = \sqrt{\frac{c}{\beta}} \quad \text{equatorial deformation radius.}$$

We will see that L_e plays a role for the equator that the Rossby radius L_R played for the midlatitudes. Local wave phenomena will only reside inside the L_e length scale. Finally, we write down the full Kelvin wave solution,

$$(9.15) \qquad \vec{v} = \begin{pmatrix} \left(\frac{g}{H}\right)^{1/2} G(x - ct) e^{-\frac{\beta}{c}\frac{y^2}{2}} \\ 0 \end{pmatrix}, \qquad h = G(x - ct) e^{-\frac{\beta}{c}\frac{y^2}{2}}.$$

9.1.2. Mixed Rossby-Gravity Waves. These waves go by several names and are sometimes called Yanai waves. Motivated by the manipulations that arose for Kelvin waves, we transform the u- and h-variables in the linear equatorial shallow water equations (9.9) according to the characteristics (9.12):

$$(9.16) \qquad \begin{aligned} \frac{\partial q}{\partial t} + c\frac{\partial q}{\partial x} + c\frac{\partial v}{\partial y} - \beta y v &= 0, \\ \frac{\partial r}{\partial t} - c\frac{\partial r}{\partial x} - c\frac{\partial v}{\partial y} - \beta y v &= 0, \\ \frac{\partial v}{\partial t} + \beta y \frac{q - r}{2} + c\frac{\partial}{\partial y}\frac{q + r}{2} &= 0. \end{aligned}$$

We look for solutions with $r \equiv 0$ again, which implies

$$c\frac{\partial v}{\partial y} + \beta yy = 0.$$

The Kelvin waves were the trivial $v = 0$ solution, but we see there is a general solution with nonzero meridional velocity,

$$(9.17) \qquad v = A(x, t) e^{-\frac{\beta}{c}\frac{y^2}{2}}.$$

Plugging this into the remaining two equations in (9.16) gives two coupled equations for u and q. To cancel the y-dependence, we need q of the form

$$(9.18) \qquad q = B(x, t) y e^{-\frac{\beta}{c}\frac{y^2}{2}}.$$

This leads to a 2×2 dispersive wave system

$$(9.19) \qquad \frac{\partial B}{\partial t} + c\frac{\partial B}{\partial x} = (2\beta)A, \qquad \frac{\partial A}{\partial t} = -\frac{c}{2}B.$$

Ignoring momentarily the coupling terms on the right-hand side, we see two characteristic wave speeds, 0 and c. The former we associate with planetary waves, and the latter with gravity waves. But here they are mixed by the coupling. We need to find the eigenmodes of the system. For Fourier eigenfunctions

$$(9.20) \qquad B(x, t) = B(k)e^{i(kx - \omega t)} \quad \text{and} \quad A(x, t) = A(k)e^{i(kx - \omega t)},$$

the dispersion relation is given by the roots of the characteristic equation $\omega(\omega - kc) - \beta c = 0$,

$$(9.21) \qquad \omega_{\pm} = \frac{ck}{2} \pm \left(\frac{c^2 k^2}{4} + c\beta\right)^{1/2}.$$

In terms of the equatorial length $L_e = (\frac{c}{\beta})^{1/2}$, this can be written

$$(9.22) \qquad \frac{\omega_{\pm}}{\frac{c}{2}} = k \pm \left(k^2 + \frac{4\beta}{c}\right)^{1/2} = k \pm (k^2 + 4L_e^{-2})^{1/2}.$$

The k term represents advection, which is added onto a term that looks like the gravity wave dispersion relation

$$\frac{\omega_{\pm}}{c} = \pm\left(|\vec{k}|^2 + \frac{f_0^2}{c^2}\right)^{1/2} = \pm(|\vec{k}|^2 + L_R^{-2})^{1/2} \quad \text{midlatitude gravity.}$$

This is the first connection that was hinted at earlier between the roles of the Rossby and equatorial deformation radii. The group velocity

$$(9.23) \qquad \frac{1}{\frac{c}{2}}\frac{d\omega}{dk} = 1 \pm \frac{k}{(k^2 + \frac{4\beta}{c})^{1/2}} > 0$$

always points to the east. This is important for wave packets, short-wavelength waves with large-scale modulation. However, the phase velocity $\frac{\omega}{k}$ points, for example, to the west for ω_+ for all k with $k < 0$ and for ω_- for all k with $k > 0$. Let's write down the velocity field for a mixed wave:

$$(9.24) \qquad \vec{v} = \begin{pmatrix} -\frac{\omega y}{c}e^{-\frac{\beta y^2}{2c}} \sin(kx - \omega t) \\ e^{-\frac{\beta y^2}{2c}} \cos(kx - \omega t) \end{pmatrix}.$$

In contrast to Kelvin waves, the mixed planetary-gravity waves are dispersive, and they are not in geostrophic balance. The fluid velocity on the equator is solely north-south, with u taking its maximum value off the equator. The waves are still trapped in the equatorial waveguide.

EXERCISE 9.1. Using the techniques of Section 5.4 from Chapter 5, describe the long-time behavior of mixed planetary-gravity waves with arbitrary, compactly supported initial data. How do they disperse? The answer involves dispersion over wave speeds from 0 to c (the gravity wave speed) in the eastward direction. They disperse over a wider range of characteristics than gravity waves for linear rotating shallow water; discuss this in detail. Also study the dispersion relations for these mixed waves in the limits $k \to \pm\infty$ and show that they look like a gravity wave at one end of the spectrum and like a planetary Rossby wave at the other. Unlike the midlatitude waves, there is no separation of time scales, as the frequencies take both large and small values.

9.1.3. Weakly Nonlinear Kelvin Waves. We consider the nonlinear equatorial shallow water equations,

$$(9.25) \qquad \begin{aligned} \eta_t + [(1+\eta)u]_x + [(1+\eta)v]_y &= 0, \\ u_t + uu_x + vu_y + \eta_x - yv &= 0, \\ v_t + uv_x + vv_y + \eta_y + yu &= 0, \end{aligned}$$

written in nondimensional form through the length scale $L_e = (\frac{c}{\beta})^{1/2}$, time scale $T_e = (c\beta)^{-1/2}$, velocity scale c, and unit for height $\frac{c^2}{g}$. As shown in the formula from (9.15) above, the linear Kelvin waves are nondispersive but equatorially trapped. It is natural to see whether nonlinearity can play an important role in the self-interaction of these waves at longer times. Here we assess this nonlinear interaction through formal asymptotic methods similar to those utilized in Section 5.7.

We consider solutions to these equations in the form of power series in small parameter ϵ (weakly nonlinear expansion),

$$(9.26) \qquad \begin{pmatrix} \eta \\ u \\ v \end{pmatrix} = \begin{pmatrix} \epsilon K(x-t,\tau)e^{-\frac{y^2}{2}} \\ \epsilon K(x-t,\tau)e^{-\frac{y^2}{2}} \\ 0 \end{pmatrix} + \epsilon^2 \vec{u}_2 \,,$$

where the parameter $\tau = \epsilon t$ is a long-time scale. We seek solutions that are valid uniformly for long times with τ of order 1. This particular form of the first-order solution is drawn from the linear theory of Kelvin waves in Section 9.1.1 (recall that $q = u + \eta$, $r = -u + \eta$, and $v = 0$ for a linear Kelvin wave). The first-order solution $\vec{u}_1$ is the Kelvin wave solution of the following linear system:

$$(9.27) \qquad \begin{aligned} \frac{\partial q}{\partial t} + \frac{\partial q}{\partial x} + \frac{\partial v}{\partial y} - yv &= 0 \,, \\ \frac{\partial r}{\partial x} - \frac{\partial r}{\partial x} + \frac{\partial v}{\partial y} + yv &= 0 \,, \\ \frac{\partial v}{\partial t} + y\frac{q-r}{2} + \frac{\partial}{\partial y}\frac{q+r}{2} &= 0 \,. \end{aligned}$$

As discussed in Sections 5.7 and 8.3 above, in order for the asymptotic expansion in (9.26) to be valid formally, we must require that solutions at the next order grow sublinearly in t, i.e.,

$$(9.28) \qquad |\vec{u}_2| = o(t) \,.$$

Now consider the forced linear system for $\vec{u}_2$, which appears at the next order:

$$(9.29\text{a}) \qquad \frac{\partial \eta_2}{\partial t} + \frac{\partial u_2}{\partial x} + \frac{\partial v_2}{\partial y} = A \,,$$

$$(9.29\text{b}) \qquad \frac{\partial u_2}{\partial t} + \frac{\partial \eta_2}{\partial x} - yv_2 = B \,,$$

$$(9.29\text{c}) \qquad \frac{\partial v_2}{\partial t} + \frac{\partial \eta_2}{\partial x} + yu_2 = 0 \,.$$

The last equation in the system above has no forcing, since the y-component of first-order velocity satisfies $v_1 = 0$. The components of forcing, A and B, have terms arising from two different sources: from a slow time derivative and from nonlinearity. They have the following structure:

$$(9.30) \qquad \begin{aligned} A &= -e^{-\frac{y^2}{2}} K_\tau - e^{-y^2}(K^2)_\theta \,, \\ B &= -e^{-\frac{y^2}{2}} K_r - e^{-y^2}\left(\frac{1}{2}K^2\right)_\theta \,, \end{aligned}$$

where θ is the phase variable, $\theta = x - t$. Now, we add the equations in (9.29a) and (9.29b) to get

$$(9.31) \qquad \frac{\partial(u_2 + \eta_2)}{\partial t} + \frac{\partial(u_2 + \eta_2)}{\partial x} + \frac{\partial v_2}{\partial y} - y v_2 = (A + B) .$$

In order to compute the nonlinear Kelvin wave response to forcing, we project the above equation onto the Kelvin mode by multiplying it by $e^{-y^2/2}$ and noticing that

$$(9.32) \qquad e^{-\frac{y^2}{2}} \left(\frac{\partial v_2}{\partial y} - y v_2 \right) = \frac{\partial}{\partial y} \left(e^{-\frac{y^2}{2}} v_2 \right)$$

is a perfect derivative. If we assume that v_2 does not grow too much in the y-direction in order to have a valid asymptotic expansion (which is a natural requirement for equatorial waves), then

$$(9.33) \qquad \lim_{L \to \infty} \int_{-L}^{L} \frac{\partial}{\partial y} \left(e^{-\frac{y^2}{2}} v_2 \right) dy = \lim_{L \to \infty} \left. \left(e^{-\frac{y^2}{2}} v_2 \right) \right|_{-L}^{L} = 0 .$$

In order to formulate the nonresonance condition and to shorten the notation, we introduce the Kelvin wave projection of the second-order solution $\tilde{q}_2$ by

$$(9.34) \qquad \tilde{q}_2 = \int_{-\infty}^{\infty} e^{-\frac{y^2}{2}} (u_2 + \eta_2) dy .$$

The resulting projected equation for the Kelvin wave response to forcing has the form

$$(9.35) \quad \frac{\partial \tilde{q}_2}{\partial t} + \frac{\partial \tilde{q}_2}{\partial x} = -2 \int_{-\infty}^{\infty} e^{-y^2} K_\tau(x-t, \tau) dy - \frac{3}{2} \int_{-\infty}^{\infty} e^{-\frac{3y^2}{2}} K_\theta^2(x-t, \tau) dy .$$

This equation presents a specific example of an "auxiliary problem" discussed earlier in Section 5.7 and arising in a number of forced dispersive systems,

$$(9.36) \qquad \frac{\partial \tilde{q}}{\partial t} + \frac{\partial \tilde{q}}{\partial x} = f(x - t, \tau) ,$$

where the forcing can depend on a long-time scale. The equation above has an exact solution, vanishing at $t = 0$,

$$(9.37) \qquad \tilde{q} = t f(x - t) .$$

This solution is obviously secular for long times of order $\tau = \epsilon t$. In order for a solution to be valid, we must require a sublinear growth of the solution $\tilde{q}$. The only way to satisfy this requirement is to set $f \equiv 0$. Imposing this condition in (9.35) yields the

NONLINEAR KELVIN WAVE EQUATION:

$$(9.38) \quad c_1 K_\tau + c_2 K_\theta^2 = 0 , \quad c_1 = 2 \int_{-\infty}^{\infty} e^{-y^2} dy , \quad c_2 = \frac{3}{2} \int_{-\infty}^{\infty} e^{-\frac{3}{2} y^2} dy .$$

This is the famous inviscid Burgers equation, which has the typical feature of breaking, shock formation, and dissipation of energy stored in the propagated wave. It arises in many contexts as the canonical asymptotic equation for nondispersive waves; see [**16**, chap. 1].

In the equatorial atmosphere and ocean, Kelvin waves interact with many other processes. Novel asymptotic equations involving the coupling of Kelvin waves with mixed Rossby-gravity waves through topography yield the Burgers equation in (9.38) coupled at large scales to ODEs. These equations are derived and some of their remarkable properties are discussed by Rosales, Tabak, Turner, and the author in [**22**]. In particular, the asymptotic equations have smooth traveling waves; they should be studied further analytically.

9.2. The Equatorial Primitive Equations

Motivated by the elementary solutions of the equatorial shallow water equations in Section 9.1, here we show how to obtain a complete family of waves for the linear equatorial primitive equations. An expansion technique as in Section 1.5 plays a central role. We consider the linearized hydrostatic primitive equations in the troposphere with source terms. We assume the equatorial β-plane approximation that covers the whole of the tropics up to 30° latitude with the maximum percentage error of only 14% (see Gill [**11**]),

$$
\begin{aligned}
\frac{\partial \vec{v}_H}{\partial t} + \beta y \vec{v}_H^\perp &= -\nabla_H p \,, \\
-p_z + B &= 0 \,, \\
\frac{\partial B}{\partial t} + N^2 w &= \mathcal{S}(x, y, z, t) \,, \\
\operatorname{div}_H \vec{v}_H + w_z &= 0 \,.
\end{aligned}
$$

(9.39)

Here $\vec{v}_H = (u(x, y, z, t), v(x, y, z, t))$ is the horizontal velocity field with $\vec{v}_H^\perp = (-v, u)$, w is the vertical velocity, B is the buoyancy with N the constant buoyancy frequency, β is the Coriolis parameter coefficient, $\beta = 2\Omega/R$ with $\Omega = (2\pi)/24$ hours the angular velocity of the earth, $R = 6378$ km its radius, and $\mathcal{S}$ a forcing term that may represent heating from evaporation or cooling by radiation effects, etc. We assume rigid lid boundary conditions at top and bottom of the troposphere, namely,

(9.40)
$$
w(x, y, z, t)\big|_{z=0, H} = 0 \,,
$$

where H is the height of the troposphere, and $z = 0$ represents the surface of the earth.

As shown already in Section 1.5, under the rigid lid boundary conditions in (9.40), the system in (9.39) decouples into a barotropic equation and an infinite number of shallow water equations. Since the wave properties of the barotropic mode have been developed in Sections 1.2, 1.5, and 5.1 above and involve Rossby waves, we will not repeat this discussion here and concentrate on the other equatorial modes.

We assume the following general expansion in vertical normal modes for the source term,

(9.41)
$$
\mathcal{S}(x, y, z, t) = \sum_{q=1}^{+\infty} \mathcal{S}_q(x, y, t) \frac{d G_q(z)}{dz} \,,
$$

with the following expansions for the variables:

$$\vec{v}_H(x, y, z, t) = \sum_{q=1}^{+\infty} \vec{v}_H^q(x, y, t) G_q(z) ,$$

$$p(x, y, z, t) = \sum_{q=1}^{+\infty} p^q(x, y, t) G_q(z) ,$$

(9.42)

$$B(x, y, z, t) = \sum_{q=1}^{+\infty} B^q(x, y, t) \frac{d G_q(z)}{dz} ,$$

$$w(x, y, z, t) = \sum_{q=1}^{+\infty} w^q(x, y, t) \frac{d G_q(z)}{dz} .$$

As in Section 1.5, the vertical modes $G_q(z)$ will be determined by the above procedure.

By plugging in the ansatz in (9.42) into (9.39) and using (9.41), we get

(9.43)
$$\frac{\partial \vec{v}_H^q}{\partial t} + \beta y \vec{v}^q{}_H^\perp = -\nabla_H p^q ,$$

(9.44)
$$p^q = B^q ,$$

(9.45)
$$\frac{\partial B^q}{\partial t} + N^2 w^q = \mathcal{S}_q(x, y, t) ,$$

(9.46)
$$\operatorname{div}_H \vec{v}_H^q G_q(z) + w^q \frac{d^2 G}{dz^2} = 0 .$$

Also combining (9.44) and (9.45), we get

$$w^q = \frac{1}{N^2} \left(-\frac{\partial p^q}{\partial t} + \mathcal{S}_q(x, y, t) \right) ,$$

so that substituting back into (9.46) leads to

(9.47)
$$\operatorname{div}_H \vec{v}_H^q G_q(z) + \left(-\frac{\partial p^q}{\partial t} + \mathcal{S}_q(x, y, t) \right) \frac{1}{N^2} \frac{d^2 G}{dz^2} = 0 .$$

Hence, separable solutions for (9.39) will be possible only if the factor $d^2 G/dz^2$ is a constant times $G(z)$, which with the boundary conditions in (9.40) gives the vertical structure equations

(9.48)
$$\frac{d^2 G_q}{dz^2} + \lambda_q^2 G_q = 0 , \quad \frac{d G_q}{dz} = 0 , \quad z = 0, H, \quad q = 1, 2, \ldots ,$$

where $0 < \lambda_1 < \lambda_2 < \cdots$ are the separation constants. They are inversely proportional to the depths associated with each baroclinic mode. The solution of (9.48) is given by

$$G_q(z) = \cos\left(\frac{q\pi z}{H}\right) \quad \text{with } \lambda_q = \frac{q\pi}{H} .$$

We set $p = p_q/c_q$, with $c_q = N/\lambda_q$, $q = 1, 2, \ldots$. Notice $q = 0$ corresponds to $G \equiv 1$, which gives a solution independent of z with a zero forcing term and is associated with the barotropic mode. Also note that a general source term S for (9.39), with zero vertical average, has a complete expansion through (9.41). Combining (9.48) and (9.47) and adding the equations in (9.43) gives the following system of forced shallow water equations:

$$(9.49) \qquad \frac{\partial u}{\partial t} - \beta y v + c_q \frac{\partial p}{\partial x} = 0,$$

$$(9.50) \qquad \frac{\partial v}{\partial t} + \beta y u + c_q \frac{\partial p}{\partial y} = 0,$$

$$(9.51) \qquad \frac{\partial p}{\partial t} + c_q \frac{\partial u}{\partial x} + c_q \frac{\partial v}{\partial y} = \frac{1}{c_q} S_q(x, y, t).$$

The constants c_q, $q = 1, 2, \ldots$, are the speeds of the equatorial Kelvin waves associated with each one of the baroclinic (vertical) modes.

We introduce the Riemann invariant variables via the orthogonal transformation

$$(9.52) \qquad q = \frac{1}{\sqrt{2}}(p + u), \qquad r = \frac{1}{\sqrt{2}}(p - u).$$

Writing the equations in (9.49), (9.50), and (9.51) through the variables q, r, and v and dropping the q subscript to avoid confusion, we obtain the

SYMMETRIC FORM FOR LINEAR EQUATORIAL SHALLOW WATER EQUATIONS:

$$(9.53) \qquad \frac{\partial q}{\partial t} + c \frac{\partial q}{\partial x} + \frac{c}{\sqrt{2}} \left(\frac{\partial v}{\partial y} - \frac{\beta}{c} y v \right) = \frac{1}{\sqrt{2}c} S(x, y, t),$$

$$(9.54) \qquad \frac{\partial r}{\partial t} - c \frac{\partial r}{\partial x} + \frac{c}{\sqrt{2}} \left(\frac{\partial v}{\partial y} + \frac{\beta}{c} y v \right) = \frac{1}{\sqrt{2}c} S(x, y, t),$$

$$(9.55) \qquad \frac{\partial v}{\partial t} + \frac{c}{\sqrt{2}} \left(\frac{\partial q}{\partial y} + \frac{\beta}{c} y q \right) + \frac{c}{\sqrt{2}} \left(\frac{\partial r}{\partial y} - \frac{\beta}{c} y r \right) = 0.$$

Note that in the variables q, r, and v, the equatorial shallow water equations are a symmetric hyperbolic system. The Kelvin wave and mixed Rossby-gravity waves in Section 9.1 involve combinations of low-order parabolic cylinder functions in the y-variables. This fact and the symmetric form in (9.53)–(9.55) suggest that parabolic cylinder functions might be useful for a complete eigenfunction expansion of the equatorial shallow water equations. Indeed, this is the case as shown below.

9.2.1. Parabolic Cylinder Functions, Hermite Polynomials, and Recursive Relations. We first introduce the parabolic cylinder functions and discuss some properties related to them. The parabolic cylinder functions are given by

$$(9.56) \qquad D_m(\eta) = 2^{-m/2} H_m \left(\frac{\eta}{\sqrt{2}} \right) e^{-\eta^2/4} \qquad H_m(\xi) = (-1)^m e^{\xi^2} \frac{\partial^m e^{-\xi^2}}{\partial \xi^m}$$

where H_m, $m \geq 0$, are the Hermite polynomials with the first few given by

$$H_0(\xi) = 1, \quad H_1(\xi) = 2\xi, \quad H_2(\xi) = 4\xi^2 - 2,$$
$$H_3(\xi) = 8\xi^3 - 12\xi, \quad H_4(\xi) = 16\xi^4 - 48\xi^2 + 12.$$

The parabolic cylinder functions are the well-known solutions of the harmonic oscillator

(9.57) $$f''(\eta) + \frac{1}{2}\left(2m + 1 - \frac{\eta^2}{2}\right)f(\eta) = 0.$$

This can be easily shown by using the following recursive formulas for Hermite polynomials:

(9.58) $$H_{m+1}(\xi) - 2\xi H_m(\xi) + H_m'(\xi) = 0$$

(9.59) $$H_m'(\xi) = 2\xi H_m(\xi).$$

By using (9.57), it is easy to check that the D_m's form an orthogonal set of $L^2(\mathbb{R})$ with respect to the inner product $\langle f, g \rangle = \int f(\eta)g(\eta)d\eta$; given n and m, multiply the ODE for D_m by D_n and the ODE for D_n by D_m and sum. Hence, for each constant c, $q = 1, 2, \ldots$, in (9.53)–(9.55), we consider the orthonormal basis

(9.60) $$\phi_m^q(y) = \left(m!\sqrt{\frac{\pi c}{\beta}}\right)^{-1/2} D_m\left(\left(\frac{2\beta}{c}\right)^{1/2} y\right).$$

Note that the length scale $(\beta/c)^{-1/2}$ represents the equatorial Rossby deformation radius. Also from (9.58) and (9.59), we can derive the following properties related to the raising and lowering operators of quantum mechanics:

$$\mathcal{L}_\pm = \frac{\partial}{\partial \eta} \pm \frac{1}{2}\eta.$$

We have

(9.61) $$\mathcal{L}_+ D_m(\eta) = m D_{m-1}(\eta),$$

(9.62) $$\mathcal{L}_- D_m(\eta) = -D_{m+1}(\eta).$$

We introduce the change of variables

$$\eta = \left(\frac{2\beta}{c}\right)^{1/2} y$$

so that

(9.63) $$\mathcal{L}_\pm = \left(\frac{2\beta}{c}\right)^{-1/2}\frac{d}{dy} \pm \left(\frac{\beta}{2c}\right)^{1/2} y = \left(\frac{2\beta}{c}\right)^{-1/2}\left(\frac{d}{dy} \pm \frac{\beta}{c}y\right),$$

and we rewrite the raising and lowering operators associated with each constant c in the variable y as

(9.64) $$L_\pm^q = \frac{d}{dy} \pm \frac{\beta}{c}y.$$

By using (9.60)–(9.64), we get the following identities:

$$(9.65) \qquad L_-^q \phi_m^q(y) = -\left(\frac{2\beta}{c}\right)^{1/2} (m+1)^{1/2} \phi_{m+1}^q(y),$$

$$(9.66) \qquad L_+^q \phi_m^q(y) = \left(\frac{2\beta}{c}\right)^{1/2} (m)^{1/2} \phi_{m-1}^q(y).$$

9.2.2. The Complete Expansion for Equatorial Shallow Water. Note that in terms of the raising and lowering operators $L_\pm$ and the variables (q, r, v), the equatorial shallow water equations from (9.53)–(9.55) can be rewritten as

$$(9.67) \qquad \frac{\partial q}{\partial t} + c\frac{\partial q}{\partial x} + \frac{c}{\sqrt{2}}L_-v = \frac{1}{\sqrt{2}c}\mathcal{S}(x, y, t),$$

$$(9.68) \qquad \frac{\partial r}{\partial t} - c\frac{\partial r}{\partial x} + \frac{c}{\sqrt{2}}L_+v = \frac{1}{\sqrt{2}c}\mathcal{S}(x, y, t),$$

$$(9.69) \qquad \frac{\partial v}{\partial t} + \frac{c}{\sqrt{2}}L_+q + \frac{c}{\sqrt{2}}L_-r = 0.$$

Now, we assume the following expansions:

$$(9.70) \qquad \mathcal{S}(x, y, t) = \sum_{m=0}^{+\infty} \overline{\mathcal{S}}_m(x, t)\phi_m(y)$$

and

$$(9.71) \qquad \begin{pmatrix} q \\ r \\ v \end{pmatrix}(x, y, t) = \sum_{m=0}^{+\infty} \begin{pmatrix} \overline{q}_m \\ \overline{r}_m \\ \overline{v}_m \end{pmatrix}(x, t)\phi_m(y).$$

By plugging (9.70) and (9.71) into (9.67)–(9.69) and using the identities in (9.65) and (9.66), we have, after taking the inner product with ϕ_m, for each m, the constant-coefficient PDEs

$$\frac{\partial \overline{q}_m}{\partial t} + c\frac{\partial \overline{q}_m}{\partial x} - \frac{c}{\sqrt{2}}\left(\frac{2\beta}{c}\right)^{1/2}(m)^{1/2}\overline{v}_{m-1} = \frac{1}{\sqrt{2}c}\overline{\mathcal{S}}_m,$$

$$\frac{\partial \overline{r}_m}{\partial t} + c\frac{\partial \overline{r}_m}{\partial x}\phi_m^q + \frac{c}{\sqrt{2}}\left(\frac{2\beta}{c}\right)^{1/2}(m+1)^{1/2}\overline{v}_{m+1}\phi_m^q = \frac{1}{\sqrt{2}c}\overline{\mathcal{S}}_m,$$

$$\frac{\partial \overline{v}_m}{\partial t} + \frac{c}{\sqrt{2}}\left(\frac{2\beta}{c}\right)^{1/2}(m+1)^{1/2}\overline{q}_{m+1} - \frac{c}{\sqrt{2}}\left(\frac{2\beta}{c}\right)^{1/2}(m)^{1/2}\overline{r}_{m-1} = 0,$$

where the coefficients of negative index are zero. This system of equations decouples as follows:

- a single PDE for $\overline{q}_0$,

$$(9.72) \qquad \frac{\partial \overline{q}_0}{\partial t} + c\frac{\partial \overline{q}_0}{\partial x} = \frac{1}{\sqrt{2}c}\overline{\mathcal{S}}_0(x, t),$$

- a 2×2 system coupling $\overline{v}_0$ and $\overline{q}_1$,

$$(9.73) \quad \frac{\partial \overline{v}_0}{\partial t} + (\beta c)^{1/2}\overline{q}_1 = 0, \quad \frac{\partial \overline{q}_1}{\partial t} + c\frac{\partial \overline{q}_1}{\partial x} - (\beta c)^{1/2}\overline{v}_0 = \frac{1}{\sqrt{2}c}\overline{\mathcal{S}}_1(x,t),$$

- three equations coupling $\overline{r}_{m-2}$, $\overline{v}_{m-1}$, and $\overline{q}_m$ for $m \geq 2$,

$$
\begin{aligned}
& \frac{\partial \overline{q}_m}{\partial t} + c\frac{\partial \overline{q}_m}{\partial x} - (m\beta c)^{1/2}\overline{v}_{m-1} = \frac{1}{\sqrt{2}c}\overline{\mathcal{S}}_m(x,t), \\
(9.74) \quad & \frac{\partial \overline{r}_{m-2}}{\partial t} - c\frac{\partial \overline{r}_{m-2}}{\partial x} + (\beta c(m-1))^{1/2}\overline{v}_{m-1} = \frac{1}{\sqrt{2}c}\overline{\mathcal{S}}_{m-2}(x,t), \\
& \frac{\partial \overline{v}_{m-1}}{\partial t} + (\beta cm)^{1/2}\overline{q}_m - (\beta c(m-1))^{1/2}\overline{r}_{m-2} = 0.
\end{aligned}
$$

Notice for the 3×3 systems that if we introduce

$$\vec{u}_m = \begin{pmatrix} \overline{q}_m \\ \overline{r}_{m-2} \\ \overline{v}_{m-1} \end{pmatrix} \quad \text{for } m \geq 2,$$

we have

$$\frac{\partial \vec{u}_m}{\partial t} + A_m\frac{\partial \vec{u}}{\partial x} + B_m\vec{u}_m = \mathcal{S}_m$$

where

$$A_m = \begin{bmatrix} c & 0 & 0 \\ 0 & -c & 0 \\ 0 & 0 & 0 \end{bmatrix},$$

$$B_m = \begin{bmatrix} 0 & 0 & -(m\beta c)^{1/2} \\ 0 & 0 & ((m-1)\beta c)^{1/2} \\ (m\beta c)^{1/2} & -((m-1)\beta c)^{1/2} & 0 \end{bmatrix},$$

and

$$\mathcal{S}_m = \frac{1}{\sqrt{2}c}\begin{pmatrix} \overline{\mathcal{S}}_m(x,t) \\ \overline{\mathcal{S}}_{m-2}(x,t) \\ 0 \end{pmatrix}.$$

Thus A_m is symmetric and B_m is skew-symmetric, so the systems in (9.74) provide rich examples for the dispersive wave theory developed in Chapter 5.

To summarize, we have shown through the above series of expansions and changes of variables that the linearized equatorial shallow water equations reduce to a countable number of constant-coefficient 3×3 dispersive wave systems through a judicious use of parabolic cylinder functions.

9.2.3. Free Equatorial Waves. The solutions of (9.72), (9.73), and (9.74) generalize the explicit equatorial waves from Section 9.1. Suppose that the forcing terms $\overline{\mathcal{S}}_{q,m}$ in the recursive relations (9.72), (9.73), and (9.74) are zero. Looking for plane-wave-like solutions, we set

$$(9.75) \quad \begin{pmatrix} \overline{q}_m \\ \overline{r}_m \\ \overline{v}_m \end{pmatrix} = \mathrm{Re}\begin{pmatrix} q_m \\ r_m \\ v_m \end{pmatrix}e^{i(kx-\omega t)}.$$

By plugging in this ansatz into the recursive relations we get the *dispersion relations* linking the frequencies ω to the wavenumber k and the expressions of the eigenvectors for each one of the systems, from which we deduce the different families of equatorial waves.

Kelvin Waves. The first equation (9.72) gives

$$(9.76) \qquad \omega = ck$$

and the associated solution in terms of the (c-rescaled) pressure p, and the horizontal velocity field (u, v) is given by

$$(9.77) \qquad \begin{pmatrix} p \\ u \\ v \end{pmatrix}(x, y, t) = \begin{pmatrix} 1 \\ 1 \\ 0 \end{pmatrix} \cos(k(x - ct))\phi_0^q(y).$$

This solution is known as the *Kelvin wave* from Section 9.1.1. The Kelvin wave can also be expressed in a general fashion as

$$(9.78) \qquad \begin{pmatrix} p \\ u \\ v \end{pmatrix}(x, y, t) = \begin{pmatrix} 1 \\ 1 \\ 0 \end{pmatrix} G(x - ct)e^{-(\frac{\beta}{c})\frac{y^2}{2}},$$

so that the Kelvin wave associated with the baroclinic mode q travels eastward with the speed $c = \frac{NH}{q\pi}$. Higher modes have slower Kelvin waves. With a typical value for the buoyancy frequency in the troposphere of $N \approx 0.01 \text{ s}^{-1}$, and a typical depth of the troposphere of $H \approx 16$ km, the speed of the Kelvin wave associated with the first baroclinic is fixed to $c_1 = 50 \text{ ms}^{-1}$, $c_2 = 25 \text{ ms}^{-1}$ for the second, and so on. The meridional velocity v is zero so that the Kelvin wave is always parallel to the equator with the strength of the wave diminishing as one moves away from the equator.

Mixed Rossby-Gravity Waves. When plugging (9.75) into the system in (9.73), as in Section 9.1, we obtain a 2×2 linear eigenvalue problem with constant coefficients; then, by substituting $v_0 = -i(\beta c)^{1/2}q_1/\omega$ in the second equation, we get

$$(9.79) \qquad \omega_{\pm} = \frac{1}{2}ck \pm \frac{1}{2}\sqrt{(ck)^2 + 4\beta c},$$

and the solution is

$$(9.80) \qquad \begin{pmatrix} p \\ u \\ v \end{pmatrix}(x, y, t) = \begin{pmatrix} \frac{1}{\sqrt{2}} \cos(kx - \omega_{\pm}t)\phi_1(y) \\ \frac{1}{\sqrt{2}} \cos(kx - \omega_{\pm}t)\phi_1(y) \\ ((\beta c)^{1/2}/\omega_{\pm}) \sin(kx - \omega_{\pm}t)\phi_0(y) \end{pmatrix}.$$

These are the two mixed Rossby-gravity (MGR) waves discussed in Section 9.1.

Equatorial Rossby and Gravity Waves. With (9.75), the system in (9.74) is equivalent to the 3×3 eigenvalue problem

$$
\begin{bmatrix}
-i\omega + ick & 0 & -(m\beta c)^{1/2} \\
0 & -i\omega - ick & ((m-1)\beta c)^{1/2} \\
(m\beta c)^{1/2} & -((m-1)\beta c)^{1/2} & -i\omega
\end{bmatrix}
\begin{pmatrix}
q_m \\
r_{m-2} \\
v_{m-1}
\end{pmatrix} = 0
$$

where the characteristic equation is

(9.81) $$\omega(\omega^2 - c^2 k^2) - \beta c((2m-1)\omega + ck) = 0,$$

and the eigenvectors are given by

$$q_m = i\frac{(m\beta c)^{1/2}}{\omega - ck} v_{m-1}, \qquad r_{m-2} = -i\frac{((m-1)\beta c)^{1/2}}{\omega + ck} v_{m-1},$$

where ω is any solution of (9.81). Note that the solution $\omega(k) \equiv 0$ is not possible in (9.81), so it can be rewritten as

$$\left(\frac{\omega}{c}\right)^2 - k^2 - \frac{\beta}{\omega} k = \frac{\beta}{c}(2M+1) \quad \text{with } M = m - 1.$$

This equation has three distinct solutions that can be approximated as follows: When $\frac{\omega}{c}$ is small, we have one single solution,

(9.82) $$\omega_0(k) \approx -\frac{\beta k}{\frac{\beta}{c}(2M+1) + k^2},$$

and when $\frac{\beta}{\omega} k$ is small, we get

(9.83) $$\omega_{\pm}(k) \approx \pm\sqrt{c^2 k^2 + c\beta(2M+1)} \quad \text{for } M = 1, 2, \ldots.$$

The maximum errors in these simplified approximations are only a few percent (see Gill [**11**, chap. 11]). From Section 5.1 and (9.2), recall that the dispersion relation for the midlatitude Rossby waves in the β-plane approximation, $f = f_0 + \beta y$ with $f_0 \neq 0$, is given by

$$\omega_r(k, l) = -\frac{\beta k}{k^2 + l^2 + L_R^{-2}}.$$

Also from Section 4.4 for the midlatitude Poincaré waves in an f-plane approximation $f = f_0$, it is given by

$$\omega_p(\vec{\mathbf{k}}) = \pm c\sqrt{|\vec{k}|^2 + L_R^{-2}}.$$

Recall from (9.2) that $L_R = \sqrt{gH}/f_0$ is the midlatitude Rossby radius. The waves associated with the dispersion relation in (9.82) are known as the equatorial *Rossby waves* or equatorial *planetary waves* because the form of the dispersion relation resembles the one of the midlatitude Rossby waves while the waves corresponding to (9.83) are called the equatorial *inertio-gravity waves* or simply equatorial *gravity waves*; their dispersion relations are approximately the same as those of the midlatitude Poincaré waves. Recall that the equatorial deformation radius is $L_e = (\frac{c}{\beta})^{1/2}$ and in both (9.82) and (9.83), $L_e^{-2}(2M+1)$ plays the same role as L_R^{-2} for midlatitudes.

Furthermore, notice the solution in (9.76) corresponds to $M = -1$, and the solutions in (9.79) correspond to $M = 0$. For this reason, regarding the dispersion relations, the Kelvin wave is asymptotic to the gravity waves; also, the constant c, which plays the role of the midlatitude gravity wave speed in the dispersion relations as discussed above, is the actual Kelvin wave speed as mentioned earlier. The behavior of the mixed Rossby-gravity waves is related to both the Rossby and gravity waves but in a more complex fashion.

Precisely, we have the following remark:

REMARK. When k is large, (9.79) can be Taylor-expanded as follows:

$$\omega_\pm = \frac{1}{2}ck \pm \frac{1}{2}c|k|\left(1 + \frac{2\beta}{ck^2} + o\left(\frac{1}{k}\right)\right)$$

so that when $k > 0, k \gg 1$,

$$\omega_+ = ck + o\left(\frac{1}{k}\right),$$

and when $k < 0, k \ll 1$,

$$\omega_+ = -\frac{\beta}{k} + o(1),$$

i.e, when $k > 0$ the MRG wave associated with ω_+ is asymptotically equivalent to the gravity waves in (9.83) and when $k < 0$ it behaves like the Rossby waves in (9.82). Also notice for $k > 0$ the MRG wave is traveling eastward like a Kelvin wave while for $k < 0$ it moves westward like a planetary or Rossby wave. We have the same behavior for the MRG wave associated with ω_- in the opposite directions.

We conclude this section by plotting the dispersion relations for the equatorial Kelvin, Rossby, and gravity waves for $M = 1, 2, 3$ and the structures of the solutions for the MRG, Rossby, and gravity waves with $M = 1$ for the first baroclinic mode ($c_1 = 50$ ms^{-1}). In Figure 9.1 we plot the dispersion relations. We see from this figure that there is scale separation between the gravity and Rossby waves; the gravity waves travel faster than the Rossby waves. In Figures 9.2, 9.3, 9.4, and 9.5 are shown the structures of the Kelvin, MRG, symmetric ($M = 1$), and antisymmetric ($M = 2$) equatorial Rossby and equatorial gravity waves, respectively. The filled contours represent the pressure at the bottom of the troposphere while the horizontal velocity profile at the bottom of the troposphere is represented by the arrows.

The wavenumber is fixed to its positive value represented by the wavenumber 5, except for the MRG wave, for which the negative wavenumber -5 is also considered, since we know, from the above remark, that the MRG wave behaves differently when the wavenumber is negative or positive. It has eastward phase velocity when $k > 0$ and westward phase when $k < 0$. Note that the wavenumber 1 corresponds to a wavelength equal to the circumference of the earth, 40,000 km; hence, the wavelength considered here is 8000 km. The waves displayed here are normalized such that

$$\max \sqrt{u^2 + v^2 + p^2} = 1 \text{ m}^2\text{s}^{-2}.$$

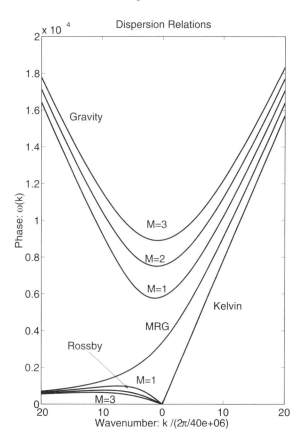

FIGURE 9.1. Phase versus the wavenumber normalized by the wavenumber 1 for the first baroclinic $c_1 = 50 \text{ ms}^{-1}$.

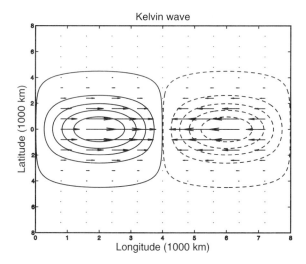

FIGURE 9.2. Filled contours of the pressure and velocity profile (arrows). Here and below, dotted contours are low pressure.

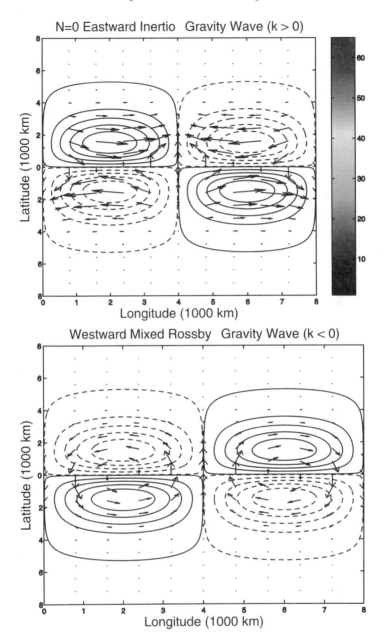

FIGURE 9.3. Filled contours of the pressure and velocity profile (arrows).

We have identified the equatorial Rossby, gravity, and MRG waves in the above analysis by the similarity of the dispersion relations for these equatorial waves with those for midlatitudes. In fact, the quantitative structure of these waves in Figures 9.2, 9.3, 9.4, and 9.5 reveals a much closer correspondence. First note that the Kelvin wave in Figure 9.2 is an eastward propagating equatorially trapped gravity wave with westward (eastward) near-surface winds in regimes of low (high)

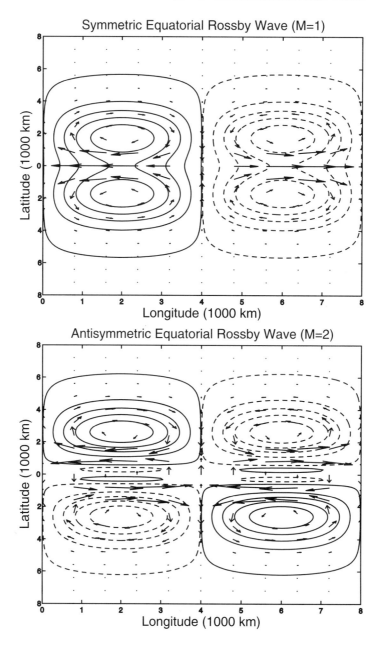

FIGURE 9.4. Filled contours of the pressure and velocity profile (arrows).

near-surface pressure. On the other hand, both the symmetric ($M = 1$) and anti-symmetric ($M = 2$) equatorial Rossby waves in Figure 9.4 show similar behavior as midlatitude Rossby waves. Recall that cyclonic flow is counterclockwise in the Northern Hemisphere and clockwise in the Southern Hemisphere with the opposite behavior for anticyclones; also, our experience reading weather maps in midlatitudes associates near-surface cyclonic (anticyclonic) flow with near-surface low

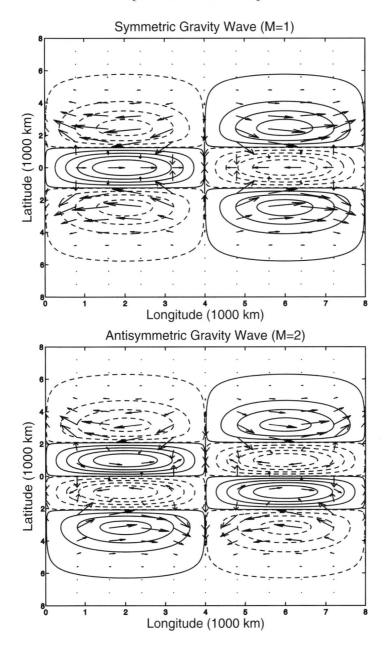

FIGURE 9.5. Filled contours of the pressure and velocity profile (arrows).

(high) pressure. With this information the equatorial Rossby waves in Figure 9.4 look exactly like equatorially trapped arrays of cyclones and anticyclones with the required corresponding pressure structure. An eastward propagating MRG wave is depicted in the upper panel of Figure 9.3 with a spatial scale of 8000 km. Note that just like the Kelvin wave in Figure 9.2, this wave has a predominantly west-ward (eastward) velocity component in the regions of low (high) pressure; thus, the structure of even large-scale eastward propagating MRG waves strongly resembles an antisymmetric equatorially trapped gravity wave. On the other hand, the westward propagating large-scale MRG wave in the lower panel in Figure 9.3 shares the same fluid dynamic features with the equatorial Rossby waves from Figure 9.4; this MRG wave has cyclonic (anticyclonic) flow in regions of low (high) pressure with interesting cross-equatorial flow in these waves. The waves in Figure 9.5 clearly have the features of more complex equatorially trapped gravity waves.

9.3. The Nonlinear Equatorial Long-Wave Equations

In Chapters 4 and 7, we heavily exploited the nondegenerate character of the Coriolis terms from rotation to get rigorously derived, simplified dynamic equations, the quasi-geostrophic equations, in the low Rossby number limit where overall geostrophic balance is a key component of the reduced dynamics. In Chapter 8 we established that such behavior persists in this limit, even with arbitrary unbalanced initial data because there is strict time-scale separation between the quasi-geostrophic modes and the vortical modes when the Coriolis terms do not vanish.

For equatorial geophysical flows, the vanishing of the Coriolis terms at the equator is the main new feature; we have shown in Sections 9.1 and 9.2 above that new equatorial trapped waves emerge as a consequence of this degeneracy such as Kelvin waves that have geostrophic balance only in the meridional or y-direction and behave like gravity waves in the zonal or x-direction and mixed Rossby-gravity waves that directly exhibit no separation of time scales between the gravity modes and vortical modes. Thus, it is a challenging and important problem to develop simplified reduced dynamical equations for equatorial fluid motion. One version of such simplified dynamics is discussed below; this is an active contemporary research topic in AOS with an interesting menagerie of reduced dynamics; the reader can consult the recent paper of Majda and Klein [21] and the references therein.

To derive the simplified reduced dynamics, first we recall the basic nondimensional units for the equatorial shallow water equations, already utilized at the beginning of Section 9.1.3. These are the following:

(9.84)
$$\text{unit length scale:} \quad L_e = \left(\frac{c}{\beta}\right)^{1/2}, \quad \text{unit time scale:} \quad T_e = (c\beta)^{-1/2},$$

$$\text{unit velocity scale:} \quad V_e = c, \quad \text{unit height scale:} \quad H_e = \frac{c^2}{g}.$$

With these units, the nondimensional equatorial shallow water equations with dissipation and thermal (height) forcing have the form

(9.85)
$$u_t + uu_x + vu_y + h_x - yv = -du,$$
$$v_t + uv_x + vv_y + h_y + yu = -dv,$$
$$h_t + (u_x + v_y) + (hu)_x + (hv)_y = \widetilde{Q}(x, y, t) - \tilde{d}h.$$

To motivate the typical physical scales of motion, as in Section 9.2, we use the representative physical value for $c = 50$ m/s with

$$(9.86) \qquad L_e \cong 1500 \text{ km}, \quad T_e \cong 8 \text{ hours}.$$

Since the circumference of the earth is $\bar{L} \cong 40{,}000$ km, clearly $L_e/\bar{L}$ satisfies $L_e/\bar{L} \ll 1$, so large-scale variations in x are allowed. Nonlinear equatorial long-wave equations (NLELWE) are the simplified dynamical equations that emerge by systematically exploiting the long-wave scaling in the zonal or x-direction, occurring for flows near the equator.

In equatorial long-wave scaling, the nondimensional variables in (9.85) are rescaled anisotropically by

$$(9.87) \qquad x' = \delta x, \quad t' = \delta t, \quad y' = y, \quad u' = u, \quad \delta v' = v, \quad h' = h.$$

With the scalings in (9.87), we are tacitly assuming a special regime of solutions for (9.85) with the special form

$$(9.88) \qquad u = u'(\delta x, y, \delta t), \quad v = \delta v'(\delta x, y, \delta t), \quad h = h'(\delta x, y, \delta t).$$

To be compatible with these scalings in (9.87) and (9.88), we assume that the dissipation and forcing in (9.85) satisfies

$$(9.89) \qquad \tilde{d} = \delta d_\theta, \quad d = \delta \bar{d}, \quad \widetilde{Q} = \delta Q(\delta x, y, \delta t).$$

Changing space and time variables in (9.85) according to (9.87) together with (9.88) and (9.89) and dropping all primes yields the

LONG-WAVE SCALED EQUATORIAL SHALLOW WATER EQUATIONS:

$$(9.90) \qquad \begin{aligned} h_t + (u_x + v_y) + (hu)_x + (hv)_y &= Q(x, y, t) - d_\theta h, \\ u_t + uu_x + vu_y + h_x - yv &= -\bar{d}u, \\ h_y + yu + \delta^2(v_t + uv_x + vv_y) &= \delta(-\bar{d}v). \end{aligned}$$

The simplified dynamics in NLELWE emerges immediately from (9.90) by dropping the terms of order δ and δ^2 in the meridional or y momentum equation; this implies that geostrophic balance is enforced in the meridional or y-direction alone but not in the x-direction. The resulting reduced dynamics are the

NONLINEAR EQUATORIAL LONG-WAVE EQUATIONS:

$$(9.91) \qquad \begin{aligned} h_t + (u_x + v_y) + (hu)_x + (hv)_y &= Q(x, y, t) - d_\theta h, \\ u_t + uu_x + vu_y + h_x - yv = -\bar{d}u, \quad h_y + yu &= 0. \end{aligned}$$

The equations derived formally in (9.91) are expected to apply for a large regime of equatorial motions. With the representative small value of $\delta = 0.1$ and the typical values in (9.86), variations in x of order 10,000 km and y of order 1500 km are allowed over times of several days with zonal velocities of 50 m/s and meridional velocities of 5 m/s; these bounds permit values easily within the regime of interesting observed equatorial motions. The same reasoning applies to the dissipation and forcing mechanisms assuming (9.89), but this won't be discussed here (see Majda and Klein [21]). There is a transparent similarity between the anisotropic scalings utilized in (9.87) and (9.88) and similar anisotropic scalings utilized in Prandtl's boundary layer theory for high Reynolds number incompressible flows (see Chorin and Marsden [2]); here the equatorial waveguide plays the role of the boundary layer.

9.3.1. Conserved Quantities for NLELWE. Here, we ignore forcing and dissipation for the NLELWE in (9.91) and display the analogues of conservation of potential vorticity and energy for NLELWE that emerge from the corresponding conservation principles established in Sections 4.2 and 4.3 for rotating shallow water with the anisotropic scalings in (9.87).

With the nondimensional form in (9.85) and the scalings in (9.87), Ertel's potential vorticity conservation for rotating shallow water, derived in Section 4.2, becomes

$$\frac{D}{Dt}\left(\frac{y + \delta^2 v_x - u_y}{1 + h}\right) = 0\,.$$

Thus, in the limit $\delta \ll 1$, we have

CONSERVATION OF POTENTIAL VORTICITY FOR NLELW:

(9.92)
$$\frac{D}{Dt}\left(\frac{y - u_y}{1 + h}\right) = 0\,.$$

Similarly, with (9.85) and (9.87), the local conservation of energy principle for rotating shallow water from Section 4.3 becomes the following:

$$E_0 = \frac{1}{2}\left[(1 + h)(u^2 + \delta^2 v^2) + h^2\right]$$

and

$$\frac{\partial E_0}{\partial t} + \mathrm{div}\left(\vec{v}\left[\left(h + \frac{1}{2}h^2\right) + E\right]\right) = 0\,.$$

Thus, in the limit $\delta \ll 1$, we have

LOCAL CONSERVATION OF ENERGY FOR NLELW:

(9.93)
$$E = \frac{1}{2}\left[(1 + h)(u^2) + h^2\right]$$

and

$$\frac{\partial E}{\partial t} + \mathrm{div}\left(\vec{v}\left[\left(h + \frac{1}{2}h^2\right) + E\right]\right) = 0\,.$$

Note that only zonal kinetic energy plus potential energy is involved in the energy conservation principle. Proceeding in the same fashion from other conservation laws for equatorial shallow water, we derive the infinite number of local conservation laws for potential vorticity,

(9.94) $$\frac{\partial}{\partial t}(1 + h)G(Q_0) + \mathrm{div}\left(\vec{v}(1 + h)(G(Q_0))\right) = 0\,, \quad Q_0 = \frac{y - u_y}{1 + h}\,,$$

for an arbitrary smooth nonlinear function $G(q)$. Detailed derivations of (9.91), (9.92), and (9.93) are left as an exercise for the interested reader.

9.3.2. Linear Theory for Equatorial Long Waves. Which of the equatorial waves discussed in detail in Sections 9.1 and 9.2 are retained by the NLELWE and how are these waves approximated? The simplest way to gain insight into these issues is to linearize the equations in (9.91) at the trivial background state $(u, h) = (0, 0)$ and neglect both forcing and dissipation. Following this procedure, the linear equatorial long-wave equations (LELWE) are given by

$$(9.95) \qquad u_t + h_x - yv = 0, \quad h_t + u_x + v_y = 0, \quad h_y + yu = 0.$$

As in Sections 9.1 and 9.2, we introduce characteristic variables associated with the right- and left-moving waves, r and l, with

$$(9.96) \qquad r = \frac{1}{2}(u + h), \qquad l = \frac{1}{2}(-u + h),$$

and also recall the raising and lowering operators

$$(9.97) \qquad L_\pm = \frac{d}{dy} \pm y.$$

When the LELWE in (9.95) are rewritten using (9.95) and (9.96), we obtain the
SKEW-SYMMETRIC FORM FOR LELWE:

$$(9.98) \qquad r_t + r_x + \frac{1}{2}L_- v = 0, \quad l_t - l_x + \frac{1}{2}L_+ v = 0, \quad L_+ r + L_- l = 0.$$

Now, as in Section 9.2, it is natural to expand r, l, and v in parabolic cylinder functions,

$$r(x, y, t) = \sum_{N=0}^{\infty} r_N(x, t)\phi_N(y),$$

$$(9.99) \qquad\qquad l(x, y, t) = \sum_{N=0}^{\infty} l_N(x, t)\phi_N(y),$$

$$v(x, y, t) = \sum_{N=0}^{\infty} v_N(x, t)\phi_N(y).$$

As in Section 9.2.2 above, by utilizing the properties of the raising and lowering operators $L_\pm$ and (9.99) inserted into (9.98), we obtain the equations

$$(9.100) \qquad \frac{\partial}{\partial t}r_0 + \frac{\partial}{\partial x}r_0 = 0 \ \text{Kelvin wave}, \quad r_1 = 0, v_0 = 0 \ \text{MRG wave},$$

and the following three coupled equations for the triad l_N, v_{N+1}, and r_{N+2} for $N = 0, 1, 2, \ldots,$

$$\frac{\partial}{\partial t}r_{N+2} + \frac{\partial}{\partial x}r_{N+2} - \frac{1}{\sqrt{2}}(N + 2)^{1/2}v_{N+1} = 0,$$

$$(9.101) \qquad \frac{\partial}{\partial t}l_N - \frac{\partial}{\partial x}l_N + \frac{1}{\sqrt{2}}(N + 1)^{1/2}v_{N+1} = 0,$$

$$(N + 1)^{1/2}l_N = (N + 2)^{1/2}r_{N+2}.$$

Note that the last algebraic equation in (9.101) represents meridional or y geostrophic balance; eliminating v_{N+1} from the first two equations in (9.101) and utilizing this constraint results in the single equation

$$(9.102) \qquad \frac{\partial}{\partial t} l_N - \frac{1}{2N+3} \frac{\partial}{\partial x} l_N = 0$$

with

$$r_{N+2} = (N+1)^{1/2}(N+2)^{-1/2} l_N$$

for $N = 0, 1, 2, \dots$. Clearly v_{N+1} is determined afterwards from (9.101) and (9.102). Thus, the formulas in (9.99), (9.100), and (9.102) completely solve the LELWE.

Now we can address the important issue of the waves retained in the dynamics by the equatorial long-wave approximations. From (9.99) and (9.100), the linear Kelvin wave is retained by the equatorial long-wave dynamics, but clearly the MRG wave is completely absent. Recall from the discussion in Section 9.2.3 that in the nondimensional units utilized here with $c = 1$ and $\beta = 1$, the equatorial Rossby waves have the dispersion relations

$$(9.103) \qquad \omega_N(k) = -\frac{k}{k^2 + 2N + 3}, \qquad N = 0, 1, 2, \dots.$$

Note that N from this section and M from Section 9.2.3 satisfy $N = M + 1$ to clarify the correspondence in (9.103). With (9.103), the zonal group velocity of the equatorial Rossby waves at large scales is given by

$$(9.104) \qquad \left. \frac{d\omega_N}{dk} \right|_{k=0} = -\frac{1}{2N+3}.$$

Notice from (9.102) that this large-scale group velocity for equatorial Rossby waves coincides exactly with the propagation speed for l_N and r_{N+2} in (9.102) for $N = 0, 1, 2, 3$. Thus, this suggests that the LELWE retains the large-scale equatorial Rossby waves for $M = 1, 2, 3, \dots$ and a detailed comparison of these waves, which we omit here, confirms this. To summarize the conclusions of this analysis, the NLELWE retains the equatorial Kelvin wave and the large-scale features of the equatorial Rossby waves as well as their nonlinear interaction. These equations filter out the higher-frequency equatorial gravity waves for $M = 1, 2, 3, \dots$ (see Figure 9.1) as well as the lower-frequency mixed Rossby-gravity wave.

9.3.3. The Analogy Between NLELWE and Incompressible Flow. The fashion in which we have solved the LELWE demonstrates that the meridional velocity field v has the role of a Lagrange multiplier that can be determined afterward from the solution. The skew-symmetric form for the LELWE in (9.98) displays this clearly. To develop an analogy between NLELWE and incompressible flow, consider the equations for inviscid two-dimensional incompressible flow [**2, 19**]

$$(9.105) \qquad \frac{D\vec{v}}{Dt} = -\nabla p, \qquad \text{div } \vec{v} = 0.$$

In the standard projection method formulation due to Leray (see [**19**, chap. 1, sect. 1.7]) the equation for the pressure,

$$(9.106) \qquad \Delta p = \sum_{i,j} \frac{\partial v_i}{\partial x_j} \frac{\partial v_j}{\partial x_i},$$

is inverted explicitly and inserted back into the first equation in (9.105) so that the velocity is determined completely by this nonlocal equation in this formulation; the incompressibility constraint is automatically satisfied and the pressure is determined afterwards through (9.106). This formulation highlights the role of the pressure as the Lagrange multiplier that guarantees the incompressible constraint $\operatorname{div} \vec{v} = 0$. Note that the vector operations div and $-\nabla$ are adjoints in the L^2 inner product.

The NLELWE in (9.91) have a very similar mathematical structure; this is readily demonstrated through the skew-symmetric form of LELWE in (9.98). I claim the following analogies are useful to establish this:

	NLELWE	Two-dimensional incompressible fluid
Lagrange multiplier	meridional velocity v	pressure p
Constraints on dynamics	meridional geostrophic balance $L_+ r + L_- l = 0$	incompressibility $\operatorname{div} \vec{v} = 0$
Normal operator to constraints	$v \longmapsto -\begin{pmatrix} L_- v \\ L_+ v \end{pmatrix}$	$p \longmapsto -\nabla p$

Notice that as in the case of two-dimensional incompressible fluid flow, the two operators

$$(9.107) \qquad (r, l) \longmapsto L_+ r + L_- l \quad \text{and} \quad v \longmapsto -\begin{pmatrix} L_- v \\ L_+ v \end{pmatrix}$$

are adjoints in the standard L^2 inner product. The detailed verification is an elementary exercise for the reader. With the above table and (9.107), it is evident that LELWE in (9.98) can be regarded as solving two linear hyperbolic equations with a constraint where the velocity v is a Lagrange multiplier. The same remarks apply to NLELWE in (9.91); compared with incompressible flow, these equations have the novel feature that the Lagrange multiplier v also enters nonlinearly as a coefficient in the dynamics. The equations in (9.92) or (9.98) also have a projection formalism analogous to Leray's formulation for incompressible flow. The role of the Laplacian for incompressible flow is played by the Hermite operator

$$(9.108) \qquad Hf = \frac{1}{2}(L_+ L_- + L_- L_+)f = \frac{d^2}{dy^2}f - y^2 f,$$

which is the familiar spatial operator for the quantum mechanical harmonic oscillator.

In the AOS community the LELWE in (9.95) have an important role in very successful simplified models for El Ninõ phenomena in the equatorial ocean (see

[**11, 29**]). These equations for LELWE are often derived in a different ad hoc fashion from formal considerations (see Gill's book [**11**, chap. 11]). Here we have presented a new derivation that includes balanced nonlinear effects in NLELWE from (9.91). It would be very interesting to decide the well-posedness and existence theory for NLELWE. An important application of LELWE to the steady tropical circulation is developed next.

9.4. A Simple Model for the Steady Circulation of the Equatorial Atmosphere

Effects of heating play an enormous role in tropical meteorology. Heating triggers deep penetrative convection, which occurs in the layer 3-7 km deep and which is responsible for significant high-altitude cloud formation. Accumulated heat is then released in the middle of the troposphere and affects the equatorial circulation. Here we use the material developed in Sections 9.2 and 9.3 to make a very simple model for these processes (see chapter 11 of Gill [**11**]).

As a model for these processes, we use the linearized primitive equations with forcing:

$$(9.109a) \qquad \frac{\partial \vec{V}_H}{\partial t} + \beta y \vec{V}_H^{\perp} = -\nabla_H P - d\vec{V}_H \,,$$

$$(9.109b) \qquad \frac{\partial P}{\partial z} = \theta \,,$$

$$(9.109c) \qquad \frac{\partial \theta}{\partial t} + N^2 W = \widetilde{Q}(\vec{x}_H, z, t) - d_\theta \theta \,,$$

$$(9.109d) \qquad \mathrm{div}_H \vec{V}_H + \frac{\partial W}{\partial z} = 0 \,.$$

We introduce the potential temperature θ, normalized as the buoyancy, and Brunt-Väisälä frequency N^2. We have added a frictional force $d\vec{V}_H$ into the momentum equations and the heating $\widetilde{Q}(\vec{x}_H, z, t) \geq 0$ into equation (9.109c) in (9.109) along with the radiative damping term $d_\theta \theta$. We assume the rigid lid approximation with the following boundary conditions in the vertical:

$$(9.110) \qquad W\big|_{(z=0,H)} = 0 \,.$$

As in Section 9.2, we separate the vertical and time-dependent horizontal structure for $\vec{V}_H$ and P:

$$(9.111) \qquad \begin{pmatrix} U \\ V \\ P \end{pmatrix} = \begin{pmatrix} u(\vec{x}_H, t) \\ v(\vec{x}_H, t) \\ p(\vec{x}_H, t) \end{pmatrix} G(z) \,,$$

where $G(z) = \cos(\frac{\pi z}{H})$. Here we make the basic assumption that all the forcing due to heating occurs on the first baroclinic mode. This is consistent with (9.111). Scaling out dimensional parameters so that $\beta = \frac{1}{2}$, we reduce the momentum equations in (9.109a) and (9.109b) to

$$(9.112) \qquad \frac{\partial u}{\partial t} - \frac{1}{2}yv = -\frac{\partial p}{\partial x} - d_1 u \,, \qquad \frac{\partial v}{\partial t} + \frac{1}{2}yu = -\frac{\partial p}{\partial y} - d_2 v \,.$$

The first baroclinic mode, therefore, assumes the following vertical structure:

$$(9.113) \qquad G(z) = \cos\left(\frac{\pi z}{H}\right), \quad G'(z) = -\sin\left(\frac{\pi z}{H}\right)\frac{\pi}{H}.$$

It follows from the equation in (9.109c) that

$$(9.114) \qquad \theta = G'(z)p(\vec{x}_H, t).$$

Since $G' < 0$, positive heating means that

$$(9.115) \qquad \tilde{Q} = -Q(x, y, t)G'(z), \quad Q \geq 0.$$

In the same spirit we separate the vertical and time-dependent horizontal structure for the vertical velocity $W(x, y, z, t)$:

$$(9.116) \qquad W(x, y, z, t) = -w(x, y, t)G'(z),$$

so that

$$(9.117) \qquad W > 0 \quad \text{for } w > 0.$$

Thus, upward motion in the full model corresponds to upward motion in the reduced equations. From equation (9.109c) we derive that

$$(9.118) \qquad W = \frac{\tilde{Q}}{N^2} - \frac{d_\theta \theta}{N^2} - \frac{1}{N}\frac{\partial \theta}{\partial t},$$

or, after separation of the vertical structure,

$$(9.119) \qquad w = \frac{Q}{N^2} + \frac{dp}{N^2} + \frac{1}{N}\frac{\partial p}{\partial t}.$$

To complete the formulation, we extract the dynamic equation for pressure from the incompressibility constraint in (9.109d),

$$(9.120) \qquad \frac{\partial p}{\partial t} + u_x + v_y = -Q(x, y, t) - dp.$$

We are ready now to write down the

REDUCED EQUATIONS WITH HEATING AND RADIATIVE DAMPING:

$$(9.121) \qquad \begin{aligned} \frac{\partial u}{\partial t} - \frac{1}{2}yv &= -\frac{\partial p}{\partial x} - d_1 u, \\ \frac{\partial v}{\partial t} + \frac{1}{2}yu &= -\frac{\partial p}{\partial y} - d_2 v, \\ \frac{\partial p}{\partial t} + u_x + v_y &= -Q(x, y, t) - dp. \end{aligned}$$

To obtain the physical interpretation of solutions of the reduced system in (9.121), we use the following formulas:

$$\begin{pmatrix} U(x, y, z, t) \\ V(x, y, z, t) \\ P(x, y, z, t) \end{pmatrix} = \begin{pmatrix} u(x, y, t) \\ v(x, y, t) \\ p(x, y, t) \end{pmatrix} G(x), \quad G(x) = \cos\left(\frac{\pi x}{H}\right),$$

(9.122) $$W(x, y, z, t) = -w(x, y, t)G'(z),$$

$$w(x, y, t) = \frac{Q(x, y, t)}{N^2} + \frac{dp}{N^2} + \frac{1}{N}\frac{\partial p}{\partial t},$$

$$\theta(x, y, z, t) = G'(z)p(x, y, t).$$

The coefficients of damping d_1 and d_2 are roughly the same size in (9.121). The goal here is a simple model for the steady (time-dependent) circulation of the equatorial atmosphere as a response to known heating. The equations in (9.121) have equatorial gravity waves, Kelvin waves, and Rossby waves. Next we follow Section 9.3 and make the long-wave approximation so that the v-component of momentum is replaced by geostrophic balance and equatorial gravity and MRG waves are filtered out. We also assume equal coefficients ε for x momentum and radiative damping, and a time-independent solution. Under these assumptions the equatorial shallow water equations in (9.121) become

(9.123a) $$\epsilon u - \frac{1}{2}yv = -\frac{\partial p}{\partial x},$$

(9.123b) $$\frac{1}{2}yu = -\frac{\partial p}{\partial y},$$

(9.123c) $$\epsilon p + u_x + v_y = -Q(x, y, t).$$

From (9.122), the vertical velocity $w(x, y, t)$ has the following form:

(9.124) $$w(x, y, t) = \epsilon p + Q.$$

The equation in (9.123b) has a simple meaning of meridional geostrophic balance.

In order to solve the equations in (9.123), we introduce the characteristic variables

(9.125) $$q = u + p, \qquad r = -u + p,$$

and rewrite the equations in the "characteristic" form by adding and subtracting the equations in (9.123a) and (9.123c),

(9.126a) $$\epsilon p + \frac{\partial q}{\partial x} + \frac{\partial v}{\partial y} - \frac{1}{2}yv = -Q,$$

(9.126b) $$\frac{\partial q}{\partial y} + \frac{1}{2}yq + \frac{\partial r}{\partial y} - \frac{1}{2}yr = 0,$$

(9.126c) $$\epsilon r - \frac{\partial r}{\partial x} + \frac{\partial v}{\partial y} + \frac{1}{2}yv = -Q.$$

In familiar fashion, we introduce the following expansion of our solution (q, r, v),

(9.127)
$$\begin{pmatrix} q(x, y) \\ r(x, y) \\ v(x, y) \end{pmatrix} = \sum_{N=0}^{\infty} \begin{pmatrix} yq_N(x) \\ r_N(x) \\ v_N(x) \end{pmatrix} D_N(y),$$

where $D_N(y)$ is the Nth-order parabolic cylinder function. We assume that the forcing $Q(x, y)$ can also be decomposed in the same way,

(9.128)
$$Q = \sum_{N=0}^{\infty} Q_N(x) D_N(y).$$

As in Sections 9.2 and 9.3, under this decomposition, the PDEs in (9.126) decouple into sets of three ODEs for r_{N-1}, v_N, and q_{N+1},

(9.129)
$$\epsilon q_{N+1} + \frac{dq_{N+1}}{dx} - v_N = -Q_{N-1},$$

$$\epsilon r_{N-1} - \frac{dr_{N-1}}{dx} + N v_N = -Q_{N-1},$$

$$(N + 1)q_{N+1} = r_{N-1},$$

for $N = 1, 2, \ldots$. The two special cases $N = -1$ and $N = 0$ are exceptional and yield the *Kelvin wave*:

(9.130)
$$\epsilon q_0 + \frac{dq_0}{dx} = -Q_0$$

and the *MRG wave*:

(9.131)
$$q_1 = 0, \quad \epsilon q_1 + \frac{dq_1}{dx} - v_0 = -Q_1.$$

The equation in (9.126b) and the sequence of ODEs originating from it represents meridional geostrophic balance.

Next, we consider two particular examples of forcing Q in the first baroclinic mode, namely, a symmetric and antisymmetric forcing.

9.4.1. Symmetric Forcing. In the first example we assume that the forcing has a symmetric structure across the equator,

(9.132)
$$Q(x, y) = F(x)e^{-\frac{1}{4}y^2},$$

where $F(x)$ is some smooth function with localized support (a good example of such a function gives a segment of $\cos x$ over $-\frac{\pi}{2} \leq x \leq \frac{\pi}{2}$). For the atmosphere such heating may be thought of as an intensive heating over the Indonesian marine continent; for example, in terms of our notation, we have

(9.133)
$$Q_0(x) = F(x), \quad Q_j = 0, \ j \geq 1.$$

Since only the Q_1 mode of forcing is excited, only the Kelvin ($N = -1$), MRG ($N = 0$), or the first triad ($N = 1$) can be excited due to the linear structure of the problem.

Symmetric Response to Symmetric Heating: Kelvin Wave. The Kelvin mode response satisfies the equation in (9.130). The only physically relevant solution to this equation with localized forcing must be zero to the west of the support of Q_0 and decays on the scale of order ϵ to the east. It therefore has the steady structure of an eastward-propagating Kelvin wave.

Antisymmetric Response to Symmetric Heating. The MRG wave has trivial structure for this case:

$$(9.134) \qquad q_1 = 0, \quad -v_0 = -Q_0 \equiv 0,$$

which is a natural response of the antisymmetric mode to symmetric forcing.

First Triad Response to Symmetric Heating. The only triad with a nonzero solution corresponds to $N = 1$ and satisfies the following equations:

$$
(9.135) \qquad
\begin{aligned}
\epsilon r_0 - \frac{dr_0}{dx} + v_1 &= -Q_0, \\[2mm]
\epsilon q_2 + \frac{dq_2}{dx} - v_1 &= 0 \Rightarrow \epsilon r_0 - \frac{dr_0}{dx} + \epsilon q_2 + \frac{dq_2}{dx} = -Q_0, \\[2mm]
r_0 &= 2q_2.
\end{aligned}
$$

These equations give rise to

$$(9.136) \qquad 3\epsilon q_2 - \frac{dq_2}{dx} = -Q_0 \quad \text{with} \quad v_1 = \epsilon q_2 + \frac{dq_2}{dx}.$$

The solution of this equation represents a westward-moving equatorial Rossby wave with spatial decay rate of 3ϵ. We are now ready to summarize the results in the form of the following:

PROPOSITION 9.1 *For the symmetric steady forcing $Q(x, y) = F(x)e^{-(1/4)y^2}$ and homogeneous initial conditions a steady response of the linearized primitive equations in (9.123) can be represented in terms of only four parabolic cylinder functions, $(D_0, D_1, D_2, D_3) = (1, y, y^2-1, y^3-3y)e^{-(1/4)y^2}$ with nonzero coefficients. The solution has a symmetric component,*

$$
(9.137) \qquad
\begin{aligned}
\epsilon q_0 + \frac{dq_0}{dx} &= -F & &\text{eastward Kelvin wave,} \\[2mm]
3\epsilon q_2 - \frac{dq_2}{dx} &= -F & &\text{westward equatorial Rossby wave,} \\[2mm]
r_0 = 2q_2, \quad v_0 &= 0 & &\text{meridional balance,}
\end{aligned}
$$

and an antisymmetric component,

$$(9.138) \qquad v_1 = \epsilon q_2 + \frac{dq_2}{dx}.$$

All other coefficients are zero, i.e.,

$$(9.139) \quad v_0 = 0, v_j = 0, j \geq 0, \quad r_j = 0, j \geq 1, \quad q_1 = 0, q_j = 0, \ j \geq 3.$$

Returning to the original variables u, v, w, and p in section 11.14 of his book [11], Gill analyzes the physical meaning of the solution. In particular, he deduces that there are two symmetrically located cyclones on both sides of the equator. He shows that there is a double Walker cell with air ascending near the heating region and then being transported zonally along the equator and then descending back and completing the circulation in the zonal direction. There is also a pair of Hadley cells, with air ascending near the equator and then moving to midlatitudes, where it descends. The Walker cell east of the heating source associated with the eastward Kelvin wave has an extent roughly three times as large as the westward cell associated with the equatorial Rossby wave. This structure follows from the formulas in (9.137).

9.4.2. Antisymmetric Forcing. In the second example, consider antisymmetric forcing of the form

$$(9.140) \qquad Q(x, y) = Q_1(x)ye^{-\frac{1}{4}y^2}, \qquad Q_0 = 0, \ Q_1 = G(x).$$

In this situation, again only a few modes of the expansion in (9.127) will be excited. In particular, we consider the same three cases as in the previous section.

Symmetric Response to Antisymmetric Heating. This time the Kelvin component will be trivial, since

$$(9.141) \qquad \epsilon q_0 + \frac{dq_0}{dx} = 0$$

has only a trivial physically significant solution.

Antisymmetric Response to Antisymmetric Heating. The MRG mode also has a simple structure,

$$(9.142) \qquad v_0 = Q_1 = G(x).$$

Second Triad Response to Antisymmetric Heating. The equations for $N = 2$ triad have the form

$$(9.143) \qquad \epsilon r_1 - \frac{dr_1}{dx} + 2v_2 = -Q_1, \qquad \epsilon q_3 + \frac{dq_3}{dx} - v_2 = 0, \qquad 3q_3 = r_1,$$

which yields

$$(9.144) \qquad 5\epsilon q_3 - \frac{dq_3}{dx} = -Q_1.$$

This is a westward steady equatorial Rossby wave with the strong decay rate 5ϵ, so the antisymmetric response is strongly confined to the forcing region. The only nonzero components are v_0, v_2, r_1, and q_3. Physically, we find that there is a "cyclone-anticyclone" pair located around the equator near the perturbation region. There is also a single dominant Hadley cell, spanning the equator, with rising air in the heating hemisphere. Obviously, these two symmetric and antisymmetric steady linear responses to prescribed heating can be superimposed to generate a more complete elementary picture of the tropical circulation. See section 11.14 of Gill's book [11] and the references there to his earlier papers. What we have attempted to do in this section is to provide all the necessary technical background for reading that work. One of the central issues in tropical meteorology is that the heating distribution is a subtle time-dependent nonlinear function of the tropical circulation itself; many other ideas beyond these simple models are needed to make progress (see Smith [35], Majda and Klein [21], and the references given there).

Bibliography

[1] Anile, A. M., Hunter, J. K., Pantano, P., and Russo, G. *Ray methods for nonlinear waves in fluids and plasmas*. Pitman Monographs and Surveys in Pure and Applied Mathematics, 57. Longman Scientific & Technical, Harlow; copublished in the United States with Wiley, New York, 1993.

[2] Chorin, A. J., and Marsden, J. E. *A mathematical introduction to fluid mechanics*. Third edition. Texts in Applied Mathematics, 4. Springer, New York, 1993.

[3] Craik, A. D. D. *Wave interactions and fluid flows*. Cambridge Monographs on Mechanics and Applied Mathematics. Cambridge University Press, Cambridge, 1985.

[4] Cushman-Roisin, B. *Introduction to geophysical fluid dynamics*. Prentice Hall, Englewood Cliffs, N.J., 1994.

[5] Duistermaat, J. J. *Fourier integral operators*. Translated from Dutch notes of a course given at Nijmegen University, February 1970 to December 1971. Courant Institute of Mathematical Sciences, New York University, New York, 1973.

[6] Embid, P. F., and Majda, A. J. Averaging over fast gravity waves for geophysical flows with arbitrary potential vorticity. *Comm. Partial Differential Equations* 21(3-4): 619–658, 1996.

[7] ———. Averaging over fast gravity waves for geophysical flow with unbalanced initial data. *Theoret. Comput. Fluid Dynam.* 11: 155–169, 1998.

[8] ———. Low Froude number limiting dynamics for stably stratified flow with small or finite Rossby numbers. *Geophys. Astrophys. Fluid Dynam.* 87(1-2): 1–50, 1998.

[9] Fincham, A. M., Maxworthy, T., and Spedding, G. R. The horizontal and vertical structure of the vorticity field in freely-decaying stratified grid turbulence. *Dynam. Atmos. Ocean* 23: 153–169, 1996.

[10] Folland, G. B. *Introduction to partial differential equations*. Mathematical Notes. Princeton University Press, Princeton, N.J., 1976.

[11] Gill, A. E. *Atmosphere-ocean dynamics*. Academic Press, New York, 1982.

[12] Hochstadt, H. *Differential equations. A modern approach*. Republication, with minor corrections, of the 1964 original. Dover, New York, 1975.

[13] John, F. *Partial differential equations*. Reprint of the fourth edition. Applied Mathematical Sciences, 1. Springer, New York, 1991.

[14] Klainerman, S., and Majda, A. Singular limits of quasilinear hyperbolic systems with large parameters and the incompressible limit of compressible fluids. *Comm. Pure Appl. Math.* 34(4): 481–524, 1981.

[15] ———. Compressible and incompressible fluids. *Comm. Pure Appl. Math.* 35(5): 629–653, 1982.

[16] Majda, A. *Compressible fluid flow and systems of conservation laws in several space variables*. Applied Mathematical Sciences, 53. Springer, New York, 1984.

[17] ———. Vorticity and the mathematical theory of incompressible fluid flow. *Comm. Pure Appl. Math.* 39(S): suppl., S187–S220, 1986.

[18] ———. Real world turbulence and modern applied mathematics. *Mathematics: frontiers and perspectives*, 137–151. American Mathematical Society, Providence, R.I., 2000.

[19] Majda, A., and Bertozzi, A. *Vorticity and incompressible flow*. Cambridge Texts in Applied Mathematics, 27. Cambridge University Press, Cambridge, 2002.

[20] Majda, A. J., and Grote, M. J. Model dynamics and vertical collapse in decaying strongly stratified flows. *Phys. Fluids* 9(10): 2932–294, 1997.

[21] Majda, A. J., and Klein, R. Systematic multi-scale models for the tropics. *J. Atmospheric Sci.*, in press.

[22] Majda, A, J., Rosales, R.; Tabak, E. G., and Turner, C. V. Interaction of large-scale equatorial waves and dispersion of Kelvin waves through topographic resonances. *J. Atmospheric Sci.* 56(24): 4118–4133, 1999.

[23] Majda, A. J., and Shefter, M. G. Elementary stratified flows with instability at large Richardson number. *J. Fluid Mech.* 376: 319–350, 1998.

[24] ———. The instability of stratified flows at large Richardson numbers. *Proc. Natl. Acad. Sci. USA* 95(14): 7850-7853, 1998.

[25] ———. Nonlinear instability of elementary stratified flows at large Richardson number. *Chaos* Special issue on mixing. 10(1): 3–27, 2000.

[26] Majda, A. J., and Wang, X. *Nonlinear dynamics and statistical theories for basic geomphysical flows*. Cambridge University Press, in preparation.

[27] McComas, C. H., and Bretherton, F. P. Resonant interaction of oceanic internal waves. *J. Geophysical Research* 83(9): 1397–1412, 1977.

[28] Muller, P., Holloway, G., Henyey, F. S., and Pomphrey, N. Nonlinear interactions among internal gravity waves. *Reviews Geophysics* 24(3): 493–536, 1986.

[29] Pedlosky, J. *Geophysical fluid dynamics*. Springer, New York, 1979.

[30] Philander, S. G. *El Niño, La Niña, and the southern oscillation*. Academic Press, San Diego, 1990.

[31] Riley, James J., and Lelong, M.-P. Fluid motions in the presence of strong stable stratification. *Annual review of fluid mechanics, Vol. 32*, 613–657. Annual Review of Fluid Mechanics, 32. Annual Reviews, Palo Alto, Calif., 2000.

[32] Ripa, P. On the theory of nonlinear wave-wave interactions among geophysical waves. *J. Fluid Mech.* 103: 87–105, 1981.

[33] Salmon, R. *Lectures on geophysical fluid dynamics*. Oxford University Press, New York, 1998.

[34] Schochet, S. Fast singular limits of hyperbolic PDEs. *J. Differential Equations* 114(2): 476–512, 1994.

[35] Smith, R. K., ed. *The physics and parameterization of moist atmospheric convection*. NATO Advanced Study Institute Series C. Mathematical and Physical Sciences, 505. Kluwer, Norwell, Mass., 1997.

[36] Temam, R. *Navier-Stokes equations*. Theory and numerical analysis. Studies in Mathematics and Its Applications, 2. North-Holland, Amsterdam–New York–Oxford, 1977.

[37] *Theoretical and computational fluid dynamics* 11(3/4): the entire issue, 1998.

Titles in This Series